The problem of evolution

Contemporary problems in philosophy
George F. McLean, Editor

Situationism and the New Morality, Robert L. Cunningham
Human and Artificial Intelligence, Frederick J. Crosson

The problem of evolution

A study of the philosophical repercussions of evolutionary science

by **JOHN N. DEELY** *Institute for Philosophical Research*
and
RAYMOND J. NOGAR

New York
APPLETON-CENTURY-CROFTS
Educational Division
MEREDITH CORPORATION

73 74 75 76 77 / 10 9 8 7 6 5 4 3 2 1

Library of Congress Card Number: 78–163701

PRINTED IN THE UNITED STATES OF AMERICA

390–25998–5

"The intellectual attrition of the sea of the timely crashing against the craggy rocks of the timeless may result in a polished marble of great value. The strength of both the tradition and contemporary thought may emerge in harmony and counter-balance the grave weaknesses in each view."

RAYMOND J. NOGAR, O.P.

To whose memory
this book is dedicated.

Preface

For a long time, philosophy of science has been roughly synonymous with the philosophy of physics. There are good reasons why this has been so. The revolution in man's conception of the world that took place in the 16th and 17th centuries, and the technology that has given us unprecedented control over the material order, have both been in large part results of the detailed development and extensive application of the mathematical schema definitive of physics in the modern sense of the word.

But however good the reasons, this preoccupation of the philosophers with the physical sciences has been something of a mixed blessing. For it has led many to believe that the method of physics is the paradigm of scientific explanation without qualification, the implication being that other efforts of the understanding are rational only to the extent that they admit of subsumption under the mathematical formulae which link explanation with prediction control.

Evolutionary science, by its success, proves the falsity of the view that every explanation as such must conform to the model or type of modern physics. In so doing, it does not in any way detract from the power and importance of mathematical physics, but simply demonstrates the limits of such an approach to the understanding of nature. Explanation thus proves reducible to two diverse formal schemes, one in which the rule of explanation is conformability to the requirements of mathematical formulation, and one in which the rule of explanation is the identification and isolation of proper causes. In the former explanatory scheme, deduction—and hence prediction—is a primary value; in the latter, deduction is a secondary value.

The transition from Classical Antiquity to Darwin's world is the subject of the First Part of this book, "Historical Perspective: The Impact of Evolution on Scientific Method," and hence on our conception of the organization and growth of human understanding. We define "Classical Antiquity" as that period of human thought from roughly the 5th century B.C. to the 16th century A.D., during which the mathematical method of the physical sciences, though known, was little appreciated and in a rudimentary state, having been applied rigorously only to the fields of astronomy, music, and optics. "Darwin's world" is the present period in which we see established, side by side, two quite different schemas within which the facts of nature can be placed for the purposes of understanding. Both these schemas are thoroughly rational and scientific, and hence usable in tandem, despite their formal differences.

The Second Part of the book, "Contemporary Discussions," attempts to isolate and develop the main areas relevant for humanistic thought in which the discovery of evolution has had a critical impact. We have identified six such "zones of crisis." The first is man's understanding of his own origins and ties with the animal world. The second is his understanding of the distinctively human environment provided by cultural traditions. The third is the region of man's moral sensibilities, and the justifiability or unjustifiability of our sense of "right" and "wrong." The fourth is man's understanding of understanding itself, what it is, how it develops, and what its limits are. The fifth is the zone of man's religious sensibilities, and his conception of his relation—or lack of relation—to something greater than this world. The sixth is man's attempt to put together the various and disparate pieces of insight and feeling into a coherent and overall picture.

These six areas of crisis correspond to the six sections of the Second Part. In each section, readings have been carefully introduced and organized so as to indicate the key controversies and implications engendered by taking an evolutionary view of the six topics—human origins, cultural life, moral responsibility, metaphysics, religion, and the future of man.

The Third Part of this book is a two-part Bibliography. Part I is the bibliography of the sources actually cited in the essay, "The Impact of Evolution on Scientific Method," in the First Part of this volume. Part II of the bibliography is a listing of sources, both books and articles, that the interested reader might profitably consult for further understanding of the themes and issues developed in the Second Part of this volume, "Contemporary Discussions." This part of the bibliography is accordingly itself divided into six sections, keyed to the six sections of readings or "contemporary discussions."

The extensive bibliography and overall scope of this volume, it is hoped, will make it eminently suitable for purposes of instruction and research, both at undergraduate and graduate levels, within the university.

Finally, I want here to say a few words *in memoriam*.

The challenge this book poses the reader is simply that of thinking long and hard about the evidence indicating that the universe and all of its parts, man and his spirit included, are unfolding in an evolutionary fashion, and of facing up to the implications of this evidence.

In doing this, the book stands, with all its inadequacies and as best it can, as a final memorial to my old professor and dear friend, Raymond J. Nogar, who died suddenly, unexpectedly, and tragically in the course of our collaboration on this volume. "Those who breathe easily about evolution today, as if it were 'old hat,' easily forget that within their own lifetime the 'evolution of man' was still an objectionable phrase which seemed contrary to *Genesis* and the traditional philosophy and theology."[1] Raymond Nogar made that observation in connection with the life-work of Père Teilhard de Chardin and "what he has done for Christian philosophers and theologians in this area." But for us of the English speaking world, the same observation could be made in calling to mind Ray Nogar's own life work. As late as 1963, with the publication of his major study of *The Wisdom of Evolution* (since translated into French, Spanish, and German), Raymond Nogar was in my estimation the first thinker of the Anglo-Saxon tradition who faced up publicly, while respecting integrally the requirements of each explanatory level, not only to the experimental detail and scientific theory of a world in evolution (this indeed had been done by many since Darwin), but also to the philosophical and theological implications of the scientific fact. It was precisely because he did not take the scientific foundations of evolutionary thought for granted (a common attitude among both philosophers and theologians); because he did not dismiss the Western philosophical heritage without further ado, on the grounds that it had once proposed a doctrine of unchanging specific types (a common attitude among scientists and those theologians content to take scientific fact for granted as long as it fits their personal biases); because he did not believe in a theology that presents simple and pat answers immune to philosophical criticism and safer still from the particular researches of natural science; it was because, in short, he was a very large and careful thinker, attentive always to truth, that Raymond Nogar was able to make some of the most reluctant and fundamentalist Christian thinkers acknowledge the power of evolutionary thought; while on the other hand, he was able to bring some of the most ardent and convinced evolutionists to acknowledge its limits.

He did nothing less than map the surest route to the decisive issues. His approach remains the most appropriate one for any thoughtful man:

---

[1] Raymond J. Nogar, *The Lord of the Absurd* (New York: Herder, 1966), p. 114

Not only must we have the basic facts upon which to build our case for evolution, we must also apply rigorous logic to the inferences which are often drawn from evolutionary statements. We must penetrate the facts and explanations with the sharpened mind. Our thinking must not be flabby, or our general view of the meaning of evolution will be excessive, either on the side of its power or on the side of its limits.[2]

Those who are already familiar with some of the history of biology and the controversies surrounding the rise of evolutionary science will perhaps best grasp the essential direction and distinctive spirit of Raymond Nogar's thought by meditating on the fact that it was he who first applied, with a characteristic and peculiar amalgam of poetry and precision, the term *epigenesis* to the overall process of evolutionary unfolding, in a sense in every way parallelling the contemporary use of the word "teleonomy" to describe individual development.

His death is a severe and untimely loss for the American intellectual community. In his own thought, as I well know (as do those who are familiar with his more recent writings), he had advanced well beyond certain of the positions sketched in *The Wisdom of Evolution*. He had intended to set forth in the Introductory Essay (the First Part) of this volume his new insights and clearer realization of the power of evolutionary thought. Illness at first and death irrevocably frustrated this intention. Writing the essay in his stead has been a melancholy task. The result is very different and for the purpose very inferior to what he would have written. But his was a style of mind impossible to duplicate. One does what one can. I have at least reason to think what I have written sheds new light on the problem that weighed on Father Nogar to the end, and does so in the spirit of what my friend early taught me must be the axiom of the philosopher who respects the traditions of his profession: "Non nova, sed noviter."

John N. Deely

[2] "Preface" to *The Wisdom of Evolution* (New York: Doubleday, 1963), p. 15.

Contents

List of charts and tables

I

Historical perspective

Historical perspective

The impact
of evolution on scientific method

Any effort to understand what is now taking place in human consciousness must of necessity proceed from the fundamental change of
view which, since the sixteenth century, has been steadily exploding
and rendering fluid what had seemed to be the ultimate stability—
our concept of the world itself. To our clearer vision the universe is
no longer a State but a Process. The Cosmos has become a Cosmogenesis. And it may be said without exaggeration that, directly or
indirectly, all the intellectual crises through which civilization has
passed in the last four centuries arise out of the successive stages
whereby a static *Weltanschauung* has been and is being transformed,
in our minds and hearts, into a *Weltanschauung* of movement.

Pierre Teilhard de Chardin

THEMATIC REMARKS

When philosophy undertakes to assess the cultural state of intellectual
development at any given historical moment, it is obliged principally
to clarify basic problems and to indicate the manner in which one problem
bears on another, by discriminating the distinctions and interrelations
of the basic areas of human concern and by imparting a grasp of the distinctive methods appropriate to dealing with each. To discharge this
obligation is no mean task. It requires an appreciation of the past and
an understanding of the present in the light of that appreciation. Doubtless this explains why the great teachers of the race have always mani-

fested a philosophic spirit, for the great teachers are those who cherish the wisdom of the past and make it relevant to the present; and it is through teaching and learning that civilization is sustained, disseminated and developed.

In dealing with the question of evolution, the successful discharge of the philosopher's synthetic obligation depends on his disposal of a distinctive difficulty. For the idea of evolution is the product of the "Darwinian revolution," which, in series with the "Copernican revolution" and the "Freudian revolution," is generally understood to mark a radical rupture with the past in man's consciousness both of his own nature and of the nature of the physical world.

In what possible sense does an appreciation of Classical Antiquity contribute to an understanding of Darwin's world? And on the practical side, how, in the limited space of a general essay, could one express that contribution, assuming there to be such?

An answer to these questions was suggested to me by taking together a contention implicit in Bernard J. F. Lonergan's book, *Insight*, and the architectonic around which W. T. Jones developed his admirable *History of Western Philosophy*.

Jones points out that a philosophy is simply a set of propositions dealing with the "big" questions about the ultimate value and meaning of life, about the structure of reality. By various criteria taken together, it is not difficult to discriminate in the flux of history the main lines of the dominating systems of human thought. First of all, to be a dominating system, it is necessary that a given philosophy constitute a perspective which satisfies a large number of people for a relatively long time. By this criterion, "the influence of a philosopher may be found in forms of speech, distinctions, and information anonymously imbedded in a later civilization as truly as in the explicitly labelled doctrines which a given age attributes to him,"[1] i.e., once the distinctions, terms, and guiding principles taken over from a philosophy are transmuted by familiar use into the accustomed materials of a culture and tradition, a philosophy can endure and provide a satisfactory perspective for millions who, as likely as not, have never heard the name of the authors of the perspective.

Secondly, a philosophy, to be such, must endure not only statistically, but also critically. If it is not to be a mere part of the cultural pre-suppositions of an age, indistinguishable from superstition and naive belief, a philosophy must be defensible before the bar of reason. To qualify as a reasoned set of beliefs, the various answers afforded by a philosophy to the various "big" questions must be mutually consistent. To the extent that what a philosopher says in psychology, for example, undermines what he says in ethics, the quality of his philosophy will be impaired.

[1] Richard McKeon, "Introduction" to *The Basic Works of Aristotle*, ed. by Richard McKeon (New York: Random House, 1941), p. xi.

Finally, over and above freedom from contradiction, the main lines of a philosophy must exhibit integration before they can dominate human thought. If a philosopher's teaching reflects no more than the perspective of one society, the philosopher's teaching will become irrelevant when that society changes in an important way. Conversely, in the measure that any philosophy contains a rationalization of the social conditions that exist at any given time, it becomes obsolete as the conditions of social life undergo transformation. Thus, on the one hand, a philosophy is adequate just to the degree that its principles are capable of coming to grips with all of the implications of all of the big questions; while, on the other hand, since, as the economic basis of life shifts, the big questions are posed with a differing emphasis and insistence in different ages, it is impossible for any philosopher ever to formulate once and for all a world-view satisfactory on all counts in all periods. But the essential point contrasting this criterion of integration against the criterion of freedom from contradiction rests on the difference between a philosophy which treats problems in isolation, be it a logical or socio-cultural isolation (an eclectic philosophy), and a philosophy in which the underlying principles or organizing parts are each one already virtually the whole (an organic or dialectical philosophy).

Using these criteria, Jones finds the architectonic for any study of Classical Antiquity blueprinted in advance:

Now, even the most cursory survey of the history of Western thought shows two—and only two—periods in which a really great philosophy, in the sense in which we have defined that term, was developed. These periods were the fourth century B.C., when Plato and Aristotle worked out views which on the whole satisfied the classical world, and the thirteenth century A.D., when St. Thomas performed the same function for medieval man. . . . Compared with such enduring syntheses as these, the modern mind has produced as yet only a variety of tentative solutions. . . . One reason for our failure is the complexity of modern life. It was literally possible for a man like Aristotle to take all knowledge for his province. When Bacon made this claim in the seventeenth century, it already seemed bombastic. Today it would be fantastic. . . . Thus, what an Aristotle or a Thomas could achieve in the way of a real synthesis of all knowledge is becoming increasingly difficult, if only because of the sheer quantity of knowledge.[2]

[2] W. T. Jones, *A History of Western Philosophy* (New York: Harcourt, Brace & World, 1952), "Introduction," pp. xi-xii. Jones goes on to say in a later passage: "With Kant [1724-1804] we come, in fact, to another major synthetical effort, comparable, at least in scope, to the Platonic-Aristotelian and Thomistic syntheses. Though it failed to establish itself, as they did in their time, it nevertheless has had its own form of endurance" (p. 813). Some of the reasons for my own negative answer to the question of whether Kant represents a third alternative explanatory mode in addition to what will be called in this essay the Platonic and Aristotelian modes of explanation are indicated in fns. 55, 119, 131, 132, and 166 below. Cf. also John N. Deely, "The Philosophical Dimensions of the Origin of Species," Part II, Sec. VII, *The Thomist*, XXXIII (April, 1969), pp. 290-304, which contains a discussion of what may be termed, after Democritus (c. 460 B.C.), the Democritean, mechanistic, or atomistic explanatory mode. The same matter is discussed in the second reading of Section IV of this volume, B. Ashley's "Change and Process."

It is recognition of this last fact, commonly referred to today as the "knowledge explosion," that makes the philosopher's pedagogic task seem at first glance practically insuperable. Even granting that an appreciation of Classical Antiquity depends principally on a grasp of the main systematic lines of just three thinkers, and supposing further that an appreciation of these lines would lead to a greater understanding of Darwin's world (which has yet to be shown), from what point of view or within what perspective could one adequately delineate that understanding within the limits of a single essay?

If it is true that the content of mankind's knowledge is "so extensive that it mocks encyclopedias and overflows libraries," "so difficult that a man does well devoting his life to mastering some part of it," and "so incomplete and inadequate that it is subject to endless additions and repeated revisions," does it not follow that any cherishing of the "wisdom" of the past is hopeless antiquarianism, and that any attempt to relate such wisdom to the present is not merely arduous, but impossible?

It is here that Lonergan's approach to the problem of human understanding suggests a solution to our difficulty. As a matter of fact, there is not only the experimental level of knowing, wherein knowing is specified according to the various sciences of the phenomena of nature, and within these sciences endlessly diversified by the discovery of new things and new aspects of things; there is also another level of knowing wherein knowledge itself is universalized and unified, the sphere of a critique of knowledge or *critica* in the most proper sense, a pure reflexion upon the knowledge of things "outside" the mind (and in that sense a cognitive activity secondary by its very nature as well as in time), but a sphere or level in which it becomes possible to show "the organic diversity and essential compatibility of those zones of knowledge through which the mind passes in its great movement in search of being, to which each one of us can contribute only tiny fragments, and that at the risk of misunderstanding the activity of comrades devoted to other enterprises equally fragmentary, the total unity of which, however, reconciles in the mind of the philosopher, almost in spite of themselves, brothers-in-arms who knew not one another." [3]

To assess the post-Darwinian era in the light of the experimental sciences of Classical Antiquity (a very dim light to work by) would indeed be a hopeless and dubious task. But if we take as our primary concern not the *known* of the respective eras but the *knowing itself* exercised throughout them, and if we deal with this activity of knowing, so far as it is immanent in the Darwinian revolution, against the main epistemological lines traced within the systems of Plato, Aristotle, and Aquinas, then the synthetic

---

[3] Jacques Maritain, *The Degrees of Knowledge,* trans. from the 4th French edition under the general supervision of Gerald B. Phelan (New York: Scribner's, 1959), "Preface," p. xi.

obligation imposed on philosophy when it turns its attention to an exposition of the idea of evolution no longer seems impossible to discharge—even within the limits of a general essay.

The known is extensive, but the knowing is a recurrent structure that can be investigated sufficiently in a series of strategically chosen instances. The known is difficult to master, but in our day competent specialists have laboured to select for serious readers and to present to them in an adequate fashion the basic components [the typically distinctive noetic features, let us say] of the various departments of knowledge. Finally, the known is incomplete and subject to revision, but our concern is the knower that will be the source of the future additions and revisions.[4]

Thus, while an account of knowing cannot disregard the content of knowledge ("the task of critique cannot for one single instant dispense with the knowledge of reality without having recourse to an illusory autophagic process"[5]), to provide a discriminant or determinant of cognitive acts, that content need be treated only in a "schematic or incomplete fashion."[6] And by grounding ourselves in the principal reasons and modalities of the movement of the mind in quest of truth, and of those phases through which it passes as, step by step, starting with sense experience, it enlarges, deepens, and transforms its own life, we find ourselves at about the one point of departure from which it is reasonable within the limits of a general essay to survey the landscape of Darwin's world from the summit of Classical Antiquity: "Thoroughly understand what it is to understand, and not only will you understand the broad lines of all there is to be understood but also will you possess a fixed base, an invariant pattern, opening upon all further developments of understanding."[7] "From this point of view it may also be said that the work which metaphysics is called upon to do today is to put an end to that kind of incompatibility of temper which the humanism of the classical age had erected between science and wisdom."[8]

This brings me to my final thematic observation. While agreeing with Jones that "we today are still in an era of experiment and preparation like those eras which preceded the Greek and medieval synthesis, and [that] it would be optimistic to expect any sudden condensation of our modern diversities into a satisfactory view,"[9] still, by assessing the passage

---

[4] Bernard J. F. Lonergan, *Insight: A Study of Human Understanding* (New York: The Philosophical Library, 1965), "Introduction," p. xviii. Cf. Aristotle's *Treatise on the Parts of Animals (De partibus animalium)*, Bk. I, ch. 1, 639a1–15.
[5] J. Maritain, *The Degrees of Knowledge*, p. 75.
[6] B. Lonergan, *Insight*, p. xvii.
[7] *Ibid.*, p. xxviii and p. 748.
[8] J. Maritain, *The Degrees of Knowledge*, p. xi. See Maritain, "The Conflict of Methods at the End of the Middle Ages," *The Thomist*, III (October, 1941), esp. Sec. V, pp. 536–538.
[9] W. T. Jones, *A History of Western Philosophy*, p. xii.

from Classical Antiquity to Darwin's world from the standpoint of the logic of the epistemological types of rational knowledge—which is essentially the standpoint of modern philosophy, i.e., of philosophical inquiry since the Renaissance, preoccupied as it is with questions concerning the nature of knowledge and its limitations—we may hope to contribute in some measure to an understanding both as to why "such positive concepts as have survived modern philosophy's own critical barrage are physico-mathematical in character," [10] and as to why "these have yet proved too narrow for a world-view of the kind Aristotle or St. Thomas attained." [11]

The pages that follow therefore cannot be read as an attempt to provide a handbook of answers, but only as an attempt to extend an invitation for personal reflection on the import of evolutionary science as regards the overall outlook on life and knowledge, and to provide a guide for further research.

In view of the scope and importance of the issues involved, the reader who wishes to pursue the points at issue further will find in the footnotes occasional illustration and development of key conceptual points likely to be misunderstood or open to dispute, together with what I hope are adequate references to the differing views on the key issues; but it has been impossible within the limits an essay must respect to indicate all the sources that have been necessary to document and establish the thesis that follows. I must rely on the reader's own general background in hoping that the restricted documentation finally incorporated within the notes to this essay shall prove sufficiently extensive.

These last remarks, of course, apply principally to interested readers themselves actively engaged in research. But it has also, and principally, been my intention and hope to write an essay sufficiently straightforward and clear that the ordinary undergraduate student with an interest in his world might come by reading it to sense the movement of history and the point at which today we stand in the cultural struggle toward a balanced differentiation of the inquiries restlessly pursued by the human mind.

---

[10] *Ibid.*

[11] *Ibid.* Jones goes on to observe summarily that "the outstanding fact about modern philosophy . . . is its inability to achieve anything remotely like a satisfactory synthesis of the historic past with the contemporary world-view, which is largely based on the findings of modern science" (pp. xii–xiii). According to Mortimer J. Adler (in *What Man Has Made of Man* [New York: Ungar, 1957], p. 242), the only contemporary exception to this generalization is Jacques Maritain. Be this as it may, one contribution I hope to have made in the present essay is a disengagement of some of Maritain's key insights into the history and nature of modern science from the expressly polemical and apologetic contexts in which he inclines to set them, often erecting thus—as even Adler openly admitted (in *St. Thomas and the Gentiles* [Milwaukee: Marquette University Press, 1938], p. 15)—an unfortunate barrier to the sharing of his insight beyond "the narrow circle of those who share with him, initially, a common ground." In this same respect, I hope to have contributed a clarification of the true sense of the notion of "empiriological" science—see Section III-A below of this present essay, and Secs. III and VII of my assessment of "The Philosophical Dimensions of the Origin of Species," *The Thomist*, XXXIII (January and April, 1969), pp. 93–102 and 290–304, respectively.

Let us proceed then to our consideration of the importance of evolution for the account of what science itself is, by indicating in a summary way the difficulty and scope of the issues involved.

To some thinkers it might seem superfluous to call attention to the very special nature of the problem of origins and development of natural entities. Yet if the term "evolution" is not reduced to the mere concept of observable change, it must be so used as to involve a problem more difficult and obscure than history itself. The origins of nebulae, physical and chemical elements, life, the multitudes of organic species including man, mind, and human culture, are essentially questions of *prehistory*. They are events and processes unwitnessed and unrecorded, events which took place millions and billions of years ago. This very obvious dimension of our problem imposes a less obvious and sometimes completely overlooked condition of its solution. The problem of prehistory places serious limitations on the expectancy of a completely satisfactory solution; and yet, prehistory has turned out to be the ultimate mystery of nature so far as human existence is concerned, for it has become increasingly clear that human history cannot be properly understood, let alone interpreted, unless it is regarded as in some way an *extension* of natural history, i.e., of prehistory.

One should be clear on the formidable difficulties this entails in the effort to achieve a view of the development of mankind, and particularly of the growth of civilization, which is to serve as the adequate (as opposed to ideological) basis for an integral humanism. If many problems of historical science, which has the great advantage of some written documentation, remain inaccessible to the human mind and admit of only highly conjectural resolution, how much more the problems of cosmological origins and organic prehistory!

To speak exactly, evolution does not pose a *problem* for contemporary thought so much as it presents itself as the necessary *background* concept for issues crucial in all areas of humanistic thought. In its essential import, as we shall see, evolution is not so much a philosophical concept as it is a summary expression of the realization, gradually secured within human experience, that nothing in the universe seems exempt from radical transformation. That is the sense and justification of C. H. Waddington's comment that "however they may be thought about, the facts of evolution and development simply cannot be omitted from any discussion of the human condition which hopes to carry conviction at the present time."[12] Thus evolution is less a particular theory (neo-Darwinian biology, for example) than a fundamental and absolutely primary datum, in the very

[12] C. H. Waddington, *The Ethical Animal* (New York: Atheneum, 1961), p. 74.

precise sense binding on philosopher and scientist alike, of exhibiting itself as a physical (i.e., observable) record of past occurrences for which no logical construction can be substituted and upon which all the logical constructions in our understanding of nature finally rest.

Consequently, from the point of view of the philosopher, evolution does not constitute a philosophical study in its own right. There is no such thing as "evolution in general," no study of "evolution as such," of "evolution inasmuch as it is evolution." There are only the changes which take place in the interaction of the diverse kinds of cosmic entities, and the relations of interdependence which obtain between the various levels of their cosmic interaction. There is nothing in the known evidence to warrant the assumption that evolution is the expression or product of a single, harmonious plan or law, rather than of a multitude of lines of causality in a universe full of chance and accident.[13] This may seem to be an obvious point, but obvious or not, its importance cannot be overstressed. For the fact of evolution too easily becomes, in the mind which has no eye for essential distinctions, an ideology of evolutionism.

All is not change, but all is changing: the whole of nature exists in and through process. That is all that the fact of evolution testifies. It is why, for the philosopher, evolution, without necessarily becoming itself the exclusive object of his investigations, nonetheless provides a kind of proving ground or "authenticity test" for diverse lines of philosophical analysis, sometimes pertaining to metaphysics, sometimes to politics or society, sometimes to the human existent as such, sometimes to moral questions, sometimes to religion, sometimes to culture, etc. The world into which Darwin led us, therefore, is not an easy world to understand, nor is it a particularly comfortable or comforting sort of world. But it is a real one; it is the world that makes us and, wittingly or not, the world in which we live. And just because, as we shall see, the evolutionary process which bears us along has an inexorable automatism about it, such that it may be influenced and guided but cannot be suppressed or halted, "it is a characteristic of this world to which Darwin opened the door that unless *most* of us do enter it and live maturely and rationally in it, the future of mankind is dim, indeed—if there is any future."[14]

To this end of making possible a rationally mature life within the "world into which Darwin led us," I hope to make some lasting contribution by the following analysis of the type or kind of explanatory structure

---

[13] Even Teilhard de Chardin, that most convinced "directionalist," acknowledged the limits of the evidence on this score: see, *inter alia*, *The Phenomenon of Man* (New York: Harper, 1959), pp. 231–233 and 284. Cf. also Theodosius Dobzhansky, *Mankind Evolving* (New Haven: Yale, 1962), p. 17; Charles de Koninck, "Réflexions sur le problème de l'indeterminisme," Part I, *Revue Thomiste*, XLII, No. 2 (juillet–septembre, 1937), p. 232ff., esp. pp. 234–5; and J. N. Deely, "The Philosophical Dimensions of the Origin of Species," Part II, pp. 298–318.
[14] George Gaylord Simpson, *This View of Life* (New York: Harcourt, Brace & World, 1964), p. 24.

evolutionary thinking represents, and of the essential observational difference between the world-view of classical antiquity and the evolutionary world-view.

I. THE EVOLUTIONARY CONCEPT

To so much as glimpse the developmental transformations that have gone into all that we are and see around us, and which are subsumed under that ultimate abstraction and dominant category of contemporary culture, "evolution," is no easy task. It requires that one not only familiarize himself with the characteristic evidences of the astronomical, geological, and biological sciences, especially paleontology or the study of the fossil forms. It demands also that one allow his imagination to quicken in order to interrelate these data after the manner of a motion picture. An actual film run at high speed can be used to reveal processes so slow as to escape ordinary observation, such as the growth of a plant, the transformations of an animal embryo. Similarly, our imaginal film of evolution, by altering our time scale, can startle us into awareness of the extent and rhythm of processes subtending the cycle of individual lives and deaths which alone imposes itself upon our everyday consciousness. Restricting ourselves for the moment to the record of living things, if we run our "film" of evolution at the rate to which our senses are accustomed, we see only the processes of individual development and destruction to which we are accustomed.

With a hundredfold speeding up, individual lives become merged in the formation and transformation of species. With our film speeded up perhaps ten thousand times, single species disappear, and group radiations are revealed. We see an original type, seized by a ferment of activity, splitting up and transforming itself in many strange ways, but all the transformations eventually slowing down and stabilizing in specialized immobility. Only in the longest perspective, with a hundred-thousand-fold speedup, do over-all processes of evolution become visible—the replacement of old types by new, the emergence and gradual liberation of mind, the narrow and winding stairway of progress, and the steady advance of life up its steps of novelty.[15]

That is one of the first and most fundamental lessons one learns from acquaintance with evolutionary studies: the simple insufficiency of a conceptual horizon restricted to the span of written history and to the data of common sense observation, for posing the question of man's place and rôle in nature. For within such an horizon it is a basically *unchanging* world which seems to confront our gaze; whereas we know it

[15] Sir Julian Huxley, *Evolution in Action* (New York: Mentor, 1953), p. 28.

fact that ours is a basically *changing* world, a world in which nothing, no individual and no sector or region, is free from radical transformations. To keep to the terms of our metaphor, "with respect to the evolution which has actually taken place in the history of the earth, an observer of only the now-living animals and plants is still in a position of judging a long movie film by only the last picture frame."[16]

This lesson comes across still more dramatically when one considers not only the history of life as revealed by the fossils—be they potsherds or bones—but also the vast forces at work in the establishment of an environment wherein life became possible in the first place, and the problem of the initial passage from the inorganic to the living.

Let us imagine, for a moment, that the 4,700 million odd years of our planet's past are represented by the distance of one hundred miles, and that we are walking from the time of the earth's origin towards the present. On the first half of our journey we encounter no living thing. After traversing a full eighty-eight miles, simplest invertebrates, resembling worms and jellyfish, begin to appear. At ninety-three miles, certain organisms—those, namely, pushed aside in the swamps and along the tideflats, the failures of the sea—begin the invasion of the land masses. Our own ancestral group, the mammals, does not appear until we have a scant two of our hundred miles left to cover. The whole of man's physical evolution since the beginning of the Pleistocene epoch will take place over the last sixty yards of our journey, and the span of written history with all its panoply of civilization will be traversed in the last half of our last stride!

Yet, our planet itself must likewise be seen in a context of process, immensely slow, it is true, even by biological standards, but equally irreversible and inexorable. So too our sun and galaxy. All reality, in fact, is in evolution, "definable in general terms as a one-way, irreversible process in time, which during its course generates novelty, diversity, and higher levels of organization. It operates in all sectors of the phenomenal universe but has been most fully described and analyzed in the biological sector."[17] It is a question first of all of learning to accept the brevity, the

---

[16] Theodosius Dobzhansky, *Evolution, Genetics, and Man* (New York: Science Editions, 1963), p. 284.

[17] Sol Tax and Charles Callendar, eds., *Issues in Evolution* (Chicago: University of Chicago Press, 1960), p. 107 (Vol. III of the University of Chicago Centennial, *Evolution After Darwin*). See John N. Deely's note on evolution as an interaction transcendental of natural philosophy, in "The Philosophical Dimensions of the Origin of Species," Part I, *The Thomist*, XXXIII (January, 1969), p. 129 fn. 108. Thus Raymond J. Nogar, in *The Wisdom of Evolution* (New York: Doubleday, 1963), p. 279, remarks: "The fundamental philosophical question which the science of Darwin, Freud, Einstein, Planck and Heisenberg raises is: *what is the relationship, in reality, between the change and stability of nature?* Evolution is an extremely generalized historical process of nature, and today natural history is grossly incomplete without an account of evolutionary history. Yet there remains the predictable stability of nature which is subjected to the structure of natural laws. Science formulates the laws of stability as well as describes the

relativity, and the dependency of human existence upon a peculiar set of segmental laws of time and space; and secondly, of understanding these laws as but special cases of more general laws giving structure to space and time.

Thus, the evolutionary reality of itself divides before the mind into three main sectors or regions in which the general process is operative in quite different ways. Following Huxley,

We may call these three phases the inorganic or, if you like, cosmological; the organic or biological; and the human or psycho-social. The three sectors of the universal process differ radically in their extent, both in space and time, in the methods and mechanisms by which their self-transformations operate, in their rates of change, in the results which they produce, and in the levels of organization which they attain. They also differ in their time relations. The second phase is only possible on the basis of the first, the third on the basis of the second; so that, although all three are in operation today, their origins succeeded each other in time. There was a critical point to be surmounted before the second could arise out of the first, or the third out of the second.[18]

In the inorganic or cosmic sector the tempo is much slower and the mode of change much different from the tempo and mode of change in the organic or biological sector of the universe. All comes about through physico-chemical exchanges, and the results must be measured by the lifespan of stars. (See Table I, p. 14.)

In the organic or biological sector the tempo is faster, measurable now by the appearance and extinction of whole new life-forms, by millions rather than billions of years. Here change is effected over the course of generations by modifications in biological heredity. (See Table II, pp. 16–17.)

With the appearance of man, still another dominant mode of change emerges: social heredity, the cumulative transmission of conscious experience. Now the tempo of change may be measured by the reorganization of the thought-patterns of human groups around new insights and values. This defines the peculiarly human, cultural or psycho-social sector of evolutionary development. (See Table III, pp. 18–19.)

Thus the anthropological problem is posed, and the idea of evolution emerges as the necessary background concept for any attempt in our day to construct an integral philosophical anthropology. For it is sufficiently clear—particularly from within the tradition of Christian thought which channels through St. Thomas Aquinas—that, as Simpson puts it, "the meaning of human life and the destiny of man cannot be separable from

---

histories of change. What is the relationship between these two great efforts of nature?" Cf. also pp. 289–290, 298.

[18] Huxley, *Evolution in Action*, p. 10.

TABLE 1: THE COSMOLOGICAL SCALE OF SPACE AND TIME

A. The Scale of Space

The Near Stars:	about 10 light years distant
Diameter of Our Galaxy (the "Milky Way" Star System):	100,000 light years plus
Diameter of Average Galaxy:	10,000 to 500,000 light years plus
Distance Between Galaxies:	1 million light years upwards (no real mean, since distance within as well as between galactic clusters would be involved)
Mass of Our Galaxy:	equal to 100 billion suns
Mass of Average Galaxy:	from 30 billion to 3 trillion suns
Number of Currently Photographable Galaxies:	1 billion (perhaps more than twice this number detectable with radio telescopes)
Number of Stars in Known Universe:	100 quintillions (10^{20})
Statistical Expectancy of Stars Supporting Planetary Systems:	1 trillion (10^{12})
Statistical Likelihood of Planetary Systems Suitable For Evolution of Organic Life:	100 million (10^8: minimal conservative expectation: some would multiply this number by a thousand, some by a million)
Extent of Visible Universe:	10 billion light years plus (and this approximates to the estimated depth of the universe in time)

B. The Scale of Time

Age of Universe:	about 14 billion years (13 to 20 billion)
Age of the Galaxies:	6.5 billion years
Age of Our Sun:	5 billion years
Age of Earth:	4.7 billion years
Age of the Continents (in cooled state):	3.5 billion years
Age of Life:	2 to 2.5 billion years
Age of Oxidizing Atmosphere:	1 billion years
Age of First Clear Fossils:	600 million years
Age of Land Animals and Plants:	405 million years
Age of Dinosaurs:	230 million years
Age of Mammals:	63 million years
Age of Man:	between 1 and 2 million years
Recorded Human History:	5 to 6 thousand years
Time of Abraham:	circa 1800 B.C.
Continued Maintenance of Environmental Conditions Capable of Sustaining Life on Earth:	another 2.5 billion years (thus we of today stand at life's high noon on earth)
Time Until Earth is Destroyed by Sun:	another 4.5 to 4.7 billion years (thus we also stand midway in the history of planet earth)
Life Expectancy of Our Sun:	another 5 billion years

Man has learned to accept the brevity, the relativity, and the dependency of his own existence upon a peculiar set of segmental features of time and space, which are only special cases of more general space-time laws that he seeks to apprehend. Natural history has achieved a perspective in time of roughly 14 billion years, and a perspective in space approaching that same depth or extent. Between the advent of *animal rationale* and the start of the Biblical record, the history of man is tenuous, uncertain, and conjectural on the side of our knowledge, even as on the side of its actual working out it was precarious in the extreme. *What was man doing through all that time?* Struggling to become human, a state of accomplishment which still lies in the not near future. Yet short as it is in the drama of cosmogenesis (if the age of the earth were equalled to one calendar year, the presence of man on earth would account for one half-hour, and recorded human history for one and a half minutes), a History of Mankind allotting one page to every hundred year period would comprise a completed work of more than twenty volumes of a thousand pages each.

the meaning and destiny of life in general."[19] No less can the meaning of life in general be separated from the meaning and mystery of matter to which, in fact, life reduces as a peculiarly organized modality.[20] In final analysis the nature of man must link up structurally with that of the universe at large, even as the phenomenon of man at once presupposes (in the naturalist's sense) and yet (in the phenomenologist's sense) constitutes the phenomenon of world.

A. The discovery of evolution

The discovery of evolution began with the great voyages of exploration which put out from Europe in the late eighteenth and nineteenth centuries. Less auspiciously and more directly, it began with a few eccentrics or "hobbyists" associated for one or another reason with the great explorers, but themselves less concerned with the extension of military and economic empires than with "idly" collecting the diversity and debris of life. What began so immersed in contingency and dependent on chance has culminated, toward the middle of our own century (and indeed, for penetrating minds, considerably earlier), in a world-view fundamentally different from that of classical or medieval times, restricted, as they were, to the conceptual horizon of common experience and written documents. And with this new vision appeared what has come to define the conceptual horizon for all efforts at synthetic thought in our time: the problematic of evolution. "To a philosopher the cosmic, biological, and cultural evolutions are integral parts of the grand drama of creation," Dobzhansky

---

[19] George Gaylord Simpson, *The Meaning of Evolution* (New Haven: Yale, 1949), p. 9.

[20] In the Aristotelian tradition, this is clear in terms of the essential analysis of life both in terms of its primary formal effect (as the principle by which anything lives, feels, moves about in its environment, or understands primarily) and of its proper subject (as the fundamental actuality of a properly organized physical unit, where "properly organized" means simply organized in such a way as to enclose the capacity for self-augmentation and reduplication, the activities most proper to life). See Aristotle's *De anima* ("Treatise on the soul"), Bk. II, chs. 1 and 2.

Moreover, this same point is only a little less clear, indeed, in some respects, clearer, in the more generally empiriological formulations of modern authors. E.g., cf. Simpson, *The Meaning of Evolution*, pp. 126 and 291; G. G. Simpson, C. S. Pittendrigh, and L. H. Tiffany, *Life* (New York: Harcourt, 1957), pp. 16–17 and 35; T. Dobzhansky, *Evolution, Genetics, and Man*, pp. 19ff.; Pierre Teilhard de Chardin, *The Phenomenon of Man* (New York: Harper, 1959), p. 300. (On the sense of the phrase "empiriological formulations," see the discussion of dialectical or hypothetical facts in Division III of this essay, "World-View as *Logos* and as *Mythos*.")

On the relation of these two lines of analysis to each other and their concurrence on the problem of the origin of terrestrial life so far as it is a question of principle, see John N. Deely, "The Philosophical Dimensions of the Origin of Species," Part II, *The Thomist*, XXXIII (April, 1969), pp. 318–326, esp. pp. 323–324. On the question of the relation of the degree of organic heterogeneity ("complexity," if you like) to the level of vital activity exhibited by an organism and to the emergence of conscious organisms, see *ibid.*, pp. 301–331, esp. pp. 318–319 and fn. 264.

TABLE II: GEOLOGIC TIMETABLE

Era	Period	Epoch	Duration in Millions of Years	Time From Beginning of Period to Present (Millions of Years)	Geologic Conditions	Plant Life	Animal Life
Cenozoic (Age of Mammals)	Quaternary	Recent	0.025	0.025	End of last ice age; climate warmer	Decline of woody plants; rise of herbaceous ones	Age of man
		Pleistocene ("most recent")	1	1	Repeated glaciation; 4 ice ages	Great extinction of species	Extinction of great mammals; first human social life
	Tertiary	Pliocene ("more recent")	11	12	Continued rise of mountains of western North America; volcanic activity	Decline of forests; spread of grasslands; flowering plants, monocotyledons developed	Man evolving; elephants, horses, camels almost like modern species
		Miocene ("less recent")	16	28	Sierra and Cascade mountains formed; volcanic activity in northwest U.S.; climate cooler		Mammals at height of evolution; first manlike apes
		Oligocene ("few of the recent")	11	39	Lands lower; climate warmer	Maximum spread of forests; rise of monocotyledons, flowering plants	Archaic mammals extinct; rise of anthropoids; forerunners of most living genera of mammals
		Eocene ("dawn of the recent")	19	58	Mountains eroded; no continental seas; climate warmer		Placental mammals diversified and specialized; hoofed mammals and carnivores established
		Paleocene	17	75			Spread of archaic mammal
Rocky Mountain Revolution (Little Destruction of Fossils)							
Mesozoic (Age of Reptiles)	Cretaceous		60	135	Andes, Alps, Himalayas, Rockies formed late; earlier, inland seas and swamps; chalk, shale deposited	First monocotyledons; first oak and maple forests; gymnosperms declined	Dinosaurs reached peak; became extinct; toothed birds became extinct; first modern birds; archaic mammals common
	Jurassic		30	165	Continents fairly high; shallow seas over some of Europe and western U.S.	Increase of dicotyledons; cycads and conifers common	First toothed birds; dinosaurs larger and specialized; insectivorous marsupials

Era	Period			Physical conditions	Plant life	Animal life
	Triassic	40	205	Continents exposed; widespread desert conditions; many land deposits	Gymnosperms dominant, declining toward end; extinction of seed ferns	First dinosaurs, pterosaurs and egg-laying mammals; extinction of primitive amphibians
	Appalachian Revolution (Some Loss of Fossils)					
Paleozoic (Age of Ancient Life)	Permian	25	230	Continents rose; Appalachians formed; increasing glaciation and aridity	Decline of lycopods and horsetails	Many ancient animals died out; mammal-like reptiles, modern insects arose
	Pennsylvanian (Carboniferous)	25	255	Lands at first low; great coal swamps	Great forests of seed ferns and gymnosperms	First reptiles; insects common; spread of ancient amphibians
	Mississippian (Carboniferous)	25	280	Climate warm and humid at first, cooler later as land rose	Lycopods and horsetails dominant; gymnosperms increasingly widespread	Sea lilies at height; spread of ancient sharks
	Devonian	45	325	Smaller inland seas; land higher, more arid; glaciation	First forests; land plants well established; first gymnosperms	First amphibians; lung-fishes, sharks abundant
	Silurian	35	360	Extensive continental seas; lowlands increasingly arid as land rose	First definite evidence of land plants; algae dominant	Marine arachnids dominant; first (wingless) insects; rise of fishes
	Ordovician	65	425	Great submergence of land; warm climates even in Arctic	Land plants probably first appeared; marine algae abundant	First fishes, probably freshwater; corals, trilobites abundant; diversified molluscs
	Cambrian	80	505	Lands low, climate mild; earliest rocks with abundant fossils	Marine algae	Trilobites, brachiopods dominant; most modern phyla established
	Second Great Revolution (Considerable Loss of Fossils)					
Proterozoic		1500	2000	Great sedimentation; volcanic activity later; extensive erosion, repeated glaciations	Primitive aquatic plants—algae, fungi	Various marine protozoa; towards end, molluscs, worms, other marine invertebrates
	First Great Revolution (Considerable Loss of Fossils)					
Archeozoic		???	???	Great volcanic activity; some sedimentary deposition; extensive erosion	No recognizable fossils; indirect evidence of living things from deposits of organic material in rock	

Reproduced from Edward O. Dodson, *Evolution: Process and Product*, by permission of Reinhold Book Corporation, a subsidiary of Chapman-Reinhold, Inc., New York, 1960. Dodson's chart was itself adapted from Villee, *Biology, The Human Approach*, 3rd ed., W. B. Saunders Co., 1957.

TABLE III: CHART OF QUATERNARY TIME

Time Scale

UPPER PLEISTOCENE
- Holocene (10,000→)
- Transition Phase (15,000–10,000)
- Fourth Glacial Phase
- 80,000
- Third Interglacial Phase

MIDDLE PLEISTOCENE
- 190,000
- Third Glacial Phase
- 240,000
- Second Interglacial Phase
- 440,000

LOWER PLEISTOCENE, OR VILLAFRANCHIAN
- Second Glacial Phase
- 480,000
- First Interglacial Phase
- 550,000
- First Glacial Phase
- 600,000
- Pre-glacial Phase
- 1,000,000

Geological Divisions			Years Ago	Main Climatic Phases
	HOLO-CENE OR RECENT EPOCH		11,000	Modern Climates
QUATERNARY PERIOD — PLEISTOCENE EPOCH	UPPER PLEISTOCENE		15,000	Transition Phase; gradual retreat of the ice-sheets
			80,000	Fourth, or Würm, Glacial Phase (Wisconsin Phase in North America)
	MIDDLE PLEISTOCENE		190,000	Third Interglacial Phase (Sangamon Phase in North America)
			240,000	Third, or Riss, Glacial Phase (Illinoian Phase in North America)
			440,000	Second, or Great, Interglacial Phase (Yarmouth Phase in North America)
	LOWER PLEISTOCENE OR VILLAFRANCHIAN		480,000	Second, or Mindel, Glacial Phase (Kansan Phase in North America)
			550,000	First Interglacial Phase (Aftonian Phase in North America)
			600,000	First, or Gunz, Glacial Phase (Nebraskan Phase in North America)
			1,000,000 to 2,000,000	Pre-glacial Phase; transition from Pliocene Epoch

This chart has been devised to give a general picture of the geological, climatic, and archaeological phases of the Quaternary Period, with their associated fossils and human cultures. Most of the dates are still speculative and recent research suggests a rather greater antiquity for the Australopithecines than is shown here. The time scale on the left gives an approximate indication of the duration of the various phases; the remainder of the chart is diagrammatic only and is not drawn to scale.

(PERIOD OF MAN'S RISE TO DOMINANCE)

Characteristic European Fauna	Archaeological Divisions	Main Cultures	Main Human and Subhuman Types	Years Ago
Existing species First domestic animals	Modern urban society AGE OF METALS NEOLITHIC	Metal cultures and civilisation. Advanced stone cultures	Modern races of *Homo sapiens*	
Steppe fauna REINDEER AGE (tundra fauna)	MESOLITHIC	Campignian Erteböllian Maglemosian Tardenoisian Azilian		10,000
AGE OF WOOLLY MAMMOTH Woolly mammoth (*Mammuthus primigenius*) Woolly rhinoceros (*Rhinoceros tichorhinus*) Cave bear (*Ursus spelaeus*) Cave lion (*Panthera leo spelaea*)	UPPER PALAEOLITHIC	Magdalenian Solutrean Gravettian Aurignacian Chatelperronian	Fossil races of *Homo sapiens*: Cro-Magnon Grimaldi and Chancelade	40,000
AGE OF THE ANCIENT ELEPHANT	MIDDLE PALAEOLITHIC	Mousterian	Neanderthal Man (*Homo neanderthalensis*)	160,000
Ancient elephant (*Palaeoloxodon antiquus*) *Mammuthus trogontherii* Merck's rhinoceros (*Rhinoceros mercki*) *Hippopotamus major* *Note:*—During this period the temperate and subtropical climatic belts with their associated faunas moved frequently north and south with the advance or retreat of the ice-sheets. The above are four characteristic mammals of the time.	LOWER PALAEOLITHIC	Levalloisian Acheulian Clactonian Abbevillian Choukoutienian Oldowan	Heidelberg Man (*Homo heidelbergensis*) Java Man and Pekin Man (*Pithecanthropus*) Australopithecinae and *Telanthropus* ?*Zinjanthropus* (exact status and date of this fossil is still speculative)	600,000
AGE OF THE SOUTHERN ELEPHANT Southern elephant (*Elephas meridionalis*) Etruscan rhinoceros (*Rhinoceros etruscus*) Sabre-toothed cat (*Machairodus*)	?Origin of tool-making tradition		?Pre-Australo-pithecine sub-hominids (no known fossil evidence)	1,000,000

Based on Richard Carrington's "Chart of Quaternary Time" in *A Million Years of Man,* New York: The World Publishing Co., 1963, pp. 50–51.

muses, since "the human race with its social, intellectual, and artistic achievements, the world of living creatures, and inanimate nature all evolved gradually and by stages from very different antecedents."[21]

Thus, while the world was for centuries conceived principally in terms of *space*, today it must be conceived with primary reference to *time*, for its characterizing structural features are not self-identical in every region or at every period in the world's history. Bit by bit, the patient probing of the geographical zones and geological strata of our planet has disclosed that the world and its layer of life have not simply a record of change but a history of development, a history with an ontological content. The paleo-sciences reveal a retrospective, that is, a temporally receding, environmental sequence. They show us stratified cross-sections of the slow process through which the world had to build itself up from its simplest beginnings before it could support such life as we are familiar with.

From indiscernible beginnings in the sea, life began to multiply in number and kind, in a word, to *ramify*. Eventually, amphibious forms moved onto land, in their turn to ramify, preparing the way for entirely new metabolic organizations and so new dominant life-forms. The Age of Amphibia gave way to the Age of Reptiles, which would in turn yield to the Age of Mammals.

As the continental upthrust cooled the atmosphere, the great reptilian forms, lacking an internal temperature-regulating mechanism, became less and less able to maintain a minimal level of mental efficiency. At the same time, the emergence of the true flowering plants provided new and concentrated sources of energy that were quickly exploited by the warm-blooded mammals. Characterized by a high metabolic rate and the maintenance of a constant body temperature (homeostasis), the mammals originally consisted of squirrel-like little creatures living in trees and underbrush, wholly incapable of competing with the massive reptiles, and so confined to nocturnal and fringe environments of a reptilian world. But with the advent of the flowering plants and the cooling of the atmosphere, all this changed.

"Whirl is king," said Aristophanes, and never since life began was Whirl more truly king than eighty million years ago in the dawn of the Age of Mammals. It would come as a shock to those who believe firmly that the scroll of the future is fixed and the roads determined in advance, to observe the teetering balance of earth's history through the age of the Paleocene. The passing of the reptiles had left a hundred uninhabited life zones and a scrambling variety of newly radiating forms. Unheard-of species of giant ground birds threatened for a moment to dominate the earthly scene. Two separate orders of life contended at slightly different intervals for the pleasant grassland—for the seeds and the sleepy burrows in the sun.[22]

[21] Dobzhansky, *Evolution, Genetics, and Man*, pp. 5 and 1.
[22] Loren Eiseley, *The Immense Journey* (London: Gollancz, 1958), pp. 7–8.

The balance of life had been tipped, and the Age of Reptiles gave way to the Age of Mammals, and this new age in turn engendered the Age of Man. But it all took some two billion years. Built up by a multitude of interacting forces, including living things, the hierarchy of life step by step prepared the way for higher forms. And the type of environment defining the possibility of human beings extends far beyond the vision or experience of the things that live there. For all the extreme difficulty of getting information from which to draw by the methods of measuring prehistoric time reliable inferences, for all the dependency of the paleo-sciences upon interpolations, assumptions, extrapolations, and analogies, "the broad, overall picture of the succession of organic forms in space and time is too heavily documented by cross-checking and convergence of materials to be rejected by the objective observer."[23]

According to Bergson, "The evolutionist theory, so far as it has any importance for philosophy, requires no more. It consists above all in establishing relations of ideal kinship, and in maintaining that wherever there is the relation, of, so to speak, *logical* affiliation between forms, there is also a relation of chronological succession between the species in which these forms are materialized."[24]

B. The explanation of evolution

What Bergson is driving at is the fact that once an overall succession of organic forms in space and time had been established, once, that is to say, it was known that the now living life-forms had been preceded not merely by *other* forms but by specifically and typically *different* forms no longer existent, and these by still others, we have only two alternatives for explaining this succession. Either there has been a continuous development of life, or a discontinuous series of creations and extinctions. On either supposition, "we should still have to admit that it is successively, and not simultaneously, that the forms between which we find an ideal kinship have appeared." On the supposition of a discontinuous process, "evolution would simply have been transposed, made to pass from the visible to the

---

[23] Raymond Nogar, *The Wisdom of Evolution*, p. 63. See Theodosius Dobzhansky, *Genetics, and the Origin of Species* (3rd rev. ed.; New York: Columbia, 1951), p. 11, J. Franklin Ewing, "Precis on Evolution," *Thought*, XXV (March, 1950), pp. 59–60; W. LeGros Clark, "The Crucial Evidence For Human Evolution," *American Scientist*, XLVII (1959), pp. 299–300; E. C. Case, "The Dilemma of the Paleontologist," in *Contributions From the Museum of Paleontology*, IX, No. 5 (Ann Arbor: Michigan, 1951).

On the question of the aspect or element of necessity in evolutionary studies, i.e., on the question of their fundamentally authentic character as *science*, see Section IV of John N. Deely's article, "The Philosophical Dimensions of the Origin of Species," Part I, *The Thomist*, XXXIII (January, 1969), pp. 102–130, esp. pp. 112–129.

[24] Henri Bergson, *Creative Evolution*, authorized trans. by Arthur Mitchell (New York: Modern Library, 1941), pp. 29–30.

invisible," from the realm of the researchable to a forever unintelligible level. Such a gratuitous transposition, however, is not a rational option, either at the level of science, philosophy, or theology. This is an absolutely essential point.

The triumph of the theory of evolution as a concept, however ambiguous, metaphorical, or equivocal, is that it provides a means of synthesizing knowledge about the cosmos within a natural continuum of explanation. The order of nature cannot be described except in natural terms, the theory asserts; there is a natural bond connecting cosmic entities in their space-time continuum. As long as there is hope of joining the prehistories of cosmic species in a natural sequence by a natural explanation, cosmic problems remain in the province of natural science. No preternatural, miraculous, or special Divine Intrusion need be postulated until the possibility of these natural causal relationships be ruled out. This frame of mind is largely due to the achievement of evolutionary theory in underscoring the continuity of natural events in time and space and in insisting on searching for natural relationships among all natural events to make them intelligible in terms of natural causes. This is excellent natural science; this is the premise of realistic natural philosophy; this is axiomatic to the natural theologian.[25]

With a differing emphasis, Professor Georges Crespy illuminates this same fundamental issue. "Evolutionism is not only a scientific *theory*," he points out, "although it is principally this; it is also a *mentality*, an attitude of mind in facing the problems posed by understanding the phenomena of matter. If one attacks the 'gaps' in the theory of evolution —and since the theory of evolution is full of 'gaps' this is not at all difficult —then, in reality, one is attacking the evolutionist mentality, without saying so, and often even without knowing it."[26]

Yet the "evolutionist mentality" amounts to no more than a recognition of the intrinsic necessity for continual causal play throughout a natural developmental process. Organic evolution, understood simply and loosely as a natural process of change of successive generations, of new forms genetically and somatically related through space and time with old forms, does occur and has occurred for the total period of life on earth; and this evolution depends in its turn on those processes which gave form

[25] Raymond J. Nogar, "Evolution: Scientific and Philosophical Dimensions," in *Philosophy of Biology*, Vincent E. Smith, ed. (New York: St. John's University Press, 1962), pp. 54–55. Cf. St. Thomas Aquinas, *Summa contra gentiles*, I, ch. 42, n. 4; III, chs. 22, 76, 77, 83, 94 (available in five-volume English trans. from Image Books); *Summa theologica*, I, Q. 22, art. 3; Q. 103, art. 6. George Gaylord Simpson, *The Meaning of Evolution*, pp. 127–128. Ernst Mayr has a curt reminder for those who might incline to the expedient of projecting imaginal explanations from the basis of subjective preferences: "The complexities of biological causality do not justify embracing nonscientific ideologies. . . ." ("Cause and Effect in Biology," in *Science*, CXXXIV, November, 1961, p. 1506.)

[26] Public lecture delivered by Professor Georges Crespy of Montpellier, France, on "Evolution and Its Problems," under the sponsorship of the Chicago Theological Seminary, January 16, 1965 (p. 2 of mimeographed text). See Michael E. Ruse, "The Revolution in Biology," *Theoria*, XXXVI (1970), pp. 11, 14–15, 16–17.

to our planetary system within the galaxy. In this statement there are present all the dangers of extrapolation, all the limits of an incomplete fossil record, all the weaknesses of indirect evidence and inference. Yet one cannot deny this natural succession of species, genera, and classes of organisms without denying the whole fabric of the science of prehistory and the converging arguments adduced from every department of biology. Everything known about the process of development of organisms in genetics, embryology, morphology, taxonomy, natural history, ecology, cytology, physiology, biogeography, indeed, in every science of life, points to natural relationship and continuity of forms.

It is doubtless a lack of understanding on this point that lies behind so much of the confusion over key implications of evolutionary science in the realms of metaphysical and theological explanation. For example, it was with the just mentioned simple, absolutely fundamental and, understood rightly, incontestable philosophical principle in mind—the principle, namely, of the intrinsic necessity for continual causal play throughout a natural developmental process or series of processes—that the great theologian Sertillanges argued in 1945 (vainly, as it turned out) against the verbal opposition between "creationism" and "evolutionism," "as if these were two contrary notions." In fact, the idea of creation and the idea of evolution in their essential notes are mutually indifferent to one another, since "there is nothing to prevent us from seeing in evolution, instead of a substitute for creation, simply another perspective on the manner in which the creative fact (itself a polyvalent, metaphysical conception, that is, a conception denoting a truth not circumscribed by time or restricted to any particular empirical facet of reality) is bound up with the facts of nature."[27] For the idea of creation poses essentially neither a question of duration (whether the universe is eternal or had a beginning) nor a question of succession (whether the specific structures of the world are fixed or labile). It poses rather the question purely and simply of the total dependence of the real with respect to its existence, *"a question of dependence in being."*[28] "The Ancients," like some moderns, "generally understood creation as an arrangement starting from a primitive chaos, the causes of which were not sought, either because of the infantile state of their general metaphysics, or because the First Necessity was envisaged by them as enveloping at the same time God and the stuff presupposed for his action."[29]

---

[27] A.-D. Sertillanges, *L'Idée de création* (Paris: Aubier, 1945), p. 128. (Translations from this work are by Simone Poirier Deely and John N. Deely.)

[28] *Ibid.*, p. 6. Cf. Jacques Maritain, *The Peasant of the Garonne* (New York: Holt, 1968), fn. 26, p. 267.

[29] *Ibid.*, p. 5. In the understanding of the idea of creation evinced in the writings of Pierre Teilhard de Chardin, for example, we seem to have a reversion to the conception of the Ancients on both counts. In terms of the Ancients' conception, such of Teilhard's writings as "La luttre contre la multitude" (1917), "L'union créatrice" (1917), "Note sur les modes de

But all that is not God is a creature of God, and so we find ourselves upon the horns of a dilemma: either there is no creation and the world is not created, or something which is not God escapes the causality of God. There is only one solution to the dilemma so posed, and it is a very poor one. Not finding an acceptable sense for the term creation, one could hold to the first horn of the dilemma just proposed and contend that the universe is not created by God, it is identical with God—which is the pantheistic [or monistic] thesis.[30]

"Does God act presupposing something or presupposing nothing at all?" This alone is the principal issue in the idea of creation.[31] Everything else is purely secondary, incidental with respect to what is constitutive of the creative fact and question, which belongs first of all to metaphysics, and to science or theology only by the way; whereas with the fact and question of evolution, this order of primary and secondary is exactly inverted.

Whoever does not see that has not grasped the essential import of the notion of creation; he has restricted and anthropomorphised it beyond what is permissible.[32] Once that has been pointed out, moreover, we are at complete liberty to return calmly to the biblical conception of an initial creation after or beyond which is a divine repose.

We henceforth know well that one can conceptualize this repose in any of three forms: as sanctioning the fixity of beings in their genus and species; as giving them over to their progressive unfolding through time; or finally, as imparting to the latent psychism with which it has endowed them the responsibility for temporal creations more and more exuberant.

One is free to choose, awaiting further evidence; but it is to be fervently hoped that after so much vain quarrelling, we Christians will cease bringing forward unjustified censures respecting this doctrine of evolution, to which—under one form

---

l'action divine dans l'univers" (1920?), "Comment je vois" (1948), "Les noms de la matière" (1919), "Contingence de l'univers et goût humain de survivre" (1953), are perfectly clear: the First Necessity is definitely envisaged as enveloping simultaneously both God and the 'stuff' presupposed for his action as Creator, which action is envisaged with equal definiteness as consisting exactly in an arrangement starting from a primitive chaos; while as far as any quest for causes is concerned, Teilhard repeats tirelessly that he seeks only "an experimental law of recurrence, not an ontological analysis of causes." (See *The Future of Man* [New York: Harper, 1964], p. 110 fn. 1; *The Phenomenon of Man*, p. 29; *et alibi*.) On all these points relating to Teilhard's metaphysical position as it bears on the problem of creation, see (in addition to "Comment je vois," Teilhard's four-page *ex professo* statement of his general metaphysics) Claude Tresmontant, "Le Père Teilhard de Chardin et la Théologie," in *Lettre*, 49–50 (septembre-octobre, 1962).

[30] *Ibid.*, p. 19.

[31] *Ibid.*, pp. 6–7.

[32] At an earlier point in his study (p. 9), Sertillanges had already appended this important note: "One must understand at this point what our intention is in speaking, as we shall do throughout this work, in the language of what is called anthropomorphism. There is an anthropomorphism of a common order entailed in the very expression of thought. With regard to this anthropomorphism, everyone must make his own adjustments. Those anthropomorphisms which we exclude from our work are first of all a doctrinaire anthropomorphism, which implies a falsification of the divine; and also, in that which concerns the philosophers, a conceptual and verbal anthropomorphism which has little relation to a precise study of problems—especially in the contemporary age."

or another [under the second form, as it has turned out]—the future seems certain to belong.[33]

That future in which Sertillanges so firmly believed has become our present. Where it is a question of thinking men, one can no longer "recognize in each the right (1) to reject the idea of evolution *en bloc* if it pleases him; (2) to choose among the diverse historical forms which it might take; or simply (3) to await solution of the difficulties that it raises and the outcome of the crisis through which the transformism of yesterday is incontestably passing."[34]

With respect to the first point, "Evolution as an historical process is established as thoroughly and completely as science can establish facts of the past witnessed by no human eyes."[35]

With respect to the second point, "there is nothing in the evidence gathered by paleontology and morphology that would warrant the assumption of autogenesis," or, as it is sometimes called, *orthogenesis*,[36] i.e., specific and transpecific evolution resulting from inherent tendencies, vital urges, or cosmic goals pursued by a "latent psychism." "On the contrary, the lack of fixed plan in detail but the tendency to spread and fill whenever possible is exactly such a picture as would result from the impulses of a random opportunism,"[37] from fundamental natural units "given over to their progressive unfolding in time." The ubiquitous irregularity in tempo and mode borne out unmistakably in the fossil record negates the postulation of any overall trend subtending the individual organisms but not adaptive in nature.[38]

---

[33] *Ibid.*, pp. 142–143.

[34] *Ibid.*, p. 127.

[35] Theodosius Dobzhansky, *Genetics and the Origin of Species* (3rd ed., rev.; New York: Columbia, 1951), p. 11.

[36] Theodosius Dobzhansky, *Mankind Evolving* (New Haven: Yale, 1962), p. 17; see pp. 16–17.

[37] George Gaylord Simpson, *The Meaning of Evolution*, p. 121. "There are two aspects of opportunism: to seize such diverse opportunities as occur, and when a single opportunity or need occurs, to meet it with what is available, even if this is not the best possible." (*Ibid.*, pp. 167–168).

[38] "If 'nature' means the whole material universe, then the hope for scientific understanding becomes very remote. If to understand anything we must understand all, then science is impossible. If however, as is obvious enough, this world is not one single 'nature', but many individuals varied in nature and forming distinct and relatively independent centers of activity, then the problem of scientific understanding is worth tackling. We may hope to arrive at some essential knowledge of this or that kind of thing, even if the whole escapes us. The history of science makes clear that scientific progress requires piecemeal procedures. . . . The strongest objection to the point I have been making is the view of some physicists that the universe is one continuous 'field'. . . . A 'general field theory' [however] would cancel the evidence for the existence of primary natural units only if it could show that all changes form a rigidly deterministic system which could be attributed to one single intrinsic principle, and not to the conflict and balance of many [relatively] independent units. The whole course of modern physics, on the other hand, has been to reaffirm the fact, which is obvious enough at the macroscopic level, that this universe is not rigidly deterministic." (Benedict M. Ashley, "Does Natural Science Attain Nature or Only the Phenomena?" in The *Philosophy of Physics*, ed. by V. E. Smith [Jamaica, New York: St. John's University Press, 1961], pp. 66 and 69, *passim*.) Cf. Yves Simon, *The Tradition of Natural Law* [New York: Fordham, 1965], pp. 55–56; and note 155 below.

With respect to the third point, as recently as a generation ago, among the many evolutionary explanations proposed to account for the almost incredible diversification and complication of living forms now existent, "there seemed so little reason to choose among some of them, so much to say against any one of them, that the non-partisan student could feel only confusion or despair." Today, however, "one theory has emerged"—the variously called synthetic, integrative, neo-Darwinian, or sometimes simply biological theory of evolution—"that is judged superior and, *as far as it yet goes*, virtually irrefutable according to a large consensus." [39]

To borrow an admirable formulation from Mortimer Adler:

Having an open mind about future possibilities should not be equated, as un-fortunately it sometimes is, with having an undecided mind about present actualities; for we are obliged, at any time, to judge in the light of the evidence that is then available. [40]

If the possibility of contrary future evidence were to disbar us from drawing conclusions from the evidence now available, we could never draw any conclusions whatsoever from the data of scientific investigation. [41]

And it is nothing less than the evidence now available which con-strains one to acknowledge forthrightly that organic evolution occurs automatically wherever there is interaction of living forms through suc-ceeding generations, and that this evolution is associated in fundamental ways with those inorganic interactions inexorably at work in the formation of the stellar and planetary systems.

C. Evolution as reality and idea

It is impossible to overemphasize the points that have been made above, in differing accents, by Bergson, Nogar, Crespy, and Sertillanges. For if it is true that "the facts upon which philosophy rests are absolutely general, primordial facts, not such facts as are observable only with more-or-less difficulty—and which, as science progresses, become more and

[39] Anne Roe and George Gaylord Simpson, eds., *Behavior and Evolution* (New Haven: Yale, 1958), p. 5. See however Sections IV and VII of J. N. Deely's "The Philosophical Dimensions of the Origin of Species," *The Thomist*, XXXIII (January and April, 1969), Part I, pp. 102–130, and Part II, pp. 290–304, respectively, for an analysis of "how far this theory yet goes."

A clear statement of the main lines of the currently ascendent "synthetic theory" may be found in Part II of this volume at two places: in Section I, The Uniqueness of Man, in the third reading, "The Emergence of Man," by John N. Deely; and in Section III, The Moral Issues, in the third reading, "Man in Evolution," by F. J. Ayala.

[40] Mortimer J. Adler, *The Difference of Man and the Difference It Makes* (New York: Holt, Rinehart, and Winston, 1967), p. 113.

[41] *Ibid.*, p. 122.

more points at which the real coincides with the (always more and more complex and refined) [mathematical] constructions previously set up by scientific reason—but absolutely general and absolutely first facts," keeping always in mind, however, that these absolutely general and primordial facts do not arise out of so-called "vulgar" or "common" experience (since "vulgar experience intervenes in philosophy only as a substitute, when no scientific experience is available");[42] if this be allowed, then it must be acknowledged that the concept of evolution as we have just delineated it, a concept immediately derived from scientific experience with no intervening mathematical construction of which it is but an explanatory image, this concept, by every criteria of judgment and criticism under the proper light of philosophy, has got to be accepted as expressing a *properly philosophical fact*.

This brings us to the fundamental and primary distinction between the *fact* of evolution and the *explanation* of that fact; between the *quod* of evolution, *that which* has taken place, and the *quo* of evolution, that *by which* evolution has taken place, or between the *phenomenology* and the *ontology* of evolution; between the evolutionary *products* and the evolutionary *process*; between in a word, the *concept* and *content* of the idea of evolution; between evolution as *world-view* and as *philosophy*, between the *observations* contributing to the realization of the evolutionary universe, and the *explanations* which bring us toward an understanding of this universe. The former pertains to knowing within the perspective of a simple certitude of fact; the latter pertains to knowing in the perspective of the reason of being, or of explanation. We shall return to this fundamental distinction between the orders of observation and explanation in considering the relation of Darwin's world to the world of classical antiquity.

What is of interest to us for the moment is Dr. Nogar's point that "the explanation of *how* the process of orderly change of successive generations through time has been accomplished must be dissociated from the

---

[42] See Jacques Maritain, *The Degrees of Knowledge*, "The Proper Conditions for Philosophy. Its Relation to Facts," pp. 57–60. See Sec. III of "The Philosophical Dimensions of the Origin of Species," on "The logic of rational understanding," pp. 93–102; and cf. M. J. Adler, *The Conditions of Philosophy* (New York: Atheneum, 1965), esp. chs. 5–12, pp. 79–230. See also note 114 below. The role of what will be called in this essay "observed facts" or "data" in the empirical sense has, so far as philosophy is concerned, been recently outlined with particular clarity in a posthumous book of Yves Simon, *The Great Dialogue of Nature and Space* (New York: Magi, 1970), pp. 139–179, esp. p. 154: "In all rigor we can say that every essential part of the philosophical edifice is built on facts of common experience . . . not however . . . that every philosophical fact is at the same time a fact of common experience. . . . There are [as in the case of evolution] some philosophical facts which can be established only through the technical elaboration of an experience. On the other hand, we should also note that the majority of vulgar facts are not philosophical facts." For the full details of this summary, the reader is referred to Simon's work.

statement *that* such an orderly succession has taken place."[43] However one wishes to draw this distinction terminologically, the matter at issue is clear, and a point of widespread agreement at the level of historical consciousness attained in our century. It was with this in view that we referred earlier to evolution as the determining problematic for all synthetic thought in our day. It is in this respect (and in this respect alone) that one must accept Teilhard de Chardin's elevation of evolution above the status of a theory, system, or hypothesis to the status of "a general condition to

[43]Raymond J. Nogar, "From the Fact of Evolution to the Philosophy of Evolutionism," *The Thomist*, XXIV (1961), p. 464. The classical position here expressed by Nogar, the position, namely, that observation and explanation are distinct in principle and analytically separable in any given case (and hence that the record of evolutionary unfolding is not tied to any particular explanation of how that unfolding transpired), is the one taken by all or most of those conversant with evolutionary science. This classical view has, however, been recently subjected to attack by a number of philosophers of science whose chief and—significantly (see fn. 132 below)—almost exclusive concern has been the explanatory pattern of modern physics, which we will define shortly as the Platonic Explanatory Mode. Outstanding among these attackers, perhaps, are: (A) T. Kuhn, *The Structure of Scientific Revolutions* (2nd ed., enlarged; Chicago: University of Chicago Press, 1970); "Logic of Discovery or Psychology of Research?" in *Criticism and the Growth of Knowledge*, ed. by I. Lakatos and A. Musgrave (Cambridge: Oxford University Press, 1970), pp. 1–23; "Reflections on my Critics," in *Criticism and the Growth of Knowledge*, pp. 231–278. (B) N. R. Hanson, *Observation and Explanation* (New York: Harper & Row, 1971); *Patterns of Discovery* (Cambridge: Oxford University Press, 1958); *Perception and Discovery*, ed. by W. C. Humphreys (San Francisco: Freeman, Cooper & Co., 1969). (C) P. K. Feyerabend, "Consolations for the Specialist," in *Criticism and the Growth of Knowledge*, pp. 197–230; "How to Be a Good Empiricist: A Plea for Tolerance in Matters Epistemological," in *The Delaware Seminar in Philosophy of Science*, ed. by B. Baumrin (New York: Interscience, 1963), Vol. II, pp. 3–39; "Reply to Criticism," in *Boston Studies in the Philosophy of Science*, ed. by R. S. Cohen and M. W. Wartofsky (New York: Humanities, 1965), Vol. II, pp. 223–261. (D) S. Toulmin, *Foresight and Understanding* (New York: Harper & Row, 1961); "Does the Distinction Between Normal and Revolutionary Science Hold Water?" in *Criticism and the Growth of Knowledge*, pp. 39–48; "Reply," *Synthese*, 18 (1968), pp. 462–463.

In the classical view, it is always possible to provide an identification of facts that is sufficiently neutral to serve as a common measure and test of two opposed theories. The recently developed anti-classical view denies precisely this possibility, and asserts that all observations are theory-laden to such an extent that holders of different theories observe different facts, i.e., what proponents of different theories see in nature is so structured by and dependent upon the theories as to provide no common measure and test. This view, as Landesman points out, derives its inspiration in part from the linguistic fact that "the description and the explanation [of any given phenomenon of nature] may both be formulated in the same vocabulary" ("Introduction" to *The Problem of Universals*, ed. by C. Landesman [New York: Basic Books, 1971], p. 6), and in part from the psychological examples of Gestalt-shifts.

The excess of this recent, anti-classical position that seeks to conflate the distinct orders of observation and explanation has been just as recently shown in a number of ways.

With respect to the particular case of evolutionary science, Michael Ruse has shown the irrelevance of the anti-classical view denying the separability of fact from theory to the historical record of how the Darwinian revolution in biology came about. In other words, Ruse has shown that on the assumption of the truth of the anti-classical position, the actual events surrounding the rise of evolutionary theory become unintelligible. See Ruse, "The Revolution in Biology," *Theoria*, XXXVI (1971), pp. 1–22. R. J. Nogar, on the other hand (in company with such biologists as Dobzhansky and Simpson), has repeatedly shown the relevance of the classical position affirming the separability of fact from theory for coming to terms with the

which all theories, all hypotheses, all systems must bow and which they must satisfy henceforward if they are to be thinkable and true."[44]

But there is no need to put the matter so grandly, and good reason not to: for by evolution are meant products as well as processes; and the discovery of a world in which *nothing is exempt* from change is still not a world in which there is *nothing but* change.

It is therefore much more desirable and philosophically exact to state the point in more measured terms. "There is the theory of evolution," let us say, "and there are theories of evolution. The theory of evolution is the fact—it may surely be called 'fact' in the vernacular—that all organisms that now live or ever lived, all they are and all they do, are the outcome of genetic descent and modification from a remote, simple, unified beginning. Theories of evolution, taking the reality of evolution as given, seek to explain how this almost incredible diversification and complication have come about."[45]

And it is necessary to understand that the root difference between the evolutionary world-view and the world-view of classical or medieval times lies within this very distinction.

II. FROM CLASSICAL ANTIQUITY TO DARWIN'S WORLD

If we pause for a moment in our consideration of the evolutionary concept, and meditate, before passing over to the explanatory dimension of the datum, on the historical emergence of this concept via scientific experience

---

Darwinian revolution. In other words, on the assumption of the truth of the classical position, the events leading to the rise of evolutionary science become eminently coherent and intelligible. See Nogar, *The Wisdom of Evolution*, Part I, pp. 27–145.

With respect to the anti-classical position considered in itself, Carl R. Kordig has recently shown that, quite apart from its inability to illuminate particular historical instances of scientific revolutions in the crucial way claimed, the position does not follow from the premises given as its basis, and "leads to unintelligible and absurd consequences." See Kordig, "The Theory-Ladenness of Observation," *The Review of Metaphysics*, XXIV (March, 1971), pp. 448–484; and *The Justification of Scientific Change* (Dordrecht-Holland: D. Reidel, 1971).

The rise of evolutionary theory, indeed, is one of the most eloquent witnesses (as we shall see in these pages) to the fact that, as Kordig insists ("The Theory-Ladenness of Observation," p. 482), "what scientists observe does change," but primarily in the sense that "it increases," and hence forces crucial transitions from less to more comprehensive theories, and even, as we shall argue here, from one to another explanatory mode, in the senses to be defined shortly.

In the terms of the present discussion, the anti-classical view of scientific and intellectual development depends for its force largely on the mistaking of theories as such for world-views, and on a misunderstanding of the role imagination and myths play in linking the two: see fn. 46 and Section III below, "World-View as *Logos* and as *Mythos*," *passim*, but in connection particularly with fns. 168, 181, and 196.

[44] Pierre Teilhard de Chardin, *The Phenomenon of Man* (New York: Harper, 1959), p. 218. See also fn. 1 p. 140.

[45] Roe and Simpson, *Behaviour and Evolution*, p. 5.

as a truly philosophical problematic, we find ourselves confronted with a residuum of factors at play in man's cognitive life which have little to do with logic or evidence, with the general tendency among men to substitute habitual patterns of thinking for evidence, and with the dependence of reason on the testimony of sensation. At the same time, the fact that the evolutionary concept has emerged shows just as definitely that man is able, recurrently and intermittently at least, to function as a scientist, "if being a scientist means asserting as true only those propositions which are based upon sound evidence in a logical manner (i.e., doing non-wishful thinking)." [46] From this point of view, even "rationalization is thus the tribute which emotion pays to reason in order to conceal the latter's deficiencies"; [47] and much of the story surrounding the changeover from an essentially static to a radically dynamic world-view concerns little more than the assessment of such tributes.

It is easy, common, and true enough to put forward such statements as the following: "In the eyes of Plato and Aristotle, and in the thought of the West guided by their vision, the universe seemed in its overall duration to be structurally given once and for all. The various kinds of living things seemed to have a fundamental permanence which remained unaffected by the passage of time." Such a statement would, moreover, be perfectly in harmony with the curious inclination on the part of most narrators recounting the intellectual history of the West (particularly where science is concerned) to read dogmatic assertions into ancient

---

[46] Mortimer J. Adler, *What Man Has Made of Man* (New York: Ungar, 1937), p. 66. In Chapter Three of this book, a discussion of "The History of Psychology" (pp. 61–93), Adler forcefully and clearly demonstrates in what sense and why whatever pertains to the creative genius and psychological condition of the individual, or to his socio-historical conditions, however indispensable for the discovery of certain truths or the arrival at and formulation of certain insights, remains irrelevant to the merits of any intellectual position taken as such. In particular, see his "Digression" on "the error of 'wishful thinking,' sometimes called 'rationalization,'" *ibid.*, pp. 62–64, and his closing remarks on pp. 122–123.

The reasons for this position are not obscure. The proper aim of rational understanding is to assign the reasons for what is given in experience and to gain an understanding of that datum, not to explain it away or indulge in random guessing for the sake of some preconceived theory or personal conviction. "True, a scientist often finds it extremely useful to give his imagination free play in the *preliminary* stages of an investigation, but however brilliant the creations of his fancy they have scientific value only to the degree that they can be reduced to facts." The scientist, and equally the philosopher, in order to avoid excursions into myth-making, "deals only with what he can observe as really existing or something whose existence can be inferred from its observed effects." (Benedict Ashley, "Does Natural Science Attain Nature or Only the Phenomena?" p. 70.) See Section III of Part I of J. N. Deely's article, "The Philosophical Dimensions of the Origin of Species," *The Thomist*, XXXIII (January, 1969), pp. 93–102; and M. J. Adler's discussion of the sense in which philosophy can give us new knowledge of the world that is experienced, which discussion runs throughout his book, *The Conditions of Philosophy* (New York: Atheneum, 1965), but most closely touches on the point just made by Ashley, perhaps, on pp. 144–146.

These observations will be taken up under another aspect in Section III-A of this present essay, "World-view as *Logos*."

[47] David Bidney, *Theoretical Anthropology* (New York: Columbia, 1953), p. 5. See Section III-B of this present essay, "World-view as *Mythos*."

thinkers where only concepts tailored to the available evidences were proffered. The effect is to make modern notions look more decisive and revolutionary than they in fact are.

There are, however, three elements which, from the philosopher's standpoint, must be distinguished in this historical statement. The first concerns the *evidence* on which Plato and Aristotle took their stand; the second and third concern the *account* or *explanation* which they respectively gave for that evidence.

With respect to the question of evidence, we have already noted that any man restricted for the most part to the conceptual horizon of common experiences is bound to acknowledge that a basically unchanging world seems to confront his gaze, not indeed existentially (life does have its ups and downs), but *essentially* or so far as the structures of nature are concerned. Specific stability is a primary datum which the philosopher within this horizon is called upon to elucidate.

A. The platonic explanatory mode

Plato, considering the existential flux and intelligible constancy, located the intelligible constituents in a world beyond the phenomena, a world of transcendent numbers and ideas of which the phenomena are but the changing shadows, "the moving shadow of eternity," as Augustine would later repeat. For the tradition of thought properly called Platonic, the bridge to this ultimate reality is mathematics, to which natural science is subordinated. Such a world of form and number, being of itself incorruptible and eternal, accounts quite neatly for the specific constancies of nature. Inasmuch as the physical world successfully participates in the ideal world, it could only present intelligible constancy; while to the extent that its participation is but participation, and so intrinsically labile, the natural world presents a picture of flux and change. Such, in its essential epistemological type, was Plato's explanation.

Here we are confronted, both historically and philosophically, with the first of the two possible ways, epistemologically speaking, in which the regularities in things noted by means of sensible experience can be accounted for rationally and subsumed within an explanatory scheme. In this type of knowledge which is materially physical but formally (with respect to object and to method of conceptualization) mathematical, the *rule of explanation* prescinds from physical principles and causes with their proper intelligible value. When observations are given an explanation in this form, however, as Simon puts it, "something entirely novel takes place."[48] The very nature of mathematical abstraction renders mathe-

---

[48] Yves Simon, "Maritain's Philosophy of the Sciences," *The Thomist*, V (1943), p. 101.

matical thought indifferent to the reality of its object, inasmuch as that object can be conceived without sensible matter. Hence, mathematical-physical science tends toward indifference respecting the reality or extramental independence in being of the data it rationalizes. It tends "to make no difference between *ens reale* and *ens rationis*" (i.e., between the sort of being which exists independently of human understanding and the sort of being which is a product of human understanding). What was observed as real is explained as preter-real. "Here the mind escapes into a world of entities which were first grasped in the bodies of nature but immediately purified and reconstructed, and on which other entities, which are indifferently real or 'of reason,' will be endlessly constructed."[49] This world frees us indeed from the sensible, just as Plato contended; but it seems to achieve this rather by sacrificing any order to existence than by putting us in touch with the ultimate reality. "Physico-mathematics is, indeed, a science of the *physical real*, but a science which knows that real only by transposing it, and *not as the physical real*."[50] But if mathematical-physical knowledge "explains" in things only that kind of formal cause which is the conformity of phenomena to mathematical law, on the other hand, just because mechanistic aspects of causality can be retained within the texture of physico-mathematical explanations, this way of rationalizing the evidences compensates for those intelligibly sensible aspects from which it prescinds by enabling us to *predict and control* those aspects of the real from which it does not prescind.[51]

The establishment of a universal science of nature informed by mathematics rather than philosophy has indeed been the great achievement of modern times, based directly on the works of Descartes and Galileo; but the essentials of this method are to be found in much earlier works.

Attribution of the title 'creator of the method of the physical sciences' has given rise to many squabbles; some have wished to give it to Galileo, others to Descartes, still others to Francis Bacon, who died without ever having understood anything about this method. Frankly, the method of the physical sciences was defined by Plato and the Pythagoreans of his day with a clarity and precision that have not been surpassed; it was applied for the first time by Eudoxus when he tried to save the apparent movement of the stars by combining the rotation of homocentric spheres.[52]

[49] Maritain, *The Degrees of Knowledge*, p. 209. Cf. Simon, *The Great Dialogue of Nature and Space*, pp. 89–111, esp. 101ff.

[50] *Ibid.*, fn. 2, p. 42.

[51] One might well raise the question here as to why, if the ancients did indeed grasp the essentials of mathematical-physics, "why did they not push open the door?" Why did technological society not bud forth prior to the sixteenth century? The answers to this question are socio-cultural. In essence, as Farrington among others has shown, it was the cultural denigration of servile work which truncated the development of the technological aspect of science in ancient times: Benjamin Farrington, *Greek Science* (Baltimore: Penguin, 1944), pp. 301, 307–8, 302, and 303.

[52] Pierre Duhem, *Le Système du monde*, new ed.; Paris, 1954, Vol. I, pp. 128–129.

B. The aristotelian explanatory mode

When we turn to Aristotle's account of the fixity of specific structures, we find an attempt at quite another sort of understanding. Aristotle attempted "to construct a natural science which would rest on observation at every point, but which would seek to explain these observations *in terms of their own intelligibility*, not in some a priori or conventional fashion, and which would proceed from general to particular questions *in an orderly but not deductive fashion*."[53] It would be, not a knowledge of the real by

[53] Benedict M. Ashley, "Aristotle's Sluggish Earth, Part I: The Problematics of the *De caelo*," *The New Scholasticism*, XXXII (January, 1958), p. 8: emphasis supplied. This point is important, but historically confused. The notion of demonstration in the classical Aristotelian tradition "does not insist on a deductive movement from the known to the unknown, but on the knowledge of the proper cause of something expressed by the middle term of a syllogism. In the case of the philosophy of nature the phenomenon is better known to us than its cause, yet demonstration consists in knowing this phenomenon in a special way, namely through its cause." (Benedict M. Ashley, *Are Thomists Selling Science Short?* River Forest, Ill.: Albertus Magnus Lyceum, n.d., pp. 12–13). This is so in view of the fact that to assign the proper causes of some phenomenon is to understand in what kind of subject it inheres (its material and formal cause), and what kind of an agent has produced it and by what steps (its efficient and final cause) (*ibid.*, p. 13); and that "the difference between a descriptive definition and an essential definition . . . is not in the content of the definition but in its order": "we are dealing, therefore, not with an absolute difference between one kind of human knowledge in which is attained a perfectly ordered knowledge of nature (dianoetic intellection) and another which knows nothing of nature except its existence (perinoetic intellection), but rather with a type of intellection proper to man by which he knows at first confusedly and then more clearly as he continues his investigation both the existence of a natural unit and its nature," and the network of interrelations it sustains (Ashley, "Does Natural Science Attain Nature or Only the Phenomena," pp. 77–78; see also pp. 70–75). Consult the key passage in St. Thomas' *Commentary on Aristotle's Posterior Analytics*, Bk. II, Lect. 13, n. 7; and Zigliara's comment on this in the Leonine edition of the works of St. Thomas, tom. I, p. 375 a–b. This view is set out in Section III of Deely, "The Philosophical Dimensions of the Origin of Species," Part I, *The Thomist* XXXIII (January, 1969), pp. 93–102; and developed in a systematic textual study by Melvin A. Glutz, *The Manner of Demonstrating in Natural Philosophy* (River Forest, Ill.: Aquinas Institute, 1956). Further to this discussion, see fn. 114 below.

Maritain himself, in distinguishing sciences of explanation from "sciences" of observation (*The degrees of Knowledge*, pp. 32–34), restricts explanation properly so-called to purely deductive schema. Since many of the key contentions of biology can claim only a strong inductive support, Maritain would exclude evolutionary theory from the category of explanation. He would regard it simply as a "likely story," or an "empiriological" ("empirioschematic") account. (See *The Degrees of Knowledge*, pp. 64–66 and 192–195.) On precisely similar grounds, T. A. Goudge, in *The Ascent of Life* (Toronto: University of Toronto Press, 1961), has argued that evolutionary science employs a "narrative explanation," i.e., tells only "likely stories" in its account of life, in sharp contrast to the "covering-law explanations" characteristic of physics and chemistry.

For differing reasons, both Maritain and Goudge are mistaken in appealing to the non-deductive aspects of evolutionary theory as ground for drawing a sharp line between two types of "science" or "explanation." Maritain's mistake lies in a failure to realize that, as Ashley has pointed out in the above-cited texts, descriptive definitions do not differ from essential definitions in their cognitive content, but only in their organization of that content. (This matter will be further developed in Section III-A of this essay, "World-view as *logos*".) Goudge's mistake lies in a failure to see that, *if one prescinds from the role mathematics plays*, the explanations of modern physics and biology alike approximate closely, as to an ideal, to the "covering-law model" of explanation, as set forth by Dray (in *Laws and Explanations in*

means of the mathematical preter-real, wherein it becomes increasingly difficult and frequently impossible to differentiate the symbol from the symbolized, but rather "a knowledge whose object, present in all things of corporeal nature, is changeable being as such and the ontological principles which account for its mutability."[54]

With this ideal of science in mind, Aristotle denied Plato's World of Forms by arguing that these incorruptible and eternal essences of things, these invariant structures, are universal natures which exist outside the mind only in things singular and perishable. The immutable types, Aristotle argued, are immanent in the physical world. They are natures which are revealed through the regularities that are observed in the very order of sensible phenomena.

Here we are confronted, historically and philosophically, with the second of the two possible ways in which, epistemologically speaking, the regularities in things noted by means of sensible experience can be accounted for rationally and subsumed within an explanatory scheme. Here, the knowledge is formally as well as materially physical. Its *rule of explanation* is to "reveal to us intelligible necessities immanent in the object," the metalogical existent; to make effects known by principles or *reasons for being*, that is, by causes, "taking this latter term in the quite general sense that the ancients gave to it."[55] It prescinds only from "what in individual cases is never equal—the particular, the contingent, the variable, the unpredictable about specific events—*what is logically incidental*."[56] When observation is given explanation or reasoned in this form, without any prescinding from the materiality that renders things both perishable and observable, what are revealed to us are intelligible necessities *immanent* in the sensible object, just as in explanations formally

---

History [Oxford: Oxford University Press, 1957]), Brodbeck ("Explanation, Prediction, and 'Imperfect' Knowledge," in *Scientific Explanation, Space, and Time*, ed. by H. Feigl and G. Maxwell, Vol. III of *Minnesota Studies in the Philosophy of Science* [Minneapolis: University of Minnesota Press, 1962], pp. 231–272), Hempel (see *Aspects of Scientific Explanation* [New York: Free Press, 1965]), and others, wherein what is to be explained is explained when its sufficient conditions can be stated. This error on Goudge's part is demonstrated by Michael Ruse, "Narrative Explanation and the Theory of Evolution," *Canadian Journal of Philosophy*, I (September, 1971), pp. 59–74.

[54] Maritain, *The Degrees of Knowledge*, p. 176.

[55] *Ibid.*, p. 32. For an indication of the manner in which the rise of evolutionary biology is forcing an expansion and analogization of the narrow, univocal notion of causality as efficient causality which comes down to us in philosophy from Locke, Hume, and Kant, and in science from the suzerainty of mathematical physics since the time of Galileo and Newton, see the analysis in "The Philosophical Dimensions of the Origin of Species," esp. Secs. III, IV, and VII; and the reading selections in this volume by Benedict M. Ashley, "Change and Process," and by C. H. Waddington, "The Shape of Biological Thought," in Section IV *infra*, The Metaphysical Issues.

[56] John Herman Randall, *Aristotle* (New York: Columbia, 1960), p. 184, emphasis supplied. On this question of what is 'logically incidental' about specific events, see Charles De Koninck, "Abstraction From Matter," in *Laval theologique et philosophique*, XIII (1957), pp. 133–197; and Yves Simon, "Chance and Determinism in Philosophy and Science," Ch. X of *The Great Dialogue of Nature and Space*, pp. 181–204.

mathematical what are revealed to us are intelligible necessities *transcendent* to the sensible object as such, and consequently indifferent to its existential status.

Because such a knowledge of nature seeks to understand the things of nature not only from the point of view of quantity or quantified being as such but from the point of view of sensible being, its predictive index is very low. In fact, it is indirect, deriving heuristic value principally "by reason of the stimulations it is capable of giving to the minds of scientists."[57] Because it discloses intelligible reasons immanent in the things of nature as physical, the universality and necessity of this science is predominantly negative, that is, retrospective.[58] The laws it declares express the order of a cause to its effect, to be sure, and in this sense are "eternal" truths, since even when "in the flux of particular events, another cause comes along to interfere with the realization of its effect, that order remains";[59] so that, in the *order of explanation* this science, even as mathematical-physical science, sets before the mind "intelligibles freed from the concrete existence that cloaks them here below, essences delivered from existence in time."[60] (Whereas, by contrast, "the other sciences, sciences of observation, do indeed tend to such truths, but they do not succeed in emerging above existence in time, precisely because they attain intelligible natures only in the signs and substitutes that experience furnishes of them, and therefore in a manner that inevitably depends on existential conditions. Thus, the truths stated by them affirm, indeed, a necessary bond between subject and predicate, but also suppose the very existence of the subject, since the necessity they evince is not seen in itself but remains tangled with existence in time—and to that extent, if I may say so, garbed in contingency."[61])

---

[57] Maritain, *The Degrees of Knowledge*, pp. 177–178. See G. G. Simpson's essay on "The Historical Factor in Science" in his book, *This View of Life* (New York: Harcourt, 1963), pp. 121–148.

[58] G. G. Simpson remarks: "The testing of hypothetical generalizations or proposed explanations against a historical record has some of the aspects of predictive testing. Here, however, one does not say, 'If so and so holds good, such and such will occur,' but, 'if so and so has held good, such and such must have occurred.' (Again I think that the difference in tense is logically significant and that a parity principle is not applicable.)" *This View of Life*, p. 144.

[59] Maritain, *The Degrees of Knowledge*, pp. 28–29. G. G. Simpson explains this same point in *This View of Life*, pp. 125–127.

[60] *Ibid.*, p. 33.

[61] *Ibid.*, pp. 33–34. On p. 34, Maritain expresses the distinction between what I have called the order of observation over against the order of explanation in the following terms: "it is plain that sciences of the second category, sciences of observation . . . since they are less perfectly sciences and do not succeed in realizing the perfect type of scientific knowledge, are not sufficient unto themselves. Of their very nature, they tend to sciences of the first category, to sciences of explanation properly so called. . . . They are necessarily attracted to them. In virtue of their very nature as sciences, they invincibly tend to rationalize themselves, to become more perfectly explanatory . . . and to that extent they are subject to the regulation of . . . either philosophy or mathematics," i.e., to formulation and expression in either the Aristotelian or the Platonic explanatory mode. (See however the qualifications placed on Maritain's use of this distinction, in fn. 53 above.) Cf. Simpson, *This View of Life*, pp. 125–127, 143–144.

Nevertheless, the "eternal truths" of the natural physicist have a thoroughly *negative* character, that is, their empirical content consists purely and simply in *what they forbid*, just because they are pure objects of understanding and not objects of sensible apprehension or imaginative representation, are, in short, "unimaginable by nature."[62] This is indeed why this natural or philosophical (as contrasted with the mathematical-physical) mode of explanation has no direct heuristic value.

Among the evolutionists, none have grasped these points more clearly than Simpson:

> The search for historical laws is, I maintain, mistaken in principle. Laws apply, in the dictionary definition, "under the same conditions," or in my amendment "to the extent that factors affecting the relationship are explicit in the law," or in common parlance "other things being equal." But in a history, a sequence of real, individual events, other things never are equal. Historical events, whether in the history of the earth, the history of life, or recorded human history, are determined by the immanent characteristics of the universe acting on and within particular configurations, and never by either the immanent or the configurational alone.[63]

On the other hand, the "eternal truths" of the mathematical physicist (i.e., axioms which are neither assessable by reference to motion or that kind of time which is motion's measure, nor false within every known mathematical system), just because they are objects capable, either directly or indirectly, of imaginative representation,[64] and are not pure objects of understanding or objects of sensible apprehension—these truths have a *positive*, prospective, or predictive character so far as the world of natural things is concerned.[65] For whatever exists in a material

---

[62] In the ontological explanation typical of philosophy of nature, "being is still considered in the order of sensible and observable data. But the mind enters that order in the search of their intimate nature and intelligible reasons. That is why, in following this path, it arrives at notions like corporeal substance, quality, operative potency, material or formal cause," *natural selection* (see the references in fn. 107 below and fn. 59 above), "etc.—notions which, while they bear reference to the observable world, do not, designate objects which are themselves representable to the senses and expressible in an image or a spatio-temporal scheme. Such objects are not defined by observations or measurements to be effected in a given way." (*The Degrees of Knowledge*, pp. 147–148). See fn. 65 *infra* for the sense of the expression "what they forbid" for the present context.

[63] G. G. Simpson, *This View of Life*, p. 128. See fn. 155 below.

[64] Inasmuch as, on the one hand, so far as the basis of arithmetic is concerned, discrete quantity is directly constructible in imaginative intuition (simply represent two unities); while, on the other hand, so far as the basis of geometry is concerned, continuous quantity is likewise directly constructible. And these two sciences of discrete and continuous non-physical quantity respectively, are the point of departure for the whole of mathematics. See J. Maritain, *The Degrees of Knowledge*, pp. 35 fn. 3, 140–146, and 165–173; Y. Simon, "The Nature and Process of Mathematical Abstraction," *The Thomist*, XXXIX (April, 1965), pp. 117–139; and Philip J. Davis' article, "Number," *Scientific American*, 211 (September, 1964), pp. 50–59.

[65] It is true, as Karl Popper has pointed out (cf. *The Logic of Scientific Discovery*, New York: Basic Books, 1959, p. 41), that there is a sense in which the empirical content of *every* physical theory, be it mathematical or physical, "consists in what it forbids." When one wishes to understand how the mathematical differs from the natural physical explanation or theory,

way has extension, including qualities; and that is all that is necessary in order that a formally mathematical law hold for any given thing, regardless of its existential state or "particular configuration," as Simpson would say. Because the properties of any given mathematical essence are not really distinct from it,[66] in this mode of explanation, just as there is no question of causal sequence, so also there is no room for chance interventions.[67]

C. Contrasts

It is necessary to insist on this distinction based on the predominantly retrospective and the predominantly prospective characters, respectively, of physical explanations formulated in the philosophical or the mathematical mode if we are to appreciate the full import of the difference

---

why a heuristic value attaches essentially to the former and only incidentally to the latter, however, it is not helpful to use an analogous formula without making explicit the difference in meaning which obtains in each case.

The point I am making, then, depends simply on the fact that whatever might be said of the physical world in the philosophical mode of explanation virtually contains whatever might in principle be said in the physico-mathematical mode, but not vice-versa. Curiously, one of the clearest and most straightforward statements of the ontological reasons for this fact may be found in Thomas Aquinas, *In libros Aristotelis de caelo et mundo expositio*, Book I, lect. 3, n. 24; and again in Book III, lect. 3, n. 560. See Deely, "The Philosophical Dimensions of the Origin of Species," Part II, Sect. VII, pp. 290–298 and fn. 220a. As Simpson puts it (*This View of Life*, Ch. 7, "The Historical Factor in Science," pp. 121–148): "Prediction is inferring results from causes. Historical science is largely involved with quite the opposite: inferring causes (of course *including causal configurations*) from results" (p. 146: emphasis supplied). For this reason, "it cannot be assumed and indeed will be found untrue that parity of explanation and prediction is valid in historical science" (p. 137). "With considerable oversimplification it might be said that historical science is mainly postdictive [what I have termed "retrospective"]," and non-historical science mainly predictive [or "prospective"]," with the asymmetry in their logical (and epistemological) types deriving from the fact that in the former instance "the antecedent occurrence [the anticipated discovery] is not always a *necessary* consequence of any fact, principle, hypothesis, theory, law, or postulate advanced before the postdiction was made" (p. 147). See fn. 170 below.

[66] "Nature, in the physics of Aristotle, signifies entity, essence, whatness, quiddity with a constitutional relation to action, operation, movement, growth, development. A nature is a way of being which does not possess its state of accomplishment instantly but is designed to reach it through a progression. (*Phys.* 2.1. 192b, *Met.* 5.4. 1014b.) . . . The formalist majority and the intuitionist minority in modern mathematics would agree that a mathematical object, whatever it may be, is not a nature in the sense defined above, and that, whereas we may call it, if we please, essence, whatness, quiddity, etc., we may not attribute to it a dynamism, a tendency to forge its way in the world of becoming. It does not grow; it is what it is by definition, by construction, instantly; it is possessed of its proper condition of accomplishment immediately and does not have to acquire it by growth." (Yves Simon, *The Tradition of Natural Law*, pp. 43–44).

[67] These remarks make clear how different the notion of determinism in nature will be from the standpoint of the mathematical physicist and from the standpoint of the natural or philosophical physicist. See Yves Simon, "Maritain's Philosophy of the Sciences," pp. 98–99. (For the bearing of this distinction on the question of human freedom, see Simon, *loc. cit.*; and Maritain, *The Degrees of Knowledge*, pp. 186–192, esp. p. 191.) See further fn. 73 below.

between the epistemological type, degree, or kind of knowing exhibited in Aristotle's approach to understanding the given, as opposed to Plato's.

The objects given in experience are certainly particular and contingent; scientific knowledge, as distinct from opinion, certainly bears on the universal and the necessary. It is true that it was this paradox which induced Plato, meditating the fact of certain knowledge, to set up a world of Divine Ideas to which mathematics is the bridge, since *as* ideas the forms of the universe, separate, eternal, and perfect', were necessary and their properties could be known accordingly through the certitude and precision of mathematics. It is also true that Aristotle followed Plato in teaching that there is science, simply speaking, only of the incorruptible and eternal.

But the incorruptible and eternal, the universal and the necessary, are such as *things* only so long as they *transcend* the physical world. Once they are regarded as essences *immanent* to that world, they cease to be *that which* exists (*id quod existit*) to become rather principles *by which* (*principia quo*) things are. In these terms, the whole dispute between Plato and Aristotle turns on the judgment as to whether forms are to be regarded as *things* or as *principles* of things. In the former case, the fixity and immutability of species would be a positive eternity: a given species would either be as it is or not be; in the latter case, the fixity and immutability of species would (in principle at least[68]) be a merely negative one, a condition attaching to specific structures once understood *as understood*. On the part of the things understood *as things*, there could be no question of an actual, i.e., positive, immutability of form as such.[69]

---

[68] "As far back as the twelfth century, the temptation of fixing natural forms with a stability they do not have was warned against. Many philosophers, in commenting on Aristotle's *Physics*, attempted to make the term nature a *nomen absolutum*, an absolute concept. Thomas Aquinas (1225–74) corrected this interpretation of Aristotle's notion by pointing out that nature is composed of both matter and form and *nature is as much the potential as the actual attributes* of a natural body. Nature is a principle, that is to say, a relation of the generator to the generated, and *cosmic natures are no more fixed than this relation*. True to the Aristotelian principle that there is no other way to know how fixed this relation is than to observe nature, Aquinas and his students repudiated, in theory, the Platonic tendency to identify temporal natures with eternal essences (see esp. Aquinas' *In II Phys.*).

"There was one difficulty. The science of cosmic prehistory was not yet in existence. . . .

"Prior to 1800, the world of nature seemed to reveal only the permanent side of her regularity. Then, the dynamic history of nature—how change, even of species, entered into natural development of the cosmos—was only a faint suggestion on the horizons of science. Consequently, it is not strange that natural philosophers and scientists of the greater period of history have tended to view the cosmos as having ageless or eternal qualities. . . .

"But the most realistic natural philosophers followed the principle that nature was not something absolutely fixed, but rather a relationship between the generator and generated having both perdurance and fluidity. They incorporated the limits of natural permanence *in their theory*." Raymond J. Nogar, *The Wisdom of Evolution* (New York: Doubleday, 1963), pp. 318–319.

[69] See *The Material Logic of John of St. Thomas*, trans. by Yves R. Simon, John J. Glanville, and G. Donald Hollenhorst (Chicago: University of Chicago Press, 1955), esp. Ch. II.

It is just this point which seems to go unaccounted for in Jonas' opinion that "the liquida-

Thus it is simply erroneous to contend that the contemporary vision of species "as a concretion of history, our belief that kinds are only snail-slow rhythms in a world forever in change, would be for Aristotle a betrayal of the very spirit of knowledge, of mind and the real."[70] For, in order to understand the contemporary vision and to contrast it with that of the ancients, "it is not, to be sure, a question of giving up that [ordinary] logic," which Aristotle systematized for the first time, "or of revolting against it."[71] It is simply a question of considering carefully the distinction between the possible and the real, in order to "see that 'possibility' signifies two entirely different things and that most of the time we waver between them, involuntarily playing upon the meaning of the word."[72] And in grasping the three terms (not two) involved in the possible/real distinction, one grasps as well the sense in which the ancients asserted—and were right in asserting—that scientific knowledge is indifferent to the singular—not absolutely, but only with respect to the negative sense of "the possible."[73]

Hamlet was doubtless possible before being realized, if that means that there was no insurmountable obstacle to its realization. In the particular sense one calls possible what is not impossible; and it stands to reason that this non-impossibility of a thing is the condition of its realization. But the possible thus understood is in no degree virtual, something ideally pre-existent. . . . Nevertheless, from the quite negative sense of the term 'impossible' one passes surreptitiously, unconsciously, to the positive sense. Possibility signified 'absence of hindrance' a moment ago: now you make of it a 'pre-existence under the form of an idea,' which is quite another thing. In the first meaning of the word it was a truism to say that the possibility of a thing precedes its reality: by that you meant simply that obstacles, having been sur-

---

tion of immutable essences . . . signifies the final victory of nominalism over realism." (Hans Jonas, *The Phenomenon of Life*, New York: Harper, 1966, p. 45). See "The Philosophical Dimensions of the Origin of Species," Part II, esp. Sections VI and VIII, in *The Thomist*, XXXIII (April, 1969), pp. 251–290 and 305–331, respectively.

[70] Marjorie Grene, *A Portrait of Aristotle* (Chicago, 1963), p. 137.

[71] Henri Bergson, *The Creative Mind*, trans. by Mabelle L. Andison (New York: Philosophical Library, 1946), p. 26.

[72] *Ibid.*, p. 21.

[73] The notion that perfectly scientific, i.e., universal and necessary, knowledge of the singular as such is possible depends on the view that whatever happens in the natural world happens of necessity. This error, as we have noted above (in fn. 67), springs principally from a false conception of the nature of causal determinism. "In the first place, it is not true that if any cause whatever is present its effect necessarily follows, for there are causal actions which do not necessarily achieve what they tend to effect, but . . . in a particular instance may be robbed of their efficacy by the conflicting influence of some other cause(s). . . . In the second place, every individual substance (omne quod est per se), that is, everything that constitutes an essence in the sense of an individual existent or is in the strict sense a being, has a cause; but that which is by accident has *as such* no cause, inasmuch as, not being truly one, it is not truly being." (Thomas Aquinas, *Summa*, I, q. 115, art. 6.) Further to this point, see fn. 38 above and fn. 155 below. The best general discussion of this topic I have come across is Yves Simon's analysis of "Chance and Determinism in Philosophy and Science," in *The Great Dialogue of Nature and Space*, pp. 181–205.

mounted, were surmountable. But in the second meaning it is an absurdity, for it is clear that a mind in which the *Hamlet* of Shakespeare had taken shape in the form of possible would by that fact have created its reality: it would thus have been Shakespeare himself.[74]

We are speaking of man's understanding of the world; it is a question of human knowledge, therefore; and since, within the framework of Aristotelian metaphysics and psychology, "reality is not referred to knowledge but the reverse,"[75] there can be no question that "the truth of those things which do not always stand in the same relation to being is not unaffected by change," for "as a thing stands with regard to being, so does it stand with regard to truth."[76] So far as human insight is concerned, "the thing and the idea of the thing" are "created at one stroke when a truly new form [was not every individual form, taken as such, though not specifically, unique—"truly new"—in the philosophy of Aristotle?] invented by art or nature is concerned."[77] One need only avoid unconsciously playing on the positive and negative senses of the word possible in assessing the evolutionary data, adhering the while to the most rigorous logic, if one wishes to understand the evolutionary as opposed to the static world-view; for what is decisive in any philosophy for which essences are *principia quo*, principles by which things are, is not the *eidos*, the Idea of the world, but the realization that "actions have to do with singular things and all processes of generation belong to singular things,"[78] because this implies that "universals are generated only accidentally when singular things are generated,"[79] so that it is necessary to say that natural process "creates, as it goes on, not only the forms of life, but the ideas that will enable the intellect to understand it, the terms which will serve to express it."[80]

Dr. Nogar has, I think, brought this point out (so far as it involves an issue at once historical and philosophical) better than anyone else. After examining all the various uses to which Aristotle put the term "nature" and discriminating among these which was the basic usage so far as Aristotle's own explanations of observed data are concerned, Dr. Nogar was able to remark:

Some of Aristotle's commentators attempted to make nature a thing intrinsically generated by which the progeny was organized and exercised its energies. For them,

[74] Bergson, *The Creative Mind*, p. 102.
[75] Thomas Aquinas, *In duodecim libros metaphysicorum Aristotelis expositio*, Bk. V, lect. 9, n. 896 (see also n. 895).
[76] *Ibid.*, Bk. II, lect. 2, n. 298.
[77] Bergson, *The Creative Mind*, p. 22.
[78] Aquinas, *In I Met.*, lect. 1, n. 21.
[79] *Ibid.*, Bk. VII, lect. 7, n. 1422.
[80] Bergson, *Creative Evolution*, p. 114. I am quite well aware that in having contextualized him thus I have given what Maritain has called the "Bergsonism of intention" precedence over the "Bergsonism of fact."

nature became an object like a motor in a car. Aristotle insisted upon the *relative* meaning of the concept: it signified the determined relationship of the generator and the generated by which the thing generated received its characteristic organization and activity, whether living or not. By inferential steps from inheritance in the wide sense of any cosmic natural generation, Aristotle formulated his physical definition of nature. *Nature is the spontaneous source and cause of activity and passivity endowed by the generator upon the natural entity generated, intrinsically determining its fundamental characteristics and attributes both structural and functional.* Nature, then, as a principle and cause of characteristic activity and receptivity received from the generator, is an empirical coordinate concept, that is to say, one based upon the investigation of the constant, stable, typical, and unique relation set up by the generator's self-replication.[81]

In short, once Aristotle made Plato's transcendent forms into universal natures existing outside the mind only in things singular and perishable, natural species could no longer be regarded as eternal and immutable (on the side of the things themselves) for epistemological reasons. Once immanent, natural species could be "no more and no less permanent, stable, unique, and constant than the relation of generator-generated manifested under the closest scrutiny."[82] Henceforward, so long as a thinker wished to remain within the order of a natural philosophical assignation of reasons for being, "a mathematical or metaphysical conception of essence as an absolutely fixed and eternal idea cannot be superimposed upon natural bodies, except in the sense of an ideal type, and one must be careful here not to drift into the idealism of Plato and imagine that the real horse is the idea, and the domestic horse but a shadow of reality. As an archetype or idea, the horse can be conceived of as free of the ravages of time, but the natural history of the horse family shows it to be about 60 million years old with an estimated evolutionary rate of 0.15 genera per million years."[83]

[81] Raymond J. Nogar, "Evolution: Scientific and Philosophical Dimensions," in *Philosophy of Biology*, ed. by V. E. Smith (New York: St. John's University Press, 1962), pp. 57–58. "Nature, therefore, as the relative relation of the generator to the generated, the parent to the progeny in organic beings, is *dynamic* and *changing*, and must be conceived as of the temporal order. It is important that the permanence and stability of natural bodies be acknowledged, for regularity, unicity, and type are evident. But the permanence and stability, even of species, is no greater than the stability of the relation of the generator to the generated." (Nogar, *The Wisdom of Evolution*, p. 319). For the sort of qualifications that would have to be appended here to complete this analysis, see John N. Deely, "The Philosophical Dimensions of the Origin of Species," esp. Secs. VIII and IX, in *The Thomist*, XXXIII (April, 1969), pp. 305–335.

[82] *Ibid.*, p. 59.

[83] R. J. Nogar, *The Wisdom of Evolution*, p. 319. Similarly, "the cat is not eternal, except in the mind of the one who conceives the cat in a set, ordered complex of characteristics. The species, even as it is maintaining itself in existence, is realizing its virtualities and potentialities. It is undergoing mutations which, in turn, effect changes in the materials of heredity. The cat family has proliferated many new species, some of which have become extinct." (*Ibid.*, p. 334). Cf. Martin Heidegger's analysis of "The Limitation of Being," Part IV of his *An Introduction to Metaphysics*, trans. by Ralph Manheim (New Haven: Yale, 1959), esp. "Being and the Ought," pp. 196–199.

That is why—in accord with the requirements of the type of explanation Aristotle essayed—in seeking to understand why Aristotle regarded the specific types of things as given once and for all and why he did in fact attribute positive eternity to species, we must turn not to his metaphysics of knowledge, but to the observations to which he sought to assign reasons for being.

D. Aristotle's eternal species

In the first place we must observe that not only does like beget like, which Eiseley rightly designates "the first fact in our experience,"[84] but that within the world of experience open to Aristotle *not everything exhibited itself as subject to radical transformation.*

Everything which exists on earth, in the "sphere below the moon," experience showed to be subject to generation and corruption; but beyond this sublunary sphere experience seemed to attest to a quite different state of things. "The reason why the primary body," i.e., the heavens, "is eternal and not subject to increase or diminution, but unaging and unalterable and unmodified, will be clear from what has been said," said (as Aristotle has already reminded his readers at the opening of the chapter of the *Treatise on the Heavens* from which this quote is taken) "in part by way of assumption and in part by way of proof." Moreover, and what is decisive for the type of natural science Aristotle has undertaken to realize, "our theory seems to confirm experience," direct experience, not merely mathematically rationalizable experience, "and be confirmed by it."[85]

The truth of it is also clear from the evidence of the senses, enough at least to warrant the assent of human faith. *For in the whole range of time past, so far as our inherited records reach, no change appears to have taken place either in the whole scheme of the outermost heaven or in any of its proper parts.* The common name, too, which has been handed down from our distant ancestors even to our own day, seems to show that they conceived of it in the fashion which we have been expressing. . . . And so, *implying that the primary body is something else beyond earth, fire, air, and water* [i.e., something else beyond the types of matter which we find in the sublunary sphere], they gave the highest place a name of its own, *aeither*, derived from the fact that it 'runs always' for an eternity of time.[86]

We are here at the heart of the matter. So far as the thinking of Aristotle guided the thought of antiquity and medieval times, it was this

[84] Loren Eiseley, *Darwin's Century* (London: Scientific Book Guild, 1958), p. 208.
[85] Aristotle, *De Caelo*, Book I, ch. 3, 270b1–5. J. L. Stocks' translation in *The Basic Works of Aristotle*, Richard McKeon, ed. (New York: Random House, 1941), p. 402.
[86] *Ibid.*, 270b 1–25, p. 403 (here I have departed from Stocks' translation slightly), my emphases.

notion of the eternal heavens which provided a seemingly ontological and not merely epistemological *ratio* for the fixity of species. Hence it is not correct to say that, within the framework of Aristotelian metaphysics and psychology, it was the theory of scientific knowledge which required the positive fixity of species. Within that framework, this assertion of Marjorie Grene has been shown to be mistaken. Most fundamentally it was the enculturated conception of the eternal heavens which deflected even the most penetrating of the classical and medieval analyses of the ontological character of essential structures.

For Aristotle and St. Thomas, it was the eternal space-time of the changeless celestial spheres which determined the place and order of sublunary bodies; and so founded the rigid necessity and formal immutability of their natures. The Aristotelian essences of material beings, including those of living organisms, did not have their cosmological (or ecological) reference to what is understood today by the physical environment. That reference was rather to the unchanging heavens which, as instruments of the separated intelligences (identified in some, though not all, schools of medieval theology as the angels of revelation[87]) were regarded as the *universalis regitiva virtus generationum et corruptionum,* the governing power regulating the interactions of the terrestrial world of natures. But this physical image of the universe—a physical image, moreover, which survived into Galilean times—originally was constructed in the mathematical rather than the philosophical mode of explanation by Eudoxus (later Ptolemy) and other astronomers of the Platonic school. Later transposed with insufficient critical care ("in part by way of assumption, in part by way of experience," as Aristotle said) into a tendentially ontological explanatory framework, it became the principal factor determining first Aristotle's and later Aquinas' attitude toward the fixity of species, and especially of biological species. For example, in Aquinas' *Commentary* on the third book of Aristotle's *Metaphysics,* we find these remarks:

. . . the Philosopher shows that the first active or moving principles of all things are the same, but in relation to a certain order of rank. For first indeed are the principles without qualification incorruptible and immobile [to wit, the separated intelligences]. There are however, following on these, the incorruptible and mobile principles, to wit, the heavenly bodies, which by their motion cause generation and corruption in the world.[88]

---

[87] See James A. Weisheipl's study, "The Celestial Movers in Medieval Physics," in *The Dignity of Science,* edited by James A. Weisheipl, D.Phil. [Oxon.], (Washington, 1961), pp. 150–190.

[88] *In III Met.,* lect. 11, n. 487. See Aristotle, *Metaphysics,* Bk. XII, 1073a14–1073b17, for the "demonstration" referred to, and Aquinas *Commentary,* Bk. XII, lect. 9, "The Number of Primary Movers." (See Maritain's critical note in this regard in *The Degrees of Knowledge,* p. 224 fn. 1.)

Similarly, in commenting on the seventh book in connection with the key problem of the possible origin of living from non-living matter, what the ancients referred to as "spontaneous generation" and what we today call "biopoesis," reference is again made "to the power of the heavens, which is the universal regulating power of generations and corruptions in earthly bodies. . . ."[89]

This state of affairs in the world-view of classical antiquity, as Jacques Maritain has taken such care to manifest, is not without irony. Plato had perceived in a very clear fashion the proper method and intrinsic requirements of the mathematical mode of reasoning about nature. "In virtue of an admirable intuition of the proper conditions of physico-mathematical knowledge and of what are called the exact sciences, when ceasing to be purely mathematical, they undertake to explain the world of experience,"[90] Plato saw clearly that the creation of scientific myths is a necessary consequence of the explanatory mode of the mathematical physicist. For in physico-mathematical formulae it is impossible to differentiate the symbolized from the symbol—what belongs to the real as reasoned from what belongs to the pre- or extra-mental consistency of reality, the real as such or as real.

Aristotle, however, set about a typically different task, a task different according to its epistemological mode and type, I mean; a task the possibility of which Plato had not seen. "He founded the philosophy of sensible nature. And to do that he had to attack the Platonic metaphysics and the theory of Ideas."[91] But although Aristotle acknowledged the existence of a knowledge of nature mathematical in mode, he seems not to have grasped clearly the consequences of employing its method, however instrumentally; so that, on the one hand, although in his astronomical theory of homocentric spheres, he had himself constructed a first-rate mathematical-physical myth, on the other hand "he seems to have accorded to these spheres a full ontological value, a reality not only fundamental (as to their foundation in the nature of things) but formal and entire (as to their formality, to their thinkable constituent itself.)"[92] Accepting as the basis of his treatise *De Caelo* the universe of concentric spheres with axes at

---

[89] Aquinas, *In VII Met.*, lect. 6, n. 1403. See also nn. 1400–1401. For the contemporary state of the question concerning the problem of the origin of life in the evolutionary process, see J. N. Deely, "The Philosophical Dimensions of the Origin of Species," Part II, *The Thomist*, XXXIII (April, 1969), pp. 321–325, and references therein cited. L. Henderson's *The Fitness of the Environment*, "An Inquiry into the Biological Significance of the Properties of Matter" (Boston: Beacon Press Paperback, 1958), remains a classic statement of the wider context of this problem; while the historical development and permutations of thought on the issue can be found in any good history of the life sciences—e.g., E. Nordenskiold, *The History of Biology* (New York: Tudor, 1935); C. Singer, *The Story of Living Things* (New York: Harper, 1931).
[90] Maritain, *The Degrees of Knowledge*, p. 162.
[91] *Ibid.*, pp. 162–163.
[92] *Ibid.*, p. 163.

diverse angles in relation to one another postulated by the mathematicians, Aristotle attempted to supply the physical explanation of this mathematical diagram. "Because the point of view of the philosopher of nature predominated in him he did not see as well as Plato did the aspect of ideality necessarily embodied in the mathematical knowledge of the phenomena of nature precisely as exact science";[93] and to this oversight, coupled with Aristotle's careful observation of the reproductive pattern of living things and near-complete ignorance (*nescience*, to speak formally) of the fossil record, we owe the attitude and opinion of classical antiquity, so far as it was shaped by Aristotelian writings, on the fixity of species.

E. The fixity of species in medieval and early modern times

Within this classical tradition, the fact that the basis for opinion on the fixity of species derived from a crossing of two epistemological types, two diverse explanatory modes, seems to have raised no doubt so far as the general physical image of the universe was concerned. Nor was there any particular reason why it should have, so long as such arguments for the stationary earth as that from stellar parallax remained as the only arguments accounting for the then known evidences, "for we are obliged, at any time, to judge in the light of the evidence that is then available." Nevertheless, the "crossing" of the two explanatory modes from which the physical image in question sprang did not itself go unnoticed:

In seeking to provide an explanation for some datum, reason can be employed in either of two ways. In the first place, it can be so employed as to establish sufficiently the reasons for the fact, as in philosophy (*in scientia naturali*) there seem to be reasons sufficient for demonstrating that the movement of the heavens is of a uniform velocity [here we see the physical image of Aristotle's universe maintained]; but reason can also be employed in another fashion which does not establish the reasons for the fact, but which rather shows that explanatory reasons proposed in advance are congruous with the fact to be understood, as instanced for example in astronomy, where the theory of eccentrics and epicycles is proposed for the simple reason that the sensible appearances of the heavenly movements can thereby be saved. This latter type of explanation cannot suffice to prove anything, however, for it may well be that [as Copernicus, Galileo, and later Newton, to be followed first by Einstein and then by . . . ? would amply demonstrate] these appearances could be equally well saved within the framework of other theories.[94]

---

[93] *Ibid.*

[94] Thomas Aquinas, *Summa*, I, q. 32, art. 1 ad 2. In lect. 17 of his *Commentary* on the second book of Aristotle's *De Caelo (Treatise on the Heavens)*, n. 451, Aquinas, tracing the attempts made first by Eudoxus and subsequently by various others to account for the occasional shifts in the velocity of the planets, summarizes with this observation: "The suppositions proposed by none of these men are necessarily true: for although by granting such suppositions the appearances would be saved, it still is not necessary to say that they are true supposi-

Nor did the possible insufficiency of the evidence leading to the view of species as fixed go unremarked. Commenting on Aristotle's conclusion that the heavens are unalterable, Aquinas is careful to point out that although this view is based on the long experience and common estimation of men, nonetheless, it is a view formed concerning matters which men are able to observe only "at intervals and from a great distance." Consequently, the immutability of the heavens (upon which, it will be remembered, the eternity of species was predicated) is "a view which can only be considered probable, not certain."

For the longer-lived anything is, the more time is required for its changes to be manifest, as for example there can be changes in a man over the course of two or three years which will not be evident, whereas similar changes in a dog, or any other animal having a more rapid metabolism than man, would be readily observable within such an interval. One could argue accordingly that although the heavens are subject to transformation, the processes of change within them are of such a scale that the entire span of recorded history is not yet sufficient for making them manifest to us.[95]

There is no doubt that all this was neglected by the self-styled Aristotelians of the seventeenth and eighteenth centuries. So easily does habit substitute itself for evidence in human reasonings that, for most of them, what Marjorie Grene says of Aristotle himself holds good. So far as their understanding went, "the round of nature imitates the round of the celestial spheres: so that while father generates son and not son father, man generates man [camels are camels are camels, Nogar delighted to say, summarizing their essential world-view], eternally. Only the fixity of each ontogenic pattern through the eternity of species makes Aristotelian nature and Aristotelian knowledge possible."[96] It must be said, however, that Miss

---

tions, because it is possible that the appearances could be saved with respect to the stars and planets according to some other explanatory scheme *not yet conceived of by men. Notwithstanding, Aristotle employed suppositions of the above mentioned sort as though they were true* so far as the character of the celestial motions is concerned." Is what Aristotle did in this case so different from what contemporary mathematical-physicists undertake? Cf. A. Einstein and L. Infeld, *The Evolution of Physics* (New York: Simon and Schuster, 1950).

[95] Aquinas, *In I de caelo*, lect. 7, n. 77. For a textual analysis of the conceptual function of the celestial bodies in the thought of classical antiquity, see Thomas Litt, *Les corps célestes dans l'univers de saint Thomas d'Aquin* (Paris: Nauwelaerts, 1963). For an analysis of the theoretical implications of the removal of this function, see John Deely, "The Philosophical Dimensions of the Origin of Species," Parts I and II, *The Thomist*, XXXIII (January and April, 1969), pp. 75–149 and 251–342, respectively; and the reading selection by Benedict Ashley, "Change and Process," which appears as the second reading in Part II, Section IV of this volume, The Metaphysical Issues.

[96] Marjorie Grene, *A Portrait of Aristotle*, p. 137. See fn. 68 above; also Maritain, "The Conflict of Methods at the End of the Middle Ages," *The Thomist*, III (October, 1941), pp. 531–533; and Ashley, "Aristotle's Sluggish Earth, Part II: Media of Demonstration," *The New Scholasticism*, XXXII (April, 1958), "Conclusion," pp. 230–234; and the references in fn. 95 above.

Grene's treatment of Aristotle himself here is wide of the mark. We may summarize this whole issue by passing on the judgment of Maritain, which is the only one that will bear the weight of the evidence we have adduced: "Despite what certain popularizers may say (and even those thinkers who attribute to the ancients their own carelessness in distinguishing the intelligible from the topographical, and metaphysics from astronomy), these charges do not stand up in the case of the philosophy of Aristotle when carried back to its authentic principles."[97] "And it is fascinating to speculate how," J. H. Randall is quick to add, "had it been possible in the seventeenth century to reconstruct rather than abandon Aristotle, we might have been saved several centuries of gross confusion and error."[98]

Such "reconstruction," however, actually amounts to no more than a clear recognition of the two typically distinct ways in which reason can function in seeking to explain our experience of natural phenomena, with an equally clear appreciation of the power and limits of each of these epistemological types.

What is really new in the achievements of the science which became predominant in the 16th and 17th centuries, of "modern science," is properly speaking a physics of the physico-mathematical type. (In other scientific domains which are not thus absorbed by mathematics, modern science doubtless owes its material or technical perfection, and an autonomous conceptual lexicon which permits infinite progress in the analysis of phenomena as such, to the attraction exerted by physico-mathematics on the other kinds of knowledge, which henceforth see in the former the exemplar of knowledge.)

In truth, the epistemological principles of the Ancients considered in their very nature could easily have adapted themselves to the new physics; the logical type to which that science corresponds, and of which *astronomy* was the best example during antiquity, theoretically had its place set down in the Scholastic synthesis of sciences. This logical type is that of a science in a sense intermediary between Mathematics and Physics, but actually *mathematical* as regards its typical mode of explanation, since what is *formal* and consequently specifying in it (its formal object and its medium of demonstration) is mathematical. The explanatory deduction is mathematical. Physical reality, although of prime importance to it as subject-matter, is basically, for it, a material reservoir of facts and verifications. And thus, while natural philosophy may be characterized as a physical knowledge properly philosophical and ontological, or metaphysical by participation,[99] the new physics, on the contrary, according to the methodological principles of the Schoolmen, must be called a physical knowledge properly mathematical, or a science *formally* mathe-

---

[97] J. Maritain, *The Degrees of Knowledge*, p. 59. See also "The Conflict of Methods . . . ," pp. 532–533; and Deely, "The Philosophical Dimensions of the Origin of Species," Part I, Sections III, IV, and V, esp. pp. 136–137 and 145–146.

[98] Randall, *Aristotle*, p. 165.

[99] On this point, see Maritain, *The Degrees of Knowledge*, "Table of the Sciences," pp. 38–46, and "Structures and Methods of the Principal Kinds of Knowledge," pp. 53–60, for indispensable clarifications.

matical and *materially* physical: a formula which . . . condenses all its properties.[100] Doubtless the conception of physico-mathematics that the French physicist Pierre Duhem (and Hirn before him) defended was too mathematical and not sufficiently physical, nevertheless Duhem was right in thinking that the phenomena can be analyzed quantitatively without the existence of qualities being denied, and that the scientific method derived from Galileo and Descartes can be used without its involving in any way philosophical incompatibility with Aristotle's metaphysics.[101]

But in point of fact knowledge of a physico-mathematical sort was limited among the Ancients and the Schoolmen to certain very particular disciplines, such as astronomy and acoustics; . . . and . . . the idea of establishing a universal mathematical interpretation of physical reality by submitting the fluent detail of phenomena to the science of number nevertheless remained foreign to most of them.

Therefore, on the day when quantitative physics, having its own specific character and possessed of its own exigencies, moved to take its place within the order of sciences and to proclaim its rights, it was to enter inevitably into conflict with the philosophy of old—not only because of the error which vitiated the latter in the experimental field, but also, and this is more remarkable, because of the radical difference which separates—with respect to what is genuine and legitimate in each one of those two manners—the old manner of approaching physical realities from the new manner of approaching them.[102]

F. The conflict of methods at the dawn of modern times: The ascension of platonism and its consequences

For the world of classical antiquity, it was the power of the philosophical mode of explanation, that is, the possibility of assigning reasons for being in terms of principles and causes, which preoccupied thinkers, to the point of inclining them to overlook the limits imposed on such explanations by the state of empirical research at any given period. This in turn led to carelessness with respect to their intrinsic dependency on the sciences of observation, and so to leaving aside the proper task of a straightforward philosophy of nature, which is to assign the reasons for what is given to it and to gain an understanding of that datum. By the time of Galileo and Descartes, custom and long habituation to one mode of explanation had converted this tendency into a veritable myopia.

For the post-classical but pre-Darwinian world, it was just the opposite inclination which carried the day. Thinkers of this period were over-

---

[100] See *ibid.*, "Knowledge of the Physico-Mathematical Type and Philosophy," pp. 60–64, and "Modern Physics Considered in its General Epistemological Type," pp. 138–140.
[101] In any way, that is, save sometimes in the order of imagination with its explanatory images (cf. Maritain, *The Degrees of Knowledge*, pp. 64, 179–180, 181, 182–184). For an interpretation of the new physics exclusively in terms of the changed image of the physical world wrought in the sixteenth and seventeenth centuries, and not at all in terms of the epistemological principles determining the topography of the mind, see I. Bernard Cohen, *The Birth of a New Physics* (New York: Anchor, 1960).
[102] J. Maritain, "The Conflict of Methods . . . ," pp. 529–531.

whelmed with the discovery of the possibility of a universal knowledge of nature which would, on the one hand, save the sensible appearances, while on the other hand it would, by recomposing those very appearances in the field of mathematical ideality, make possible a type of deduction leading to a measure of prediction and control of physical phenomena completely impossible within the order of ontological explanation. But such a reconstituted universe, to the extent its creators became fascinated with the power of their explanatory mode to the neglect of its limits, tended, for reasons we have already indicated, toward mechanism and toward a mechanistically deterministic view of nature as toward an ideal limit.

Thus it is not enough to note that the physico-mathematical method was finally evolved by anti-scholastic thinkers, whom the very excess of their confidence in the application of mathematics to sense-perceived nature led instinctively to mechanism. We also must emphasize that the natural and irrepressible drive of the intelligence towards being and causes, when it met the physico-mathematical method must almost necessarily cause this discipline to be mistaken for a natural philosophy. It is because of this almost inevitable illusion that the new scientific method found itself from the very beginning, by virtue of the historical conditions of its genesis, quite ready to undergo the contamination of extraneous philosophies, and to become dependent upon a metaphysics like the Cartesian mathematicism—an accident inversely resembling that which had linked Aristotle's metaphysics to the erroneous theories of the physics of old.[103]

In the same way that the explanatory inclination which dominated in classical antiquity was brought up short in the face of physics in the modern sense of the word, the explanatory inclination of the modern period and of the so-called (because formally mathematical) exact sciences has in its turn "been brought up short in face of biology and experimental science (to say nothing of the moral sciences which concern philosophy even more closely.)"[104] The evolutionists have become quite sensitive on this point. Thus Mayr, following Scriven, considers that "one of the most important contributions to philosophy made by the evolutionary theory is that it has demonstrated the independence of explanation and prediction," that "indeterminacy does not mean lack of cause, but merely unpredictability."[105] Randall's research has led him to concur in Mayr's conclusion in this regard, though for slightly different reasons: "We often think this is a recent modern discovery, because it was forgotten in the seventeenth century."[106]

[103] *Ibid.*, p. 535.
[104] Maritain, *The Degrees of Knowledge*, pp. 45–46. See Randall, "The Significance of Aristotle's Natural Philosophy," in his *Aristotle*, pp. 165–172.
[105] Ernst Mayr, "Cause and Effect in Biology," *Science*, 134 (10 November 1961), pp. 1504 and 1505, respectively. See Michael Scriven, "Explanation and Prediction," *Science* 130, (26 August 1959), pp. 477–482.
[106] Randall, *Aristotle*, p. 184.

It is in this light that one best appreciates Hayek's remarks concerning the contemporary theory of natural selection:

It is significant that this theory has always been something of a stumbling block for the dominant conception of scientific method. It certainly does not fit the orthodox criteria of 'prediction and control' as the hallmarks of scientific method. Yet it cannot be denied that it has become the successful foundation of the whole of modern biology.[107]

What we are witnessing in our own day seems to be that very "reconstruction" which did not come about in the seventeenth century (or eighteenth, or nineteenth), and which perhaps *could not* come about before it was possible in the actual light of history to envisage not only the explanatory power but also the limits intrinsic, in typically appropriate ways, to both of these pure epistemological types.[108] For "whoever seeks to work towards the integrating of philosophy and experimental science must be at once on his guard against both a lazy separatism and a facile concordism and re-establish a vital bond between them without upsetting the distinctions and hierarchies which are essential to the universe of knowing."[109] What is called for is nothing more or less than an adequate perspective by which an integrated synthesis of all sciences can be joined with the autonomy proper to each distinct science. To this end, it seems, "two cases should be very clearly distinguished: the case of physico-mathematical science and the sciences of which it is the type, on the one hand, and the case of sciences like biology and psychology on the other,"[110] sciences of which philosophical explanation is the type.

Like the Monday morning quarterback, we are speaking here with the wisdom of retrospect. In the immediate press of historical events, essential structures are seldom so generous in showing themselves. Nonetheless, be it only retrospective, wisdom is still wisdom, and we must distinguish two very different aspects in the revolution which took place under the impetus of Galileo's turning of his telescope toward the sun and discovery of spots moving across its face, thus spelling the ruin of that image of the physical universe which depended on the perfection and immutability of the celestial spheres. As soon as the distinction between heavenly or incorruptible and earthly or corruptible matter was disposed of, the "ruling cause" of generations and corruptions in the sublunary

---

[107] F. A. Hayek, "The Theory of Complex Phenomena," in *The Critical Approach to Science and Philosophy*, Mario Bunge, ed. (Glencoe: Free Press, 1964), pp. 340–341. In section V of this essay, "The theory of evolution as an instance of pattern prediction," pp. 340–343, Hayek sets forth very clearly the ontological, a-heuristic or "negative" and non-imaginative explanatory structure of contemporary evolutionary theory as it grounds itself in the concept of natural selection.

[108] Maritain, in "The Conflict of Methods . . . ," p. 534, expresses a similar view.

[109] Maritain, *The Degrees of Knowledge*, p. 60.

[110] *Ibid.*

region was also removed. This is the first aspect of the revolution which began with Galileo.

The consequences of this as a speculative event, and not merely as a cultural event, can only be understood speculatively.[111] Traditional analyses of essence cannot be judged adequately unless they are first thought through in the light of such observations as this one by the outstanding evolutionist, Theodosius Dobzhansky:

> If the environment was absolutely constant, one could conceive of formation of ideal genotypes [i.e., substantial structures] each of which would be perfectly adapted to a certain niche in this environment. In such a static world, evolution might accomplish its task and come to a standstill [one thinks of the *Timaeus*]; doing away with the mutation process would be the ultimate improvement. *The world of reality, however, is not static.*[112]

If not psychologically, at least logically, and if one wished to remain within the explanatory mode which seeks to assign reasons for being and not just provide freely constructed schema sufficing to "save the appearances," there was an immediate and inescapable consequence to the disappearance of the celestial spheres as *causa regitiva*, as regulative or "ruling" cause: the habits of typological thinking, which were intrinsically determined by the assumption of such a "ruling cause," had to be severely modified, and in several respects abandoned. For if the phenomena are to be comprehended in an environment where *nothing* is free from substantial change, and if the exceptionless law of causality is to be integrally respected, there must be more to grasping the nature of a species than ignoring all individual peculiarities of the members which make it up. The humanly envisaged ideal "types" no longer bear a simple and immediate relation to what the natural processes are "up to." When the immediate and invariant relation of forms to the *causa regitiva* is suppressed, in short, specific fixity is replaced by the developmental potentialities of the particular individual within the specific population having to realize themselves in unusual as well as usual circumstances. In such a milieu, it becomes a question of understanding the real event and individual, not as an instanced ideal, but as an interaction product.

This aspect of the Galilean revolution remained purely virtual, however, until the actual establishment of evolutionary theory and the develop-

---

[111] "In medieval literature, the problem of celestial movers was not created by theologians. . . . The problem of celestial movers was entirely a scientific one having many ramifications." (Weisheipl, *art. cit.*, pp. 151–152). Indeed, practically all of the neo-scholastic dilemmas before the evolutionary problematic may be directly traced to their inherited tendency to treat scientific problems, whether pure or mixed, *modo philosophico*: see Part Three of M. J. Adler's study, *The Conditions of Philosophy* (New York: Atheneum, 1965), pp. 231–294, Ch. 15 in particular, pp. 250–261, esp. p. 256.

[112] Theodosius Dobzhansky, *Genetics and the Origin of Species* (3rd ed., rev.; New York: Columbia, 1951), p. 74: emphasis supplied.

ment in our own century, through the marvellous writings of such men as Mayr, Huxley, and Simpson, of taxonomic classification techniques which gradually found ways to conform to the intricate patterns of phylogenetic descent within evolutionary lines.[113] Certainly, a good deal more reflection, especially at a purely philosophical level, is necessary in this regard. Nonetheless, it is already possible in these terms to see that the ultimate import of the Galilean revolution for philosophy derives from the sciences of observation, understanding philosophy as synonymous with the natural-physical in contrast with the mathematical-physical mode of explanation, and hence as applying to every science or art capable of understanding not just the mathematical forms but the proper causes of the things it studies.[114]

The decisive difference between the classical and contemporary world-view turns out to be neither a preference for typically distinct

---

[113] See, for example: Ernst Mayr, *Systematics and the Origin of Species* (New York: Columbia, 1942); E. Mayr, E. G. Linsley, and R. L. Usinger, *Methods and Principles of Systematic Zoology* (New York: McGraw-Hill, 1953); Julian S. Huxley, ed., *The New Systematics* (Oxford, 1940); George Gaylord Simpson, *Principles of Animal Taxonomy* (New York: Columbia, 1961); Glenn L. Jepsen, George Gaylord Simpson, and Ernst Mayr, editors, *Genetics, Paleontology, and Evolution* (New York: Princeton, 1949).

[114] "From one point of view, there can be no difference between a philosophical analysis of nature and a study of nature in the scientific sense: all explanation, when it is not mathematical, assigns reasons for being.

"Yet from another point of view, there is a sense in which natural philosophy and natural science do subdivide the order of rational knowledge of nature. The question of the difference between philosophy and science and of the relations which their respective explanations sustain is a difficult one which, perhaps more than any other, has exercised contemporary reflections. It is not even agreed among those who treat this question that philosophy constitutes a mode of rational understanding in its own right; but I think that once it is seen that, with respect to the sensible, natural world, just as there are some questions for which laboratory or field research is indispensable (e.g., how does photosynthesis take place? what is the average life-span of a star? are there extinct life forms? or why do lemmings recurrently plunge to their death in the sea?), so also there are other questions for which such research is adventitious (e.g., what is change? what is chance? what is place? what is the basis for the prior possibility of agreement? why is change possible? how do relations exist? is the existence of an order of purely spiritual beings a real possibility?), then it is necessary to admit that (a) in relation to experience science and philosophy differ between themselves according to the manner in which their respective explanations depend thereon, and that (b) both belong to the order of rational understanding, sharing formally in an identical set and sequence [formally, not at all materially, speaking] of questions. (We should recognize, too, that not all questions can be assigned preclusively to either science or philosophy; there are also, so to speak, 'hybrid' questions, questions for which laboratory or field research are superflous in certain respects and helpful in others—such questions as, how is man unique? what is good for man? what role does chance play in the constitution of the world? or . . . what are the natural species or kinds which ordinary experience indicates to be real?)

"This assertion depends, of course, on there being a real analytical distinction between knowledge and experience. That there is such a distinction is plain from the fact that experience functions both as a source and as a test of our knowledge in both philosophy and science, which would be impossible if the two were not somehow distinct. In fact, this difference between experience and knowledge lies at the base of rational understanding, inasmuch as it defines the respective spheres or orders of observation and explanation." (John N. Deely, "The Philosophical Dimensions of the Origin of Species," Part I, *The Thomist*, XXXIII [January, 1969], pp. 101–102, text and fn. 50.) See further references in fn. 42 above.

explanatory modes nor a mere transformation in the physical image of the universe, but rather a *datum*, an element of experience for which no logical construction can be substituted and upon which all the logical constructions of the science of nature finally rest, the realization, specifically, that nothing in the universe is exempt from radical transformation.

The second aspect of the revolution which began with Galileo and which we must now distinguish, however, was the aspect that was not merely virtual both historically and philosophically, but *formally constitutive* from a culturological standpoint. This aspect pertains to sciences of explanation rather than to those of observation. Indeed, the ontological implications of the spots on the sun were the last thing the revolutionaries wished to follow out; it sufficed for them that these sun-spots invalidated, or rather smashed, the traditional image of the physical world. "When the historic conflict between the Aristotelian physics and the new physics opened, both sides were equally convinced that this was a conflict between two philosophies of nature."[115] The physico-mathematical science founded by Galileo and Descartes was taken by its proponents as a philosophy of nature and indeed as the only authentic one. Moreover, the predictions it made possible were put forward as *proofs* of the validity of its explanations.[116]

Then it happened that the Cartesian mechanism achieved the obliteration of the old distinction between the philosopher of nature (*physicus*) and the mathematical interpreter of nature (*astronomus, musicus* . . .). When we re-read the great work of Newton significantly titled *Philosophiae Naturalis Principia Mathematica*, we realize that the Newtonian science, once considered by positivists as the archetype of positive knowledge, was far from having rid itself of ontological ambitions.[117]

The post-Galilean thinkers did not explore the possibility of understanding the phenomena of nature in their full physical reality as interaction products, for which it is necessary to retain the notion of natural units or substance.[118] Instead, they set about the comprehension of concrete particular cases in the quite different way of applying to the details of phenomena themselves, just as they are coordinated in time and space, the purely formal connections of mathematically expressible relations. In this way they came to seek as the ideal type of modern science a knowledge "to be at once experimental (in its matter) and deductive

---

[115] Yves Simon, "Maritain's Philosophy of the Sciences," p. 97.
[116] Already a sign of profound confusion over the nature of reason's distinct modes: see text above at fn. 94.
[117] Simon, *art. cit.*, p. 97. See the text of Newton's letter to Fr. Pardies, cited in text below at fn. 179.
[118] See esp. Ashley, "Does Natural Science Attain Nature or Only the Phenomena?"; and his "Change and Process," second reading in Part II, Section IV of this volume, The Metaphysical Issues.

(in its form, but above all as regards the laws of the variations of the quantities involved)," capable for that very reason of devising endlessly varied "means of utilizing sensible nature (from the point of view of quantity indeed, but not from the point of view of being)."[119] Within such an ideal type, to be sure, "lawfulness is no longer limited to cases which occur regularly or frequently but is characteristic of every physical event. . . . Even a particular case is then assumed, without more ado, to be lawful."[120] "The step from the particular case to law . . . is automatically and immediately given by the principle of the exceptionless lawfulness of physical events."[121] What happens to chance factors in this new world-picture? (And it is indeed a picture, a *physical image* comparable in every way to the astrological attitude toward the celestial spheres as all-determining). "The distinction between lawful and chance events" is eliminated.[122] The *actual course* of history, the "configurational" in Simpson's sense, ceases to have any causal import or consequently any explanatory potential; it becomes the accident which must be discarded if we are to understand the real determinants of any given process.[123] Propositions formed in the Aristotelian explanatory mode, it is acknowledged, "show an immediate reference to the historically given reality and to the actual course of events."[124] By contrast, it is likewise acknowledged, concepts formed in the Galilean explanatory mode "unquestionably have in comparison with Aristotelian empiricism a much less empirical, a much more constructive character than the Aristotelian concepts, based immediately upon historical actuality."[125] But we are assured (by one who has certainly never grasped the type of explanation essayed by Aristotle and its implications for a world with no eternal heavens) that any immediate reference to the historically given reality and the actual course of events "really means giving up the attempt to understand the particular, always situation-determined [=process- as

---

[119] Maritain, *The Degrees of Knowledge*, p. 45. See also Maritain, *Science and Wisdom* (New York: Scribner's, 1940), pp. 43–44; and Gavin Ardley, *Aquinas and Kant* (New York: Longmans, 1950).

[120] Kurt Lewin, "The Conflict Between Aristotelian and Galileian Modes of Thought in Contemporary Psychology," Ch. 1 of *A Dynamic Theory of Personality: Selected Papers of Kurt Lewin* (New York: McGraw-Hill, 1935), pp. 25–26. (The references in fn. 105 above are relevant here.)

[121] *Ibid.*, p. 31.

[122] *Ibid.*, pp. 6 and 25–26.

[123] *Ibid.* This is so because there is no longer any criterion for distinguishing between what occurs by nature and what occurs by chance or violence, between *event* and *encounter* (see fn. 155 below), between exceptionless *causality* and exceptionless *lawfulness*—this last particularly, because in the "Galileian mode" the determinism of nature is conceived in a formally mathematical rather than ontological sense, and moreover entirely displaces the latter: see fns. 67 and 73 above.

[124] Lewin, p. 12.

[125] *Ibid.*, pp. 12–13.

distinct from product-determined] event."[126] "Galilean concepts, on the contrary, even in the course of a particular process, separate the quasi-historical," i.e., the incidental or configurational factor, the historically actual course of events, "from the factors determining the dynamics."[127]

"Thus and in similar ways," muses Simpson, "the descent from the ideal to the real in physical science has been coped with, not so much by facing it as by finding devices for ignoring it."[128] For in very truth, if there is no distinction between the causal and the lawful, if the concept of exceptionless causality is not (as Lewin assures us it is not[129]) in any way distinct from the conception of exceptionless lawfulness, if, in short, the only difference between the epistemological type of explanation incarnated by Aristotle and that incarnated by Galileo (after the Pythagoreans and in the tradition of Plato) is that the former is illusory while the latter is explanatory, then there is no way in which biology à la Darwin can be regarded as explanatory or scientific, *for within the epistemological type of Galilean concepts, data of paleontology and of pre-history generally are not expressive "of the vectors determinative"*[130] *of the dynamics of the evolutionary process.*

Again it is necessary to revert to the conclusion of Maritain, as being the only judgment capable of supporting the weight of the evidence:[131]

There are two possible ways of interpreting the conceptions of the new physics philosophically. The one transports them literally, just as they are, on to the philosophical plane, and thereby throws the mind into a zone of metaphysical confusion; the other discerns their spirit and their noetic value, in an effort to determine their proper import.[132]

---

[126] *Ibid.*, p. 31.

[127] *Ibid.*, p. 34.

[128] George Gaylord Simpson, "The Historical Factor in Science," ch. 7 of *This View of Life*, p. 129.

[129] Lewin, pp. 6, 23, 25–26, 26 fn. 1, 31, 35.

[130] *Ibid.*, pp. 33–34.

[131] A similar consideration may have been behind Mortimer Adler's judgment that Maritain seemed, in the early twentieth century, "the only contemporary philosopher who has deeply sensed the movement of history and the point at which we stand." (*What Man Has Made of Man* [New York: Ungar, 1937], p. 242). See further Mortimer Adler, "The Next Twenty-five Years in Philosophy," *The New Scholasticism*, XXV (January, 1951), pp. 92–95; and Yves Simon, "Maritain's Philosophy of the Sciences," pp. 93–96.

[132] Maritain, *The Degrees of Knowledge*, p. 171. It is probably not far wrong to regard this "zone of metaphysical confusion" as the principal *locus* of contemporary philosophy insofar as it has drawn its exclusive inspiration from the spectacular revolutions effected in physics by relativity and quantum theory: e.g., see Lincoln Barnett's assessment of *The Universe and Dr. Einstein* (New York: Signet, 1957). (Simon delineates the epistemological architecture of this zone in *art. cit.*, 99–100, with brief lines of masterful clarity.) See fn. 43 above.

It should be noted that the peculiar intensity of this conflict in our day between the rudimentary ontology of common sense (what Bergson once called "the natural metaphysics of the human mind") and the theoretical speculations of mathematical-physics has given rise

It is to this last interpretation that one is forced by the rise of evolutionary biology.

G. Tables reversed and conflict of method renewed: The immediate obstacles to evolutionary thought

Thus, in meditating the passage of thought from classical antiquity to Darwin's world, we find ourselves returned, not indeed to Aristotle, but rather to the epistemological type of explanation immanent in his philosophy. This return has been made possible however principally by the disappearance of the celestial spheres which formed the physical image of the classical world, while it has been made necessary if we are at last to resolve "the terrible misunderstanding which, for three centuries, has embroiled modern science and the *philosophia perennis*," and which "has given rise to great metaphysical errors to the extent it has been thought to provide a true philosophy of nature." [133] Because this return has been made against the backdrop of a realized universal knowledge of nature mathematical in mode, we ought to recognize in it not the sign of a restriction and impoverishment, but of an improvement and growth within the organic structure and differentiation of thought. "Of itself, it [i.e., physico-mathematics] was an admirable discovery from an epistemological point of view, yet one to which," once freed of its ontological pretensions, "we can quite easily assign a place in the system of sciences." [134]

---

in philosophical circles not only to those who, like the Logical Positivists, take as primary reference and point of departure the conclusions of theoretical (mathematical) physics, and maintain the primacy of this theoretical network even in its subjection to constant and radical reformulation; but also, on the other side, to those who, like the Phenomenologists, take their point of departure from the "life-world" (*Lebenswelt*) of common experience, from the fact that man is human before he is scientific, from the fact that it is common or lived experience which is at once prior to and constitutive of the very possibility for the specialized research and experiences of the scientific enterprise, and who accordingly seek to ground the primacy of common sense over scientific (mathematical) theory. Two phenomenologists in particular stand out in this connection: Edmund Husserl (1859–1938), the founder of the movement; and Maurice Merleau-Ponty (1908–1961), his greatest French disciple. *Inter alia*, see Husserl's "The *Lebenswelt* as forgotten foundation of meaning for natural science," in *Husserliana* (The Hague: Martinus Nijhoff, 1963), VI, 48 (this article appeared in *Philosophia*, 1936); *Phenomenology and the Crisis of Philosophy*, trans. by Quentin Lauer (New York: Harper Torchbooks, 1965). And Maurice Merleau-Ponty, *The Primacy of Perception* (Chicago: Northwestern, 1964); *Phenomenology of Perception* (New York: Humanities Press, 1962); *The Structure of Behavior* (Boston: Beacon Press, 1963).

[133] Maritain, *The Degrees of Knowledge*, p. 41. See Simon, *art. cit.*, pp. 96–101.

[134] *Ibid.* Well, perhaps not all that easily, after all—at least not from the side of subjectivity, of human knowledge taken in its psychological dimension. "Quite easily" after Maritain: but so was the idea of geometry "quite easy" after Euclid, or of evolution after Darwin! See Simon, "Maritain's Philosophy of the Sciences," pp. 93, 95, 96–98.

In any case, it seems that a true philosophy of the progress of the physical and mathematical sciences in the course of modern times, precisely because it is its duty to set
forth by critical reflection the spiritual values with which that progress is pregnant
. . . must therefore on the one hand reveal the essential compatibility of this mathematical and empiriometrical progress with the knowledge of the ontological type
which is proper to philosophy. On the other hand, it must respect the nature of the
Experimental Sciences, which of themselves escape from complete mathematization,
and it must render justice to their working methods, which will extend to ever
larger sections of the scientific domain the more they assert their autonomy. In
effect, it would be completely arbitrary to refuse to biology, and other sciences of
the same epistemological type, the rank of authentic knowledge. This type of knowledge merits the attention of the philosopher and it is playing an ever more important,
perhaps one day preponderant, role in the progress of speculative thought.[135]

It is necessary to keep all these historical and philosophical considerations well in mind as we move toward an assessment of the causal understanding achieved by contemporary evolutionary science. "Our first step
in the effort to understand how life became natural, therefore, is to avoid,"
writes Loren Eiseley, "the commonly held impression that Darwin, by
a solitary innovation—natural selection—transformed the Western world
view."[136]

Variation, selection, the struggle for existence, were all known before Darwin. *They
were seen, however, within the context of a different world view* [and, we must add,
within the range or perspective principally of reason's second explanatory mode].
Their true significance remained obscured or muted. . . . It was not really new facts
that were needed so much as a new way of looking at the world from an old set of
data.[137]

Eiseley lists four propositions which, looking back, had to be clarified
before the theory of organic evolution would prove acceptable to science.
First of all, it was necessary to grasp the *antiquity of the earth.* Secondly,
it was necessary to establish that there had been a true geological *succession
of life-forms* on the earth. Thirdly, the extent of individual *variation* in the
living world *and its prospective significance* in the creation of novelty had
to be grasped. And finally, *a conception of a relative, dynamic equilibrium*
had to replace the conception of the absolute, permanently balanced
world-machine. In the final analysis, it was the clarification of these four
propositions that made possible Darwin's world—and ours. Let us
examine, very briefly, each of them in turn. It is of particular interest to
note that the chief difficulties obscuring these propositions stem in every
case not from classical antiquity, but from the mentality of the seventeenth

---

[135] *Ibid.*, pp. 200–201. (See fn. 114 above.)
[136] Eiseley, *The Firmament of Time* (New York: Atheneum, 1962), p. 68.
[137] *Ibid.*, p. 72: emphasis added.

and eighteenth centuries, that is, from early modern times, and in general from the scientific even more than from the religious temper of that modern Age of Enlightenment.[138]

The first proposition, the antiquity of the earth, was taken for granted in classical times. Aristotle was reasonably confident the earth had existed from eternity. The Christian Middle Ages saw the eternity of the world as a philosophical possibility, although it seemed to them a datum of Revelation that in fact it had had a temporal beginning, with nothing much more definite said on the matter. It was left for the biblical chronologists of the sixteenth and seventeenth centuries, culminating in the work of the Irish Archbishop James Ussher (1581–1656), to exorcise the last "illusions" of antiquity with the discovery that the world had been created precisely in 4004 B.C.

For the establishment of the second proposition concerning the succession of forms, there were required the prior attempts to classify animals and plants collected from every part of the globe, for only when classification was attempted on this scale did difficulties multiply to such a point that the successors of Linnaeus (the eighteenth century Swedish botanist who principally authored the form of classification of organisms still in use today[139]) were forced to abandon the idea that species are fixed. As late as 1815, Cuvier (the great French zoologist of the early 19th century, who, though he as much as any one man made paleontology a distinct biological science, was himself a firm disbeliever in evolution) could still contend that no discovery of intermediate forms had ever been made. "An orderly and classified arrangement of life was an absolute necessity before the investigation of evolution, or even its recognition, could take place";[140] yet not before the great voyages of discovery of the late eighteenth and nineteenth centuries did such a reconstruction of "that which was" become possible. Inasmuch as the first step in any science is to know that a possible subject of investigation exists, "to know one thing from another," as Linnaeus said,[141] the possibility of evolutionary science as such began with these voyages.

The third proposition, concerning the extent of variation and its prospective significance as a source of specific transformations, had been

---

[138] Cf. Loren Eiseley, *Darwin's Century* (New York: Anchor, 1961), pp. 23–24; and Michael Ruse, "The Revolution in Biology," *Theoria*, XXXVI (1970), pp. 1–22.

[139] It is necessary to add that his authorship extends "only to the *form*, the terms and names used. There have been two revolutionary changes in the *principles* of classification since Linnaeus."—Simpson *et al.*, *Life* (New York: Harcourt, 1957), p. 462. (See esp. the references given in fn. 113 above.)

[140] Eiseley, *Darwin's Century*, p. 15.

[141] *A Selection of the Correspondence of Linnaeus and Other Naturalists from the Original Manuscripts*, Sir James Edward Smith, ed. (London: Longman, Hurst, Rees, Orme, and Brown, 1821), Vol. 2, p. 460. See Secs. III and IV of "The Philosophical Dimensions of the Origin of Species" in *The Thomist*, XXXIII (January, 1969), pp. 93–130.

obscured in the classical and medieval world. It was not the conception of scientific method found in Aristotle's *Posterior Analytics* which was the source of this obfuscation; rather was Aristotle's presentation obfuscated by the idea of eternal species.[142] And this idea, as we have seen in some detail, in turn derived from the presence in the classical representation of the universe of the unchanging heavens which kept all transformations in fixed categories. In principle, the significance of variation was brought into the open by the disappearance of the celestial spheres c. 1610. In fact, the invention, by Ray and Linnaeus and other classifiers of the period 1750–1850, of species immutable in themselves, i.e., fixed without any ecological reason, kept the whole issue as much in the dark as ever.

Much of the sound and fury surrounding evolutionary theory is due to a misapprehension of sorts. Evolution initially had no pretensions to the status of a *Weltanschauung*, nor did it seek to serve as a substitute for the Christian doctrine of creation. . . . The theory of evolution actually grew out of a conflict between *two distinct and opposing biological theories*. It was a family quarrel. The dominant biological theory was that *of a fixed and immediate creation of species*. This of course has little or no reference to the theological doctrine of creation *ex nihilo*. Nor is the concept of the fixity of species a logical deduction from the philosophical doctrine of the immutability of essence, although the genus and species of Linnaeus do carry some of the logical and conventional characteristics of the Aristotelian genus and species.[143]

Here again, it was the sciences of observation which rescued theorists from the illusions consequent on confusing the second explanatory mode of reason with the first.

Theories of cosmic evolution, of suns and planets emerging from gaseous nebulae in space, appeared almost simultaneously with the first intimations of organic change. The timeless Empyrean heaven was now seen to be, like the corrupt world itself, a place of endless change, of waxing and waning worlds. Although the fact waited upon geological demonstration, the new astronomy with its vast extent of space implied another order of time than man had heretofore known. For a little while the public would not grasp what the sky watchers had precipitated. It would have to be brought home to them by the resurrection of the past.[144]

Finally, let us note that Eiseley's fourth proposition likewise, the conception of the world as being in a relative rather than an absolute, and

---

[142] Complementary to the discussion in fn. 53 above, it is worth referring the reader at this point to the remarkable conclusions concerning the history and philosophy of science which emerge at the end of Benedict Ashley's important study of Aristotle's *Treatise on the Heavens*: see "Aristotle's Sluggish Earth, Part II," *The New Scholasticism*, XXXII (April, 1958), pp. 230–234.

[143] William E. Carlo, *Philosophy, Science and Knowledge* (Milwaukee: Bruce, 1967), p. 118. (While agreeing with Dr. Carlo's historical point here, I must disagree almost equally completely with his general assessment of the evolutionary question.)

[144] Eiseley, *Darwin's Century*, pp. 35–36.

a dynamic rather than a static, equilibrium, and the idea of the possibility of unlimited organic change which is bound up with this proposition, follows in principle upon the removal of the unchanging spheres, the *causa regitiva*. For "all that the Chain of Being actually needed to become a full-fledged evolutionary theory was the introduction into it of a conception of time in vast quantities added to mutability of form. It demanded, in other words, a universe not made but being made continuously."[145] Yet not until the end of the eighteenth century were such ideas entertained on a wide scale, and even then thanks principally to the popularity of one man, Comte de Buffon (an eighteenth century French naturalist who authored a voluminous natural history "so complete and so well written that it is still a household work in France"[146]).

Just as the clarification of the proposition concerning the significance of variation laid to rest the ghost of typological thinking (more exactly, of the reification of ideal types), so the clarification of this proposition disposed of the ghost of preformism, the view that holds, with regard to the development of individual organisms, that all the physical and psychological traits and characteristics of the mature adult are present actually though in miniature in the sex cells, ready made, so to speak, except for size; and holds, with regard to the evolutionary unfolding as a whole, that evolution is the working out of a built-in plan: it does not produce genuine change, i.e., novelty, but consists in the simple maturation of predetermined patterns.[147] Mayr refers to preformism as constituting, along with typological thinking, the "two basic philosophical concepts that were formerly widespread if not universally held," the rejection of which were indispensable preconditions for the formation of contemporary evolutionary theory.[148]

But here again, although preformist thinking goes all the way back to Hippocrates (460?–377? B.C.),[149] the particular expression of it which proved an obstacle for evolutionary science dated back no farther than Leibniz (1646–1716) and other enlightenment philosophers and scientists who conveyed the impetus of Newtonian mechanism in science and often deism in philosophy.

---

[145] *Ibid.*, p. 9. It is customary whenever "the great chain of being" is mentioned to refer enthusiastically to Professor Arthur O. Lovejoy's book of that very title. However, from the standpoint of the philosopher (as has been recently pointed out by M. J. Adler in *The Difference of Man and the Difference It Makes*, p. 57), one's enthusiasm for this particular "study of the history of an idea" must be seriously dampened by the realization that, in spite of the prodigious scholarship of its author, the book is blind to essential distinctions and finally inconsistent with itself.

[146] Simpson *et al.*, *Life*, pp. 804–805.

[147] See Theodosius Dobzhansky, *Mankind Evolving* (New Haven: Yale, 1962), pp. 24–26.

[148] Ernst Mayr, *Animal Species and Evolution* (Cambridge, Mass.: Harvard, 1963), pp. 4–5.

[149] The brief discussion of "Preformism vs. Epigenesis" in Nogar's *The Wisdom of Evolution*, pp. 292–294, contains some bibliographical material.

It was, in short, neither from classical antiquity nor from the scholasticism that extended it into medieval times that the immediate obstacles to evolutionary thought came—with the exception of a tendency to typological thinking, which by right should have been abandoned with the celestial spheres which were its mainstay. They came rather from the parts of the culture which carried the second or postulationally explanatory mode of reason over in a verbal form to the exegesis of scripture, or from those peculiarly modern thinkers who attempted to substitute the second explanatory mode of reason in its exact or mathematical form for the properly philosophical or natural physical mode of explanation which assigns reasons for being. "If the preceding analyses are correct, we can see that the central error of modern philosophy in the domain of the knowledge of nature has been to give the value of an ontological explanation to the type of mechanist attraction immanent in physico-mathematical knowledge, and to take the latter for a philosophy of nature." [150]

What seems to me to be essential in this matter, is to emphasize this fact, that there is a natural and inescapable proportion between means and ends, methods and objects, and that, every time we deal with genuine kinds of knowledge, the difference between methods presupposes as its very root, a more fundamental and more enlightening difference between objects. [151]

But if physico-mathematics were a natural philosophy, if it made manifest the essences and causes at work in the corporeal world, then it would be a knowledge having the ontological essence of physical reality as its proper and specifying object; and from then on we would see subverted and destroyed the genuine structure of this science. It would no longer be a science *formally* mathematical and *materially* physical, it would become a kind of monstrosity, a science which would be at the same time *formally* physical and ontological as to its specifying objects and *formally* mathematical as to its medium of demonstration and explanation. The natural and necessary proportion between the end and the means, between the specifying object and the explanatory tools in knowledge would be broken. [152]

Moreover, if physico-mathematics were a natural philosophy, and as such the paradigm or "type" of rational understanding of nature—of rational explanation of the world, then, with Kurt Lewin (among others), it would be necessary to arbitrarily refuse to accord to biology and other disciplines of the same epistemological type (such sciences as of their nature escape integral mathematization and insofar fail to fit the "orthodox" criteria of scientific knowledge as predictive and controlling) the status of authentic and objective knowledge.

[150] Maritain, *The Degrees of Knowledge*, p. 184. See "The Conflict of Methods . . . ," pp. 534–535, and *Science and Wisdom*, pp. 45–46.
[151] Maritain, "The Conflict of Methods . . . ," p. 538.
[152] *Ibid.*, p. 535.

If, however, by contrast, Mayr and Scriven are right in contending that the organization of evolutionary research has demonstrated the independence of explanation and prediction, then, in the light of the essential considerations bound up with such a demonstration, it becomes necessary to acknowledge that the significance of Darwin consists less in any particular discovery than in a return to the ancient conception of science as reasoned facts. As Eiseley puts it: "It was not natural selection that was born in 1859, as the world believes. Instead it was *natural selection without balance*,"[153] i.e., *a returning to Aristotle's epigenetic view of individual development*—"he reasoned that the best way to explain both the repetition of type and the production of novelty was to recognize the potential factor in the reproductive material and the developmental process as the progressive education or actualization of adult form"[154]— *minus the unchanging environmental reference* of the celestial, immutable spheres, which were originally a second-mode explanation (in mathematical form) anyway, and which moreover (as *causa regitiva*) were the sole guarantee that the relation of generator-generated would be absolute and not just relative across the ages. "Aristotle's concept of nature, with the operation of chance, indeterminism, and the equivocal causality of other natural cosmic agents, easily places the space-time dimensions of natural species within the theoretical structure of his natural system."[155]

---

[153] Eiseley, *The Firmament of Time*, p. 81.

[154] Nogar, *The Wisdom of Evolution*, p. 292. See "The Philosophical Dimensions of the Origin of Species," Part I, p. 131 text and fn. 115.

[155] Nogar, "Evolution: Scientific and Philosophical Dimensions," p. 60. Moreover, it is necessary to point out what is in itself of great philosophical interest, namely, that the idea of natural selection, insofar as it presupposes the mutual interference of independent lines of causation (and this is what is essential to the notion), bears witness on the one side to an *irreducible pluralism* in nature, the plurality of causal series which meet at a given moment; and on the other side, it coincides with the classical idea of *chance* as that notion was formulated and developed in the Aristotelian tradition.

    Once this has been realized, one immediately sees that almost all of the recent controversies over "teleology" and an over-all direction or goal for the evolutionary universe have been the consequence of failure to draw the proper distinctions. So far as the theological side of these debates go, Nogar brought this out clearly in his exchange with Francoeur over the question of whether from within the universe of progressive development, with its order and disorder both cosmic and human, man can discover ultimate meaning—a finality and direction for the universe *as a whole*: see "The God of Disorder" in *Continuum*, 4 (Spring, 1966), pp. 102–113, by R. J. Nogar; "The God of Disorder II: A Response," by R. Francoeur, in *Continuum*, 4 (Summer, 1966), pp. 264–271; and "The God of Disorder III: A Postscript," by R. J. Nogar, *Continuum*, 4 (Summer, 1966), pp. 272–275. On the philosophical side, there is the underlying problem of progress in evolution, which is itself centered on the question as to the role chance plays in the constitution of the world: see J. N. Deely, "The Philosophical Dimensions of the Origin of Species," Part II, *The Thomist*, XXXIII (April, 1969), Sections VII and VIII, pp. 290–335.

    On the classical notion of chance, the following texts are adequate: Aristotle, *Physics*, Bk. II, and the *Metaphysics*, Bks. 11 and 12. (Marvellously simple explanations of Aristotle's texts on the notion of physical chance can be found in Yves Simon, Ch. X of *The Great Dialogue of Nature and Space*, pp. 181–205; and in J. H. Randall's book, *Aristotle* [New York: Columbia, 1960], pp. 172–188, 229, *inter alia*. See also Yves Simon, *The Tradition of Natural Law* [New

H. The state of the methodological question

And yet it is not strictly correct to speak here of a *return* to the ancient conception of natural, i.e., philosophical, as opposed to mathematical, science; for although Darwin's great theory does indeed force a restoration of the distinction between the philosopher of nature and the mathematical interpreter of nature, still, coming as it does after three centuries of mathematical interpretation of nature have changed the face of the world, this restoration is now achieved at a higher or more mature level of cultural development within the organic structure and differentiation of the mutually irreducible (and so irreplaceable) ways of knowing. Moreover, this restoration makes it possible for the first time for natural philosophy to achieve, beyond an assessment of the ontological disposition of this universe, an authentic understanding of the balance between the necessary, the contingent, and the fortuitous in the course of events—something completely impossible as long as the Aristotelian tradition remained "basically typological resting essentially upon fixed finalities," and so "opposed to a true evolution of nature in its very theory of scientific knowledge and understanding."[156]

A closer look at nature, provided by the very realistic empirical sciences which Aristotle founded, reveals that time and space are essential properties not only of individual substances but of species themselves. Understanding which abstracts from the essential condition of the space-time contingency of natural bodies is illusory. Man cannot know the nature of a star, a camel, a salamander or a man unless he knows its origins and development, its proper space and time which makes it to be what it is.[157]

When all is said and done, therefore, and just because it stands alone, or almost alone,[158] among the great modern scientific theories as truly natural in its explanatory structure, Darwin's theory portends a denoue-

---

York: Fordham, 1965], pp. 41–66, esp. pp. 54–58.) Thomas Aquinas, *Summa*, I, q. 115, art. 6, & q. 116, art. 1; *In II Physicorum*, lects. 7–10; *In II Met.*; and *In XII Met.*, lect. 3. J. Maritain, on "Chance," in *A Preface to Metaphysics* (New York: Mentor, 1962), pp. 113–141.

[156] Raymond J. Nogar, "The God of Disorder," *Continuum*, 4 (Spring, 1966), p. 108. (Originally presented as a paper at the 1965 ACPA under the title, "The Mystery of Cosmic Epigenesis.")

[157] *Ibid.*, p. 109.

[158] Beside it stand the depth-psychology which springs from Freud and the sociology and culturology which alike spring from Durkheim. What is at work within these currents—becoming very broad now—is nothing less than the revolution in scientific understanding which we have been discussing in terms of epistemological types. See R. M. MacIver, *Social Causation* (New York: Harper, 1964); and Gerard Radnitzky, *Contemporary Schools of Metascience* (2nd. ed., rev.; Copenhagen: Munksgaard, 1970).

ment to the "epistemological drama" of modern times, a resolution of the "tragic misunderstanding" which took place at the time of Galileo and Descartes and has for three centuries embroiled science and philosophy. If this be so, the rise of evolutionary science is not only a great advance in our *understanding of nature*, the greatest such in modern times; it is a spiritual advance as well, I mean an advance in our *understanding* of what *knowledge itself* is, "its values, its degrees, and how it can foster the inner unity of the human being."[159] That is why it seems to me that the passage from classical antiquity to Darwin's world (which suffered all the other cultural vicissitudes, of course, social, political, religious, sociological, psychological, besides the purely philosophical ones I have attempted to trace [not unaware of the gaps in the outline], but which lie beyond the compass of the logic of the epistemological types of rational knowledge) must be seen not merely as circular—a complementary rivalry between two alternative views concerning the aim and method of the study of nature—but rather as *spiral*, as an expression of that essential tendency of human intelligence to correct over the long run its own excesses, and in doing so to move toward that establishment of "the freedom and autonomy as well as the vital harmony and mutual strengthening of the great disciplines of knowledge through which the intellect of man strives indefatigably toward truth."[160] This is the only alternative to that disastrous imperialism in which now one and now another type of knowing claims at once to face the full range of reality (or at least as much of it as deserves scrutiny) and at the same stroke to absorb all "genuine" knowing into itself. It would also seem to hold the sole hope for the future of humane civilization.

But we have, for good or ill, more or less aware of its implications and willingly or no, all of us made the passage to Darwin's world. From a world in which the heavens, at least, hinted exemption from transformations, to a world in which nothing at all is exempt from process: that is the essential datum. That is what locates us in Darwin's world and no longer in the classical or medieval world.

It is by reference to that single point of observation—but how many years, how many lives, and how much anguish is compressed in this one point of our collective intellectual biography!—that we are able to refer (in the accents of the curator of culture and historian of ideas) to "the idealistic philosophy of Plato and the modifications of it by Aristotle";[161] while from the standpoint of the difference between sciences of observations and sciences of explanation, that same datum compels us—paradoxically, indeed—at one and the same time *to* "*claim* that the typological

[159] Jacques Maritain, "On Human Knowledge," in *The Range of Reason* (New York: Scribner's, 1952), p. 3.
[160] *Ibid.*, p. 11.
[161] Ernst Mayr, *Animal Species and Evolution*, p. 5.

philosophies of Plato and Aristotle are incompatible with evolutionary thinking"[162] *and* (just because "the mind, even more so than the physical world and bodily organisms, possesses its own dimensions, its structure and internal hierarchy of causalities and values—immaterial though they be"[163]) *to acknowledge* that, "surprisingly enough, the fundamental issues involved in the doctrinal differences between Plato's system and Aristotle's view are raised again today by the advances of evolutionary theory."[164] So that Darwinism, more than any other doctrine responsible for the now dominant evolutionary vision of all reality, turns out to have been a thoroughly dialectical event.

III. WORLD-VIEW AS LOGOS AND AS MYTHOS

We have already noted how mathematical-physical knowledge, obliged as it is to substitute quantitatively reconstructed entities for the sensible and qualitatively determined objects of experience, when translated into words and presented as an explanation of the world of experience, results in the creation of myths which superimpose on the universe of experience an entirely different one, a universe of provisory representations, sometimes, indeed (as in the case of Einstein's space, or the case of wave-mechanics), only reductively or analogically figurable, but always myths or fables, since their whole value derives, not from the essence of the real envisaged in itself, but from the mathematical relations they sustain; so that, if accorded an ontologically explicative value, the result of such translation is a casting of the mind into a zone of metaphysical confusion. And we have noted too that it is in this sense of transporting the conceptions of mathematical-physics literally on to the philosophical plane, that many elements of the contemporary scientific world-view belong to the order of myth rather than of understanding.[165] There is quite another

---

[162] *Ibid.*: my emphasis.

[163] *The Degrees of Knowledge*, p. ix. For discussion of this last point, see John N. Deely, "The Immateriality of the Intentional as such," *The New Scholasticism*, XLII (Spring, 1968), pp. 293–306.

[164] R. J. Nogar, *The Wisdom of Evolution*, p. 316.

[165] And since moreover "there is no other way for the Philosophy of Nature to take up into its own order the well-founded myths of physico-mathematical knowledge than to become a fabricator of myths in its turn," then "perhaps it is fitting that the Philosophy of Nature add to its philosophical knowledge, properly so called, a region of philosophical myths destined to harmonize it with the well-established myths involved in physico-mathematical theories. In this way it may complete its union with the experimental body that the sciences construct for it. And so, though there can be no continuity as to the rational explanation and the understanding of things between physico-mathematical theories and the Philosophy of Nature, a secondary continuity may be established through their common ground of imagery" (J. Maritain, *The Degrees of Knowledge*, pp. 183–184), a secondary continuity thus between "two heterogeneous rational conceptions between which mathematical formalism alone assures [primary] continuity" (p. 186). See fn. 65 above.

sense of myth as well, which we will discuss in a moment; but having discerned the spirit and proper noetic value of the role played by mathematical knowledge in our understanding of the phenomena of nature, it would be well if we delineated first the structure of rational knowledge as it pertains to the *proper explanation and understanding*—rather than simple interpretation indifferently commingling mental fictions and observable processes—of the evolutionary universe.

A. World-view as *logos*

We have seen that the scientific understanding advances by a circular process or a kind of dialectic in which the mind begins with experience or observation, works out explanations in terms of its observations, gathers new evidence, modifies previous explanations to take account of the new data, and so on endlessly, with techniques of observation more and more sophisticated giving rise to explanatory structures more and more refined.

We have seen also that in this continual going and coming between observation and explanation, the mind works in two distinct explanatory modes, one philosophical, in which the rule of explanation is to assign the reasons for being or proper causes of phenomena (that which is observed); the other mathematical, in which the rule of explanation is to construct whatever causes or hypotheses would by their assumption make possible the calculation from the principles of mathematics (especially geometry) of the conditions for the phenomena in question for the past and the future—a skillful effort "to save the phenomena," making as such no pretense to arrive at the nature of things or to explain them in terms of their true, or rather, *proper*, causes and principles of being; so that, if its procedures are transparent to itself, the scientific mind will always subordinate the second mode to the first, in which alone a true understanding, both physical and literal, of the world of nature is achieved.[166] For "not all aspects of nature can be known by a quantitative procedure, and even those which are known in this way have to be interpreted in the light of the nature of a thing which underlies its quantity."[167]

---

[166] "Undoubtedly it is this genuine distinction between the intellectual habit of natural science and the intellectual habit of mathematico-natural science that has led so many of our philosophers to believe that our knowledge of nature must be split into two levels," comments Ashley. "Nevertheless, this mathematico-natural science, important as it is, does not constitute a new level of natural science but is best conceived as an instrument used by the natural scientist, a technique like his other techniques." ("Does Natural Science Attain Nature or Only The Phenomena?" p. 81). Simon (*art. cit.*, pp. 100–101) gives an express idea "of the distinctions which should be made and of the phases which should be surveyed in order to appreciate the bearing of physical [i.e., mathematico-physical] theories with regard to the knowledge of the real."

[167] B. Ashley, *The Arts of Learning and Communication* (Chicago, 1961), p. 379.

Thus world-view as *logos* refers to our understanding of nature so far as it is achieved in the light of facts or evidence, which light must be taken in three senses. Any fact is a witness to the activity of the mind, since in order to be "given" or *datum* something must be simultaneously "received" or "actively accepted" and conceived—*factum*. Nevertheless, so far as it pertains to the *order of observation*, the notion of fact or *datum* has a fundamental sense, absolutely binding on every exercise of the speculative understanding: it is, in Simon's words, "the object of an intuition"—an element of experience, let us say—"for which no logical construct can be substituted and upon which all the logical constructions of the science of nature finally rest." [168] So far as it pertains to the *order of explanation*, on the other hand, the notion of fact divides before the mind, so to speak, under the pressure of a critical analysis of the rational processes used in each particular case:

The more the mathematical is reduced to the role of enabling one by measurement and calculation to get a surer grasp of the undiluted physical and of those causes and conditions whose character as *entia realia* [or beings independent of the considering mind] the philosopher has no reason to question, the more does the result deserve to be called a fact. But the more the physical is reduced to the role of intervening only as a mere instrument for discriminating between theoretical constructions whose proper value is constituted by their mathematical amplitude and coherence or as as a mere basis for entities which the philosopher has good reason to regard as beings of reason, the more should the result be transferred to the order, not of fact, but of explanatory image. [169]

---

[168] Simon, *art. cit.*, p. 91. Nor is this fundamental sense rendered any the less binding by the easy observation that the original description or experience is already steeped in theory (Karl Popper, *The Logic of Scientific Discovery*, pp. 59ff.) and so itself dependent on a background of presuppositions and assumptions (D. M. Emmet, "The Choice of a World-Outlook," *Philosophy*, 23 [1948], p. 211). For the distinction I have insisted upon between observation and explanation, together with this definition of *datum* on which it rests, derives its validity from its reducibility to an instance of the principle of contradiction: "Experience is a source of knowledge about the things experienced, and it provides a test for what claims to be knowledge of the things experienced. To function in these ways as a source and as a test, it must be distinct both from the things experienced and from the knowledge of those things" (Adler, *The Conditions of Philosophy*, p. 132). As Waddington might put it (cf. *The Ethical Animal*, p. 27), temporal and genetical overlap should not prevent us from recognizing that the processes of observation and those of explanation are in important ways different in kind. Otherwise, it would make no sense to maintain that "to seek objectivity in questions of fact is a primary obligation" (Emmet, p. 215), or to speak of "a discipline of accuracy in dealing with empirical evidence" (*ibid.*). It is just the primacy of this obligation and the basis of such discipline that the notion of *datum* or "observed fact" as defined above accounts for. See the discussion of the current debate over the theoretical-observational distinction in fn. 43 above.

[169] Maritain, *The Degrees of Knowledge*, p. 59. Essentially the same thesis, namely, that a "critical analysis of the rational processes used in each particular case" is required both to maintain the continuity of order between the discrete kinds of knowledge and to avoid the paradigmatic theory of epistemological monism, is expressed in Bernard Lonergan's study of human understanding, *Insight* (New York: Philosophical Library, 1965): see Lonergan's "Preface," pp. ix–xv, and "Introduction," pp. xvii–xxx.

In this way, an impassable gap will always attest the difference of order that distinguishes the philosopher of nature or the natural scientist from the mathematical interpreter of nature, for "it is the possibility of being ascertained through sense experience which gives the concept its positive meaning."[170]

Within the world-view as *logos* then everything rests upon the evidences established within the triangle of facts, let us call them *observed* (or at least *measurable*) facts, *reasoned* facts, and *mathematized* facts. To this we should have to add an intermediate sort of fact, more than a simple *datum* of observation and yet less than a *factum* of explanation, a dynamic web of interpretation not indifferently commingling mental constructs and observable processes, but tentatively or *dialectically* commingling them, what we must recognize and identify as dialectical or *hypothetical facts*, "facts" destined to become in the course of the extension of our field of observations and the elaboration of our scientific and philosophical understanding of nature (and in more or less modified and corrected form) either explanatory images in their turn—mathematized facts pure and simple—or reasoned facts in the ontological order of understanding.

Thus the notion of "empiriological discipline" brought forward by Maritain and defended by Simon (Maritain's book on *Philosophy of Nature* [New York: Philosophical Library, 1951] contains Simon's essay [cited in fn. 48 above] as the fourth and concluding chapter), insofar as it means something more than science of observation, *and contrary to the opinion of both Maritain and Simon*, designates purely and simply those dialectical extensions of our understanding of nature which characterize modern research-projects and which are from a philosophical point of view "science in the making." But in view of the fact that most contemporary philosophers base their conceptions of "positive" science almost exclusively on mathematical physics, it is not surprising that the dialectical use of highly refined experimental techniques and the flexible elaboration of mathematical hypotheses and laws is taken (mistaken) for science itself without further qualification[171]; nor that Maritain, in seeking to assimilate physico-mathematics to the principles of an Aristotelian noetic, should have been led (misled) into a confusion of the properties of the Platonic explanatory mode with the characteristics proper to the order of hypothetical facts—the register of empiriology—taken in itself:

---

[170] Simon, *art. cit.*, p. 91, See fn. 65 above.
[171] See, for example, Michael Polanyi's remarks concerning *Science, Faith, and Society* (Chicago: Phoenix, 1964); or E. A. Burtt's assessment of *The Metaphysical Foundations of Modern Science* (New York: Anchor, 1954). As Simpson remarks in another context: "This is an example of the existing hegemony of the [mathematico-] physical sciences, which is not logically justifiable but has been fostered by human historical and pragmatic factors" (*This View of Life*, p. 141).

In this very empiriological category two clearly different types of explanation can be distinguished. The empirical content (in this case the measurable) may receive its form and its rule of explanation from mathematics. Then we have an "empirio-metric" type of explanation characteristic of physico-mathematical science. Or, the empirical content (in this case the observable in general) may call for a purely experimental form and rule of explanation. Then we have an "empirioschematic" type of explanation characteristic of the non-mathematical or at least non-mathe-maticized, sciences of observation (by this we mean that experience itself is not thought or rationalized according to the law of mathematical conceptualization, but according to the experimental schemas themselves discovered by reason in the phenomena). . . . Note that *in both cases* the empiriological terminology proper to the sciences of phenomena tends to be established in a more and more perfect independence from the ontological terminology of philosophy.[172]

It is remarkable how science, encroaching, so to speak, on future possibilities and undergoing especially the exigencies of its ideal form, uses only materially, and as though without recognizing them or rendering them competent, notions which belong to less evolved strata of conceptualization. . . . That is why in the kind of knowing with which we are at present concerned, the sciences of phenomena, the formally activating value is linked up with the elimination of the ontological and the philosophical, to the profit of a wholly empiriometric or empirioschematic explanation.

It is understandable that, for a mind limited by professional habits to the intelligibility of this degree, philosophical notions can lose all significance. It is

---

[172] J. Maritain, *The Degrees of Knowledge*, pp. 148–149, emphasis supplied. That a confusion of the Platonic explanatory mode with the order of hypothetical facts taken in itself is indeed present here, is indicated by the distinctive difficulty which Maritain acknowledges as conse-quent upon his denial of the possibility in principle of empirioschematic formulations attaining to the status of reasoned facts, or proper explanations: "This sort of purification"—the success of the alleged tendency of *all* the sciences of nature, in their very structure, to free the observable as much as possible from the ontological (why not say: to reduce the conceptual to the percep-tual, the intelligible to the sensible?—see fn. 175 below)—"is particularly far advanced in physics. Either by the elaboration of new concepts or the recasting of definitions, or by a new use of common concepts (of a philosophical or pre-philosophical origin), applied exclusively to sensible verifications, sciences like biology and experimental psychology, which can be put under the empirioschematic type . . . also tend to establish a more and more autonomous notional terminology. Since they abide in a much less precarious continuity with philosophy, it is more difficult for them than for physics to isolate this terminology and to prevent its being invaded by philosophical concepts which, in this domain, would give rise to pseudo-explana-tions. They persevere in the attempt, however, and often seem even to prefer rudimentary conceptual tools . . . on condition that it assures this independence" (*ibid.*, p. 149).

In other words, "experimental but non-mathematizable schema," in order to be and remain empiriological in Maritain's sense, seem to require continuance in a rudimentary state of conceptualization. But is not that the entire meaning of a hypothetical as over against reasoned and mathematized facts alike? Surely there is no "law" against the child passing into adolescence, or against the adolescent reaching maturity? Quite the contrary! Why then postulate such a law stunting the growth of our rational understanding of nature?

A recent defense of this unfortunate position taken by Maritain is elaborated by Joseph J. Sikora, *The Scientific Knowledge of Sensible Nature* (Paris: Desclee, 1966); and by Maritain himself in Appendix 3 of *The Peasant of the Garonne*, trans. by Michael Cuddihy and Elizabeth Hughes (New York: Holt, Rinehart, and Winston, 1968), "A Short Epistemological Digres-sion," pp. 270–273.

likewise understandable that the experimental sciences have in a certain sense made progress by warring on the intellect. For the intellect has a natural tendency to introduce into the conceptual register proper to the sciences meanings which derive from another register, the philosophical register, and which consequently disturb or retard experimental knowledge as such, and prevent it from achieving its pure type.[173]

But this could be so only if concepts from "the philosophical register" were opposed to the pure type of scientific explanation—and that state of affairs in its turn would be possible *if and only if* the Aristotelian explanatory mode were quite inapplicable to the details of nature.[174] Maritain's point thus seems to rest on his conception of a type of knowledge which attains the *phenomenality* but not the essences of natural realities, what Maritain calls *perinoetic* as against *dianoetic* intellection which does penetrate to the ontological character or essence of a thing.

If we grant that there is a knowledge of things in terms of phenomena and that such a knowledge in some basic sense characterizes modern science, however, it does not follow that we must accord to it an absolute status; for it may well be that we are dealing "not with an absolute difference between one kind of human knowledge in which is attained a perfectly ordered knowledge of nature (dianoetic intellection) and another which knows nothing of nature except its existence (perinoetic intellection), but rather with a type of intellection proper to man by which he knows at first confusedly and then more and more clearly as he continues his investigation."[175] Thus the fact that a vast part of modern research relies on rudimentary conceptual tools and correlations of phenomena not known to be ontologically linked constitutes a state of

---

[173] *Ibid.*, pp. 153–154. See further pp. 139–140, par. 4; and the comments in fn. 172 above.

[174] And in fact it is precisely "here that not only Maritain but De Koninck," and many, many others, "feel that the philosophy of nature has rather narrow limits. Both grant that it is possible to have essential definitions at the broad level of the problems raised in Aristotle's *Physics, De generatione et corruptione* and *De anima*, but beyond this they are doubtful. The forms of material things are 'so immersed in matter,' that . . . strict scientific demonstration of any but the most generic properties of changeable things is impossible." (Ashley, *Are Thomists Selling Science Short?* pp. 11–12. See the reference in fns. 42 and 142 above, and fn. 175 below.

[175] Ashley, "Does Natural Science Attain Nature or Only the Phenomena?", pp. 77–78. In the non-mathematical investigations of nature, what is known "more and more clearly as the investigation progresses" is "both the existence of a natural unit and its nature. This will be recognized, I think, by any scientist as the process he goes through in any work of research, moving gradually from a dim intuition that he is dealing with a special type of thing to a clearer and clearer notion of just what makes it special." (*Ibid.*, p. 78.)

Consequently, there seems to be no real foundation for the contention "that there is any fixed limit to the discovery of the specific essences of material things, nor to the strictly scientific explanation of their properties" in terms of proper causes; while at the same time there can hardly be any doubt "that the philosophy of nature"—explanations in the Aristotelian mode, let us say—"needs a dialectical extension both through mathematical and non-mathematical reasoning," i.e., must be supplemented "with dialectical knowledge, some of which is derived from the application of mathematics, and some from probable reasoning in [non-mathematical] physical terms." (Ashley, *Are Thomists Selling Science Short?*, p. 12).

affairs which we may characterize as "empiriological," without thereby requiring ourselves to postulate with Maritain an "irreducible distinction that must be recognized between the approach, the mode of conceptualization, the kind of relation to the real (in other words, the kind of truth) which are proper to the sciences of nature (by 'sciences of nature' I mean all sciences [physics as well as biology, etc.] which deal with things pertaining to the world of matter) and those proper to the philosophy of nature,"[176] and without thereby obliging ourselves to say without exception or further qualification concerning all of the disciplines participant in this vast effort of organized research and special investigation that "it is not their business to use signs grasped in experience in order to attain, through them, the real in its ontological structure or in its being, by a type of intellection ('dianoetic' intellection) that penetrates to the very essence (not apprehended in itself, certainly, but through those of its properties that fall under experience, outward or inward),"[177] even though it remains true to say that "they all have in common this essential character of depending (whether primarily or totally) on that intellection of an empiriological order ('perinoetic' intellection) which takes hold of the real insofar and only insofar as it is observable";[178] and the sub-distinction of the empiriological or hypothetical and dialectical order into the empiriometric and empirioschematic simply further specifies various research-projects according as they are ordered to results that are primarily of the mathematical or philosophical formal type, according, that is, as they are conceived in the line of mathematical physical or natural physical explanations: and the defining feature of empiriological knowledge as such is accordingly that it is a knowledge of phenomena constituted thanks to an intelligible element *imposed* on the data of sensation by the constructive work of the intelligence, not in an arbitrary way (as in wishful thinking), nor for its own sake and in view of some transrational end (as in poetry), but simply in an attempt "to save the appearances" by a plausible hypothesis *entertained solely for the sake of and in subservience to the data of direct experience*, i.e., the sciences of observation, that is to say, entertained *tentatively*, as was explained by Sir Isaac Newton in a formulation that leaves nothing to be desired as a clarification of this point:

---

[176] Maritain, "A Short Epistemological Digression" in *The Peasant of the Garonne*, p. 270, text and fn. 1. See also p. 272.

[177] *Ibid.*, p. 272, text and fn. 7.

[178] *Ibid.*, pp. 271–272, text and fn. 6. But now it is no longer a question of a difference between philosophy and science sustained in terms of a differential kind of truth—of explanatory mode, I mean, but rather of one sustained solely in terms of a differential relation to experience, without any inevitable differentiation from a formal standpoint in the epistemological type involved—a question of the difference between what Adler designates as "investigative versus non-investigative disciplines, both of which are empirical"; and now it is a question too of a difference between the order of observation and that of explanation generally: see fns. 43, 114, and 168, above.

The best and safest method of philosophizing seems to be, first to inquire diligently into properties of things, establishing those properties by experiments, and then to proceed more slowly to hypotheses for the explanation of them. *For hypotheses should be subservient only in explaining the properties of things, but not assumed in determining them; unless so far as they may furnish experiments.* For if the possibility of hypotheses is to be the test of the truth and reality of things, I see not how certainty can be obtained in any science; since numerous hypotheses may be devised, which shall seem to overcome new difficulties.[179]

The empiriological register as such therefore is a vast zone of "science (physico-mathematical and natural) in the making," or of the explanatory determination of experiences in the direction of a mathematical network (so as to become mathematized facts pure and simple) or in the direction of a properly causal scheme (so as to become reasoned facts). This vast zone of "factual" knowledge—of tendential facts, *facta vialia*—is indeed difficult to mark out, just because it is essential to it to have always shifting boundaries, and to receive its formulation in statements which tend to overlap and even contradict. It is, as we have just indicated, "empiriological" knowledge in probably the only purely defensible sense of the term, inasmuch as it is a knowledge of natural phenomena achieved thanks to an intelligible element imposed on the data of experience—not indeed arbitrarily but plausibly—a construct which *correlates* without explaining. Such dialectical definitions, or *hypothetical facts*, and the arguments based on them, do not belong strictly to the world-view as *logos*, but either to the realm of *opinion* (more or less well-founded), or pertain to the *extension* of a science (be it biology, ethics, physics, or whatever) properly so called, thus marking the twilight region in human apprehension of reality for any given individual or any given age.[180] It is principally in respect of these "facts" that it is necessary to say that "when we come to anything as complex as a world outlook, we must take account not only of methodological postulates, but of a whole background of general assumptions about what is and what is not reasonable."[181]

A vast field of critical analysis is thereby opened up. . . . The essential thing to understand is that it would be a serious mistake to conceive science [or, equally, philosophy,] in a static fashion as something achieved, "completely made." And this is true not only from the point of view of its extension and of the objects it has to know, which is obvious, but also from the point of view of its internal noetic

[179]Newton's reply to the Second Letter of Father Pardies (1672), printed in *Isaac Newton's Papers and Letters on Natural Philosophy*, I. Bernard Cohen, editor (Cambridge, Mass.: Harvard, 1958), p. 106. This indeed is the classical view, as indicated in fn. 43 above.
[180]Cf. Charles De Koninck, "The Unity and Diversity of Natural Science," in *The Philosophy of Physics*, pp. 15, 16, 23, and *passim*; and his study of *Natural Science as Philosophy* (Quebec: Laval, 1959).
[181]D. M. Emmet, "The Choice of a World-Outlook," *Philosophy*, 23 (1948), p. 211.

morphology and of its *intension* with regard to its typical forms. By the very fact that it leaves behind its prescientific basis in common sense in order to attain more and more purely the state of science, its progressive extensive growth is accompanied by a progressive intrinsic formation which brings it into line with certain determined epistemological types which it has as yet only realized partially and to diverse degrees. But though a total and homogeneous realization of these ideal types must be regarded as an asymptotic limit, it is remarkable how science, encroaching, so to speak, on future possibilities and undergoing especially the exigencies of its ideal form, uses only materially, and as though without recognizing them or rendering them competent, notions which belong to less evolved strata of conceptualization. The formal element of scientific intelligibility is current especially in the higher strata, in notions which are most typically pure.[182]

Thus the general notion of world-view as *logos* does not imply a set of definitions wholly clear and specific, still less a positivistic concentration on the physical, or a phenomenological-existential circumincision of the mental. All it requires is the possession of some essential knowledge of nature, sufficiently well-ordered to find some of the proper causes of some of the phenomena in question, and that we do not conclude with confidence beyond our knowledge of the evidence.[183] Here then is the essence of world-view as *logos*:

Nothing about material things limits human knowledge merely to the phenomena, nor the dialectical saving of the phenomena [be it verbal and hypothetical or mathematical and predictive]. Rather the phenomena are a sufficient way to the essence, provided that we undertake all the laborious work of research and experiment which the history of science proves are necessary to know the relevant phenomena.[184]

There is however a fifth kind of "fact," the tribute which emotion pays to reason in order to compensate or sometimes conceal the latter's deficiencies: let us call it the *rationalized* fact (when it is a question of concealing) or (more properly) *poetic* fact (when it is merely a question of compensating). And on this order of fact principally, *in conjunction with hypothetical facts taken not so much in their character as opinion as at the expense of their dialectical nature as extensions of scientific and philosophical understanding*, does the mythical view and *Zeitgeist* depend. For once its ordering to intellect is sacrificed, the hypothetical construct ceases to be a logical sign, in order to become rather what Maritain has aptly named a *magical* sign:

---

[182] Maritain, *The Degrees of Knowledge*, p. 153.
[183] "There are sophisticated theories which, arguing that no truth is quite true and no falsehood quite false, and that the notion of what is meant by 'fact' is a very obscure one, explain away our simple conviction that *sometimes, at a certain level* truth means correspondence with fact, and whatever our ultimate philosophical theory of truth, this simple conviction should be taken up into it and not made meaningless." (Emmet, p. 215.)
[184] Ashley, *Are Thomists Selling Science Short?*, p. 12.

In giving the word 'logic' a very broad and rather infrequent usage, but a usage which seems to me justified, I describe sign as a 'logical sign,' or a sign in the sphere of the Logos, when it is located in a certain functional status, wherein it is a sign *for the intelligence* (speculative or practical) taken as the *dominant factor* of the psychic regime or of the regime of culture. Whether the sign in itself be sensory or intelligible, it is then definitively addressed to intelligence: in the last analysis, it is related to a psychic regime dominated by the intelligence.

I describe sign as a 'magical sign,' or as a sign in the sphere of the Dream, when it is located in another functional status, wherein it is a sign *for the imagination* taken as a supreme arbiter or *dominant factor* of all psychic life or of all the life of culture. Whether the sign in itself be sensory or intelligential, it is then definitively addressed to the powers of the imagination; in the last analysis, it relates to a psychic regime immersed in the living ocean of the imagination.[185]

B. World-view as *mythos*

In assessing the reactions of our contemporaries to the observations contributing to the realization of the natural or essentially evolutionary state of the universe, and to the very incomplete but genuine explanations that constitute our understanding of these observations, we are at once reminded of Ernst Cassirer's great thesis "that philosophy of mind involves much more than a theory of knowledge: it involves a theory of prelogical conception and expression, and their final culmination," or failure to so culminate, "in reason and factual knowledge."[186] Indeed, "intellect is not merely logical reason; it involves an exceedingly more profound—and more obscure—life, which is revealed to us in proportion as we endeavor to penetrate the hidden recesses of poetic activity."[187] "The universe of concepts, logical connections, rational deliberation, in which the activity of the intellect takes definite form and shape, is preceded by the hidden workings of an immense and primal preconscious life . . . which is specifically distinct from the automatic or deaf unconscious,"[188] the Freudian unconscious, and of which, moreover, "Plato and the ancient wise men were well aware, and the disregard of which in favor of the Freudian unconscious alone is a sign of the dullness of our times."[189]

Much has been made in contemporary philosophy, and rightly so, of the fact that there exists a common root of all powers of the soul, which is

---

[185] J. Maritain, "Sign and Symbol," in *Redeeming the Time* (London: Geoffrey Bles, 1943), pp. 199–200. See also Simon, "Ideology versus Philosophy," in *The Tradition of Natural Law*, pp. 16–27.

[186] From Susanne K. Langer's "Preface" to Ernst Cassirer's *Language and Myth*, trans. by Langer (New York: Dover, 1946), p. x.

[187] Jacques Maritain, *Creative Intuition In Art and Poetry* (New York: Pantheon, 1953), p. 4.

[188] *Ibid.*, p. 94.

[189] *Ibid.*, p. 91.

hidden in an ontological unconscious or preconsciousness, and that there is in this ontoconscious dimension of the self a root activity in which the intellect and the imagination, as well as the powers of desire, love, and emotion are suchwise engaged in common that "the powers of the soul envelop one another, the universe of sense perception . . . in the universe of imagination, which is in the universe of intelligence. . . . And, according to the order of the ends and demands of nature, the first two universes move under the attraction and for the higher good of the universe of the intellect, and, to the extent to which they are not cut off from the intellect by the animal or automatic unconscious in which they lead a wild life of their own, the imagination and the senses are raised in man to a state genuinely human where they somehow participate in intelligence, and their exercise is, as it were, permeated with intelligence."[190] This is much of the meaning of man as *Dasein*.[191]

No venture is possible without a primary gift. . . . Intuition is, as far as we are concerned, an awakening from our dreams, a step quickly taken out of slumber and its starried streams. For man has many sleeps. . . . There is a sort of grace in the natural order presiding over the birth of a metaphysician just as there is over the birth of a poet. The latter thrusts his heart into things like a dart or rocket and, by divination, sees, within the very sensible itself and inseparable from it, the flash of a spiritual light in which a glimpse of God is revealed to him. The former turns away from the sensible, and through knowledge sees within the intelligible, detached from perishable things, this very spiritual light itself, captured in some conception. The metaphysician breathes an atmosphere of abstraction which is death for the artist. Imagination, the discontinuous, the unverifiable, in which the metaphysician perishes, is life itself to the artist. While both absorb rays that come down from creative Night,[192] the artist finds nourishment in a bound intelligibility which is as multi-form as God's reflections upon earth, the metaphysician finds it in a naked intelligibility that is as determined as the proper being of things. They are playing seesaw, each in turn rising up to the sky.[193]

Thus Aristotle observes that "even the lover of myth is in a sense a lover of wisdom"[194]; and St. Thomas, commenting on this observation, turns it around, saying that the philosopher himself has a certain attachment to myths and fables, which are the proper domain of the poet:

---

[190] *Ibid.*, p. 110.

[191] See John N. Deely, *The Tradition Via Heidegger* (The Hague: Martinus Nijhoff, 1971), esp. Chs. V and VII, pp. 43–61 and 88–110, respectively.

[192] The "creative Night" to which Maritain here but refers in passing, he elsewhere defines and explains in a more exact treatment as *creative intuition*: see *Creative Intuition in Art and Poetry*, esp. Ch. III, "The Preconscious Life of the Intellect," pp. 71–105, and Ch. IV, "Creative Intuition and Poetic Knowledge," pp. 106–159; and the reference in fn. 187 *supra*.

[193] Maritain, *The Degrees of Knowledge*, p. 2.

[194] Aristotle, *Metaphysics*, Bk. I, ch. 2, 982b 18–19.

Whence the earliest philosophers, who treated of the principles of things after the manner of a kind of teller of tales, were called theologizing poets. . . . The philosopher is compared to the poet, or vice versa, however, simply in the sense that both treat of wonderous matters; for fables, with which the poet concerns himself, are put together from wonderous imaginings, whereas the philosophers themselves are moved to philosophizing by the spirit of wonder.[195]

When the "marvellous tale" of the poet, however, is substituted for the evidence on which authentic philosophy (and science) depends and given priority in the very order of understanding, just then are we confronted with the *mythos* of a world-view, with "a complex of ideas, convictions, and valuations which are ultimately derived from the social and psychological heritage of the person who holds them, and which he merely attempts to rationalize in his conscious philosophizing"[196]; so that in the end it must be said that in the order of *mythos* our outlook on the world depends finally (at worst) on irrational factors refractory to criticism or (at best) on transrational factors beyond its positive and direct reach.

What overcomes the *logos* by subordinating to the extent of subjugating it to a *mythos* in the mind of the thinker? It is not a hard subordination to achieve. One need only give the obscure longings of the Self sufficient freedom to triumph over the evidences of Things—an easy enough feat, which all of us perform often enough. "It is very easy for a speculative knowledge of things as they are to be transformed," muses William Carlo, "gradually and unawares, into an artistic knowledge, a production of things as the mind would like them to be. There is a bit of the creator in all of us, a legitimate heritage."[197] And Bergson has gone far in the work of assigning profound reasons for such psychological transformations:

The impulsive zeal with which we take sides on certain questions shows how our intellect has its instincts—and what can an instinct of this kind be if not an impetus common to all our ideas, i.e., their very interpenetrations? The beliefs to which we most strongly adhere are those of which we should find it most difficult to give an account, and the reasons by which we justify them are seldom those which have led us to adopt them. *In a certain sense we have adopted them without any reason, for what makes them valuable in our eyes is that they match the colour of all our other ideas, and that from the very first we have seen in them something of ourselves.*[198]

"The more I think of it," says the mythophile, "the less I see any other criterion for truth but to promote a maximum of universal coher-

[195] *In I Met.*, lect. 3, n. 55.
[196] Emmet, "The Choice of a World-Outlook," p. 211.
[197] William E. Carlo, *Philosophy, Science and Knowledge*, p. 123.
[198] Henri Bergson, *Time and Free Will*, F. L. Pogson, trans. (New York: Harper, 1960), pp. 134–135.

ence." [199] Thus by *mythos* is meant faith over against philosophy, a total explanation of the world over against inevitably incomplete understanding based on evidence. If the total explanation assumes a theological form, let us translate *mythos* as *myth ;* if the total explanation assumes an atheistic or non-theological form, let us translate *mythos* as *ideology.* On either translation, the essence of world-view as *mythos* is the same: it decides without evidence, or rather, in advance of and beyond the evidence, by extrapolating where there is no assurance that extrapolation is valid. The process has been very clearly described (fittingly enough) by Pierre Teilhard de Chardin:

The development of Faith consists in the adherence of our intelligence to a general view of the universe, by virtue of an option (freedom) or of affectivity (affection). The essential note of the psychological act of faith is, in my opinion, to see as possible and to accept as more probable a conclusion which, because it envelops so much in space and time, goes far beyond all its analytical premises. *To believe is to effect an intellectual synthesis.* [200]

If one recalls what was said above, it is plain that the establishment of a world-view as *mythos* belongs to the workings of reason in its second explanatory mode (understanding "reason" in the "deeper and larger" sense than usual indicated above), not this time indeed consisting in the verbal translations of mathematical interpretations of the sensible (*scientific myth* properly so called), but this time consisting from the first in a verbal rationalization of one's feelings with respect to evidence at hand. Following Aristotle, let us call it *poetic myth,* for that is what we are faced with. Yet both the scientific and the poetic myth are such by reason of being products primarily of reason's creative rather than its cognitive mode, by reason of belonging, so far as they can be put forward to account for anything, to reason's second explanatory mode.

In seeking to provide an explanation of some reality, reason can be employed in either of two ways. In the first place, it can be so employed as to establish sufficiently the reasons for a fact, as in natural philosophy . . . ; but reason can also be employed

[199] Pierre Teilhard de Chardin, *How I Believe,* his own translation of *Comment je crois.* (On the authenticity of this document, see George B. Barbour, *In the Field with Teilhard de Chardin* [New York: Herder and Herder, 1965], p. 151.) It is a curious fact that Père Teilhard, who was philosophically not an idealist, was never able to formulate a notion of truth which was other than that of a purest idealism: e.g., see *The Future of Man,* pp. 36, 182, 214; or *The Phenomenon of Man,* pp. 30, 32, 59, 219. In this instance at least Simon's thesis is verified strikingly: "Men of science, willingly or not, receive their philosophical ideas from philosophers; they could not rid themselves of idealistic prejudices while philosophers were teaching idealism as the only doctrine that may account for the unquestionable ability of the mind to treat in an orderly and causal manner the universe of phenomena." (*Art. cit.,* pp. 93–94. A reasoned basis for the truth of this thesis may be found in M. J. Adler's treatment of the distinction and relation between pure and mixed questions in *The Conditions of Philosophy,* esp. in his discussion of the fourth condition of intellectual respectability, pp. 38–42, but also on pp. 44–48, 60 fn. 13, 79–91, and 95–227.)

[200] Teilhard de Chardin, *Comment je crois,* p. 2: my trans.

in another fashion, which does not establish the reasons for the facts, but which instead shows that explanatory reasons proposed on some other grounds are congruous with the given facts. . . .[201]

No use to propose this alternate explanation as the "soul" of science,[202] still less to propose it as science itself become aware of its true dimensions, a "hyper-" or "ultraphysics,"[203] for "this latter type of explanation cannot suffice to prove anything, since it may well be that appearances, the 'whole of the phenomena,' could be equally well saved within the framework of another hypothesis," not yet dreamed of by man.[204] No doubt, "like the meridians as they approach the poles, science, philosophy [here meaning principally metaphysics] and religion are bound to converge as they draw nearer to the whole."[205] Still no use to fabricate a propaedeutic to their mutual absorption. "Hyperphysics" remains distinct from all three and must be taken for what it is: world-view as *mythos*.

Nonetheless, this struggle of the human personality beyond the always incomplete evidences of knowledge is in itself a great thing, the primal workings, as we have remarked, of reason within us, where the life of sense, imagination and intellect are engaged in common. It is the source of all intuition and greatness of vision. (Thus Maritain—rightly, in my opinion—links up with what the French philosopher Blanc-de-Saint-Bonnet called "the progressive weakening of reason in modern times" and with "a so-called reason as afraid of looking at things as it is busy digging in all the details around them, and as fond of illusory explanations as it is insistent in its claim to recognize only statements of fact, the reason of those who believe that poetry is a substitute for science intended for feeble-minded persons."[206]) In itself, as Bergson has pointed out, it remains a strictly incommunicable experience, and the ideas at work in it are adopted because "they match the color of all our other ideas" (it is in this respect that an individual can express the very spirit of an age, give voice to the *Zeitgeist*) and because "from the very first we have seen in them something of ourselves" (it is in this respect that a mythos remains inextricably bound up with the obscure longings and unique moments of subjectivity, with all that differentiates poetry from science).

Hence they do not take in our minds that common looking form which they will assume as soon as we try to give expression to them in words; and, although they bear

---

[201] *Summa*, I, q. 32, art. 1 ad 2. See Jacques Maritain, *Creative Intuition in Art and Poetry*, fn. 33, p. 180 and pp. 168–70, *passim*.

[202] The expression is Joseph Donceel's in "Teilhard de Chardin: Scientist or Philosopher?" *International Philosophical Quarterly*, V (May, 1965), p. 256.

[203] The expression, of course, is that of Teilhard de Chardin in his "Preface" to *The Phenomenon of Man*, p. 30.

[204] Aquinas, *Summa theologica*, I, q. 32, art. 1 ad 2.

[205] Teilhard de Chardin, *The Phenomenon of Man*, p. 30.

[206] *Creative Intuition in Art and Poetry*, pp. 71–72. See also "Sign and Symbol," pp. 219–220.

the same name in other minds, they are by no means the same thing. The fact is that each of them has the same kind of life as a cell in an organism: everything which affects the general state of the self affects it also. But while the cell occupies a definite point in the organism, an idea which is truly ours fills the whole of our self.[207]

Thus it is that such ideas, the intellectual components for any given individual or any given age of the *world-picture*, the world-view as *mythos*, although they are commonly hardened into a *doctrine* sure of itself and of its power to renew everything, in reality can find expression—if they are not to lose what is authentic and noble about them—only "as fragments of a vast poem." They belong to those wondrous imaginings with which the poet (the poet in each of us, too) concerns himself. However:

One doesn't expect a poem to bring us any kind of rational knowledge whatever, be it scientific, philosophical, or theological. One expects it only to give us a glimpse of what, in an obscure contact, the poet has seized in himself and in things at the same time. But we can admire such a poem for its boldness and its beauty. And it can awaken in those who love it—particularly the [kind of] poem I am speaking of [world-view as *mythos*]—fertile ideas and lofty aspirations, and can likewise serve to overcome their prejudices and defences, opening their mind to the flame of living faith [be it secular or religious] which burned in the soul of the poet. For it is the privilege of poetry to be able to transmit an invisible flame, and through the grace of God, a flame of such a nature.[208]

But if poetry becomes a doctrine, then it has ceased to be poetry in order to become instead the wishful thinking of an individual or an age.[209] Milton and Dante knew what they were about, but, in its essential episte-mological type, what they wrote belonged to the same *genre* as Plato's *Timaeus,* John Ray's *The Wisdom of God Manifested in the Works of the Creation,* Teilhard de Chardin's *The Phenomenon of Man,* or Julian Huxley's *Religion Without Revelation.*

It may be, however, that it is the evolution of the world and of life taken in its reality discernible to reason that we wish to understand, and that it is the world-view as *logos* and not only as *mythos* that occupies our attention. In this case, these cosmological syntheses succeed in teaching us nothing that all men of science did not already know, for what made them men of science in the first place was the working out of explanations so far as the evidences compelled them. Plato spoke for all to whom "faith" equals the construction of a psychological synthesis, and he spoke well, for he did not confuse *truth* with the coherence and harmony of soul-satisfying synthesis:

---

[207] Bergson, *Time and Free Will,* p. 135.
[208] Maritain, "Teilhard de Chardin and Teilhardism," in *The Peasant of the Garonne,* pp. 125–126. See "Sign and Symbol," in *Redeeming the Time,* pp. 223–224.
[209] See, in addition to the references in fn. 46 above, pp. 221–222 of "Sign and Symbol."

As being is to becoming, so is truth to belief. If, then, Socrates, amid the many opinions about the Gods and the generation of the universe, we are not able to give notions which are altogether and in every respect exact and consistent with one another, do not be surprised. Enough, if we adduce probabilities as likely as any others; for we must remember that I who am the speaker, and you who are the judges, are only mortal men, and we ought to accept the tale which is probable and make no more of it than that.[210]

That is why we must translate Plato's insight into the language of our own time, that it might serve as judgment not only on the noetic structure and value of the *Timaeus*, but on those among the evolutionist writings which belong to the same epistemological type. This has been finely done by Stephen Toulmin:

It is an excellent thing that men should think deeply about their place in the world of nature, and relate their goals and ideals to the process—and potentialities—of Nature. But any attempt . . . to find a single, unambiguous intention informing the whole course of cosmic history, must be regarded with suspicion. There may be legitimate objections to scepticism; but they are as nothing compared with the risks involved in philosophical wish-fulfillment.[211]

IV. CONCLUSION

We can observe by way of summing up, then, that the rise of evolutionary science in our own century exposes the excessive rationalistic character of the seventeenth and eighteenth centuries in its impatience with distinctions between different types of knowledge. At the same time, it counters this same habit of univocity which has swung contemporary science, with its inclination to regard mathematical-physics as the paradigm of rational knowledge, to the other extreme of anti-rationalism and a despair of providing an intelligible account of nature in its own proper reality. In short, the coming of evolutionary science to maturity compels us to recognize anew the intrinsic order of the human intelligence, with the essential distinctions it requires between the typical and mutually irreplaceable forms of knowing of which the mind is capable. In achieving this it compels us to accord natural science its proper value as a genuinely philosophical knowledge. It is neither a part nor an application of metaphysics, for its certitude does not rest on the necessity of intelligible being as such. As a type of explanation, it depends neither on deduction nor induction

---

[210] Plato, "Timaeus," in *The Dialogues of Plato*, B. Jowett translation (New York: Random House, 1920), Vol. II, p. 13. (The last seven words as cited differ from Jowett's rendition.)
[211] Stephen Toulmin, "On Teilhard de Chardin," *Commentary*, 39 (March, 1965), p. 59.

exclusively, but on the knowledge of the proper cause of something. This knowledge is not a mere projection of creative imagination, nor is it but a progressive approximation to a hidden reality that always eludes us; rather is it a very incomplete but genuine understanding of the world and of man as they are.

The areas in which our knowledge is clear and our insights successful are islands joined by bridging hypotheses. The true natural scientist is not discouraged by this fact but is determined to continue his researches until hypothesis yields to genuine insight. . . . He is humble in admitting that he knows little, but he will not be persuaded that the search is in vain.[212]

Or, as De Koninck puts it:

The bewildering progress of natural science reveals not only the bottomless depths of nature and the ineffable variety of nature's works; it shows, at the same time, the unexpected limitations of any human mind, and the devious modes of knowing it must resort to, even in the study of things immediately around us. Still, to enquire what any object of nature is, and to pursue the enquiry down to the last detail, is surely a pursuit which deserves to be called philosophy. To answer such a question, all the branches of natural science should be brought into play, and each of these remains open to infinity. At least this much we know.[213]

And since it is thus in the study of nature, at the very base and outset of our human knowledge—at the very heart of the sensible and changing multiple—that the great law of the hierarchical and dynamic organization of knowledge (on which for us the good that is intellectual unity depends) first comes into play, the hope is not unfounded that "if workers are not wanting, if unreasonable prejudices (due above all, it seems, to a morbid fear of ontological research and of all philosophy ordered to a knowledge of things—as though a philosophy of being could not also be a philosophy of mind) do not turn them back from the study of . . . philosophy that claims to face the universality of the extramental real without at the same stroke pretending to absorb all knowing into itself, it might well be hoped that we will see a new dawn break upon a new and glorious scientific era—putting an end to misunderstandings engendered in the realm of experimental research by the conflict between Aristotle and Descartes— in which the sciences of phenomena would finally achieve their normal organization, some, physics above all, undergoing the attraction of mathematics and continuing their remarkable progress along this line,

---

[212] Ashley, "Does Natural Science Attain Nature or Only the Phenomena?" p. 282.
[213] Charles De Koninck, "The Unity and Diversity of Natural Science," p. 24. In speaking thus of the more general context of the issues, however, it is necessary to bear in mind the qualifications indicated in fn. 114 above.

others, biology and psychology especially, undergoing the attraction of philosophy and finding in that line the organic order they need and the conditions for a development that is not merely material, but truly worthy of the understanding. Thus there would be a general redistribution springing from the natural growth of the sciences of phenomena"—be it remembered that there are two modes of empirical formulation, schematic and metric—"but one that would also suppose—and this point is quite clear—the supreme regulation of metaphysical wisdom.

"Thus the divine good of intellectual unity, shattered for three centuries now, would be restored to the human soul." [214]

[214] Maritain, *The Degrees of Knowledge*, pp. 66–67.

II

Contemporary discussions

I

The uniqueness of man

Rationale of this section

Today there are few who would challenge the observation that man cannot reach an adequate understanding of his own nature without a knowledge of his biological background. Yet the realization that the human race with its social, intellectual, moral and aesthetic achievements (the world of culture, in short), the world of living creatures, and the inanimate world all evolved gradually and by stages from very different antecedents—this is a rather recent realization. With its dawning, evolutionary thought has entered into the substance of humanistic thought. In some sense, the rise and development of mankind are a part of the story of biological evolution.

The question of course is, in *what* sense? For the certainty of an evolutionary descent (or ascent, to be more accurate) of man from ancestors who were not men, calls into question in a very contemporary manner the root of our humanness. Therefore our first reading selections center on the nature of man's uniqueness in terms of his biological origins. It must be realized from the very outset that any particular interpretation of the origins of all man's faculties and of his "spiritual" nature demands the support of an analysis which goes beyond the assumptions of empirical research. The materialist, the pantheist, the idealist, the positivist, the Christian—all will have, or at least *may* have, a different interpretation

for the same paleontological evidence. This is not to say that basic agreement is impossible nor even that any interpretive construct will settle
equally well on the foundation of prehistory. It is only to say that if the
assumptions guiding one's interpretation of the nature of the root of our
humanness are not to go unexamined, the issue must be discussed on
philosophical grounds, if always in terms of the evolutionary evidence.

Fortunately for all concerned, these philosophical grounds have been
stated with unprecedented formal clarity thanks to the appearance early in
1968 of Mortimer J. Adler's dialectical assessment of *The Difference of Man
and The Difference It Makes*. Although this assessment, true to its dialectical nature, examines the conflict of opinions concerning the root of
man's humanness without taking sides, nonetheless, by pointing out
that the complexity of a question derives from the range of answers
that can be given to it, and that the particular question as to how man
differs can only be answered in one of the modes in which any object that
we can consider differs from any other (since "the various possible ways in
which any two comparable things can be said to differ exhaust the ways in
which man can be said to differ from everything else on earth"[1]), Adler, by
showing that there are but three basically distinct modes of difference
(thus exhausting the range of possible answers to the general question as
to how any two comparable things can be said to differ), is able with some
justice to claim for his book "that it provides the basis for understanding
and criticizing all the writing that has so far been done on this subject, as
well as whatever remains to be written in the future as new evidence
accumulates and new theories or arguments develop."[2]

In the reading selections to follow, we will see that each of the possible
answers outlined by Adler is argued by evolutionists well acquainted with
the present accumulation of scientific data. Whether one or other of these
arguments is more consistent with the implications of the evidences and
hence more persuasive, it is the task of the reader of course to decide.

After considering in this set of readings the *root* of man's humanness

---

[1] Mortimer J. Adler, *The Difference of Man and the Difference It Makes* (New York: Holt,
Rinehart & Winston, 1967), p. 15.

[2] *Ibid.*, p. 47. "We must note at once that when the question about man ceases to be treated
as a purely philosophical question and becomes a mixed question involving science as well as
philosophy, the fundamental structure of the issue is not altered. The three answers to the
question—degree, superficial kind, and radical kind—exhaust the possibilities. Those same
three answers exhaustively represent the positions actually taken in the history of Western
philosophy. The intervention of science in the consideration of the question has not increased
the number of the answers, nor has it in any way affected the structure of their opposition."
(*Ibid.*, p. 53). That the three modes of difference here enumerated really do exhaust the possible ways in which any two comparable things can be said to differ from one another may be
seen from Adler's analysis of "The Hierarchy of Essences" in *The Review of Metaphysics*, VI
(September, 1952), pp. 3–30, the terminology of which has been explicitly correlated with that
employed in *The Difference of Man* in Sec. VIII of Part II of my analysis of "The Philosophical Dimensions of the Origin of Species," *The Thomist*, XXXIII (April, 1969), pp.
305–331 (see esp. fn 254, p. 314).

we will proceed in the second set of readings to a consideration of the *fruit*, so to speak, of man's uniqueness; and then, in our third set of readings, we will return to the root question, in terms of what difference it makes how man differs.

Before proceeding to these evolutionary discussions, therefore, it will be helpfull to envisage in formal and abstract terms the three-sided controversy about the difference of man.

Thus, he could differ according to what Adler terms an *"apparent* difference in kind," or, equivalently, a "difference in degree": "When, between two things being compared, the difference in degree in a certain respect is large, and when, in addition, in that same respect, the intermediate degrees which are always possible are in fact absent or missing (i.e., not realized by actual specimens), then the large gap in the series of degrees may confer upon the two things being compared the *appearance* of a difference in kind."[3] (An example of this mode of difference would be a chromatic spectrum in which colors separated by interference bands or blackouts *seem* to differ discontinuously, whereas the fact that *in reality* they differ continuously can be exhibited by removing the interference bands.)

Again, man could differ according to a *"superficial* difference in kind": "An observable or manifest difference in kind may be based on and explained by an underlying difference in degree, in which one degree is above and the other below *a critical threshold* in a continuum of degree."[4] Thanks to this critical threshold in the series of degrees, no intermediates are possible with respect to that property in terms of which comparison is being made. (An example of this mode of difference is afforded by the difference between ice, water, and steam, where the freezing and boiling points respectively function as critical thresholds in a continuum of degrees.)

Finally, man might differ as man according to a *"radical* difference in kind": "An observable or manifest difference in kind may be based on and explained by the fact that one of the two things being compared has a factor or element in its constitution that is totally absent in the constitution of the other; in consequence of which the two things, with respect to their fundamental constitution or make-up, can also be said to differ in kind."[5] Here it is not a question of a mere critical threshold which marks the difference, but a manifest difference in kind bespeaking an underlying one as well. (Perhaps the simplest way of illustrating this mode of difference is by borrowing an example from mathematics: odd and even numbers, or three- and four-sided figures, differ radically in kind.)

Thus the word "only" indicates a difference in kind, even as the

[3] *Ibid.,* p. 23.
[4] *Ibid.,* p. 24.
[5] *Ibid.,* p. 25.

words "more" and "less" indicate a difference in degree. When two things differ in kind, they differ discontinuously: since no intermediate is possible, the law of excluded middle is applicable in discourse concerning them. On the other hand, when two things differ in degree, they differ continuously: since intermediates are always possible, the law of excluded middle may not be employed in discourse concerning them (or at least, if it is employed, it has *only* logical, no ontological, import).

This initial question of how man is in fact set apart within the biological community will seem to many students a rather "academic" issue—interesting, perhaps, but hardly relevant to practical concerns. However, as we shall see in subsequent sections of this book, notably Readings Sections 2 and 3 following, the answer one gives to the question of what makes man specifically distinct directly determines what one must logically hold concerning the sorts of freedom and responsibility man has in the face of his future evolution. For, only if man differs radically in kind is there room in his make-up for a power of free choice (*self-determination*) of the sort requisite to justify the imputation of praise and blame and moral or personal responsibility generally. If, on the other hand, man differs in kind only superficially or merely apparently, then, though he has various "freedoms," just as do other animal forms, he does not have that freedom required to become, under favorable circumstances, the master of himself and of the course of his life.

Thus, only if man differs in kind radically, rather than merely apparently or even really but only 'superficially,' is there room for the judgment that there are better and worse ways for men to live humanly, room for a properly moral aspect in cultural evolution and individual thought and action.

With this much said, we are in a position to better appreciate the discussions to follow.[6]

---

[6] Having myself done extensive research and a little writing concerning the very problem which Adler's book so neatly circumscribes, I can well appreciate and heartily concur in Adler's "reason for presenting a purely formal picture of the possible answers in advance of documenting the possibilities by reference to positions actually taken by scientists and philosophers who have concerned themselves with the difference of man. Most, if not all, of them have approached the question with too few distinctions explicitly in mind. They use the words 'degree' and 'kind' without qualifying them by such critical modifiers as 'real' and 'apparent', 'superficial' and 'radical'. The reader will find that the philosophical and scientific literature on the subject of man's difference is simply not intelligible without these distinctions, especially the distinction between a radical and a superficial difference in kind." (*Ibid.*, p. 32).

Evolution and transcendence

Theodosius Dobzhansky

CONTEXTUALIZING COMMENTS

Born and educated in Russia, Theodosius Dobzhansky came to the United States in 1927, eventually to become a naturalized citizen. He is now a professor at Rockefeller University in New York. One of the key figures in working out the role of genetics in the origin of species, Dr. Dobzhansky is the author of several books (see bibliography), one of which, *Mankind Evolving* (1963), was described by George Gaylord Simpson as "the most interesting . . . the most judicious scientific treatise that has ever been written on the nature of man." On the central issue of human origins, Dr. Dobzhansky is convinced that as regards the root of man's evolutionary uniqueness, his cultural capacity, "this difference between man and animal is fundamentally a difference in degree," although he adds that "it is so great that it may justly be *described* as qualitative."[1] Although there is nothing simplistic about Dr. Dobz-

From *The Biology of Ultimate Concern* by Theodosius Dobzhansky. Copyright © 1967 by Theodosius Dobzhansky. Reprinted by arrangement with The New American Library, Inc., New York.
[1] Theodosius Dobzhansky, *The Biological Basis of Human Freedom* (New York: Columbia University Press, 1956), p. 99: emphasis added. In this particular, I am taking exception to the alignment of authors according to patterns of agreement and disagreement over the nature of man's uniqueness made by Adler in *The Difference of Man and the Difference It Makes*. In that book, Adler explicitly aligns Dobzhansky as agreeing with Leslie White and others that man differs from all other animals superficially in kind, and as disagreeing with Darwin's view that man differs only apparently in kind, i.e., only in degree (see pp. 81–98

hansky's views on man—it must be said that he is one of the best informed
and most balanced of the students of man—he argues persuasively that all
man has and all he does is at bottom the outcome of genetic descent with
modifications from the Primates, and that there is no need nor room for
another interpretation of the facts of human prehistory: and in so arguing,
he speaks for perhaps the majority—though by no means all—of evolutionists.

Evolution and transcendence

The idea that man and the world which he inhabits are products of an
evolutionary development seems, strangely enough, "degrading" to some

---

and 128); whereas, both in the statement I have cited from *The Biological Basis of Human
Freedom* and in the reading selection which follows, Dobzhansky explicitly agrees with
Darwin's view and—by implication—dissociates himself from White's position. Am I charging
Adler with a misinterpretation of the contemporary discussion over the uniqueness of man?
By no means, for the following reason: although Dobzhansky does explicitly align himself
with Darwin, there are elsewhere in his writings statements equally explicit which are flatly
inconsistent with the view that man differs only in degree. Dobzhansky is literally open to
two contradictory interpretations, so far as his views on the nature of man go; and since Adler
is aware of and more than once—e.g., p. 92, p. 318 fn. 24—calls attention to the inconsistency
of Dobzhansky's statements on this matter, he is certainly within his rights to accord greater
seriousness to those among Dobzhansky's statements which seem—in Adler's judgment—
to accord more fully with the present weight of evidence. On the other hand, in construing
Dobzhansky's contribution to the discussion differently than does Adler, it is not necessary
that I disagree—still less, that I agree—with Adler's contention that "it is impossible for
anyone who understands the distinction between difference in degree and difference in kind
to assert, in the face of the available evidence, that man differs only in degree from other
animals" (p. 44). It suffices merely to call attention to the psychological fact that Dobzhansky,
albeit with logical inconsistency, does consider himself as agreeing with Darwin that man
differs from other animals only in degree; acknowledging the while, even as does Adler, that
Dobzhansky's writings to date legitimize two mutually exclusive interpretations. Perhaps the
underlying reason for Dobzhansky's logical bi-location is that he belongs to that large group
comprising "most, if not all," of the "scientists and philosophers who have concerned themselves with the difference of man," who "have approached the question with too few distinctions explicitly in mind" (Adler, p. 32), and is in consequence one of those who "mistakenly
identify all differences in kind with radical difference in kind. The latter does violate the
principle of [phylogenetic] continuity. Failing to recognize superficial differences in kind as
an alternative to differences in degree, they insist that all differences must be differences in
degree" (*ibid.*, p. 106; see par. 6, pp. 79–80, *ibid.*, and the statement of the seven principal
sources of the ambiguities and uncertainties in the post-Darwinian philosophical notion of
species in J. N. Deely's article, "The Philosophical Dimensions of the Origin of Species,
Part II," *The Thomist*, XXXIII [April, 1969], pp. 251–264, esp. statement 4, pp. 254–262).
However that may be, it remains a psychological and biographical fact that Dobzhansky
espouses Darwin's view of man as different only in degree; whether that espousal is logically
supported or undermined by others of Dobzhansky's equally explicit expressions of his
evolutionary views, or even by clear thinking in terms of the present state of research and
evidence, is another question again. Adler addresses himself to the latter question, and con-
strues Dobzhansky's extensive writings in the light of it. I am concerned in making this reading
selection solely with the former fact.

people. It is, on the contrary, a prerequisite of humanism in Tillich's sense,[1] since to have potentialities to be actualized man must first of all have the potentiality of evolving. He must be, individually and collectively, not a state but a process. The cosmos must be the cosmogenesis. And, in the words of Teilhard de Chardin (1964), "We see more clearly with every increase in our knowledge that we are, all of us, participants in a process, Cosmogenesis culminating in Anthropogenesis, upon which our ultimate fulfillment—one might even say, our beatification—somehow depends."[2]

The question which presents itself is whether the cosmic, the biological, and the human evolutions are three unrelated processes, or are parts, perhaps chapters or stages, of a single universal evolution. Asking this is really another way of posing the questions, how living bodies arose from the nonliving matrix, and how the stream of consciousness and self-awareness started from the straightforward physiological processes in our animal ancestors. How difficult these questions are can be seen from the following statement by an eminent biologist:

What is still so completely mysterious as to acquire for many human beings a mystical quality, is that life should have emerged from matter, and that mankind should have ever started on the road which so clearly is taking it farther and farther away from its brutish origins.[3]

Here it is necessary to guard against two oversimplifications, opposite in sign but equally misleading. One assumes complete breaks in the evolutionary continuity between life and nonlife, and between humanity and animality. The other overlooks the differences between the cosmic, biological, and human evolutions, and thus loses sight of the origin of novelty. The best hope of making the problem manageable lies, it seems to me, in using the concept of levels, or dimensions of existence, developed by dialectical Marxists on the one side and by the great theologian Paul Tillich on the other.

Stated most simply, the phenomena of the inorganic, organic, and human levels are subject to different laws peculiar to those levels. It is unnecessary to assume any intrinsic irreducibility of these laws, but unprofitable to describe the phenomena of an overlying level in terms of those of the underlying ones. One of the Soviet high priests of Marxism, Present, expresses this fairly clearly as follows:

Wherever it arose, human society must have come from the zoological world, and it was work, the process of production, that made man human. However, what removed people from the animal way of life and gave a specificity to their [new] life, became the essence and the basis of the history that ensued. . . . Likewise, in the realm of living nature, what removed the novel form of the material motion from its nonliving prehistory necessarily became its essence, its fundamental basis,[4]

According to Tillich [in the work cited in fn. 1]

No actualization of the organic dimension is possible without actualization of the inorganic, and the dimension of the spirit would remain potential without the actualization of the inorganic. . . . All of them are actual in man as we know him, but the special character of this realm is determined by the dimensions of the spiritual and historical . . . the dimension of the organic is essentially present in the inorganic; its actual appearance is dependent on conditions the description of which is the task of biology and biochemistry.

Inorganic, organic, and human evolutions occur in different dimensions, or on different levels, of the evolutionary development of the universe. The dimensions or levels are, to be sure, not entirely sundered from one another; on the contrary, there are feedback relationships between the animate and the inanimate, and between the biological and the human. Nevertheless, the different dimensions are characterized by different laws and regularities, which are best understood and investigated in terms of the dimension to which each belongs. The changes in the organic evolution are more rapid than in the inorganic. Inorganic evolution did not, however, come to a halt with the appearance of life; organic evolution is superimposed on the inorganic. Biological evolution of mankind is slower than cultural evolution; nevertheless, biological changes did not cease when culture emerged; cultural evolution is superimposed on the biological and the inorganic. As stated above, the evolutionary changes in the different dimensions are connected by feedback relationships.

The attainment of a new level or dimension is, however, a critical event in evolutionary history. I propose to call it evolutionary transcendence. The word "transcendence" is obviously not used here in the sense of philosophical transcendentalism; to transcend is to go beyond the limits of, or to surpass the ordinary, accustomed, previously utilized or well-trodden possibilities of a system. It is in this sense that Hallowell wrote, "The psychological basis of culture lies not only in a capacity for highly complex forms of learning but in a capacity for transcending what is learned, a potentiality for innovation, creativity, reorganization and change."[5] Erich Fromm wrote that man "is driven by the urge to transcend the role of the creature," and that "he transcends the separateness of his individual existence by becoming part of somebody or something bigger than himself."[6]

Cosmic evolution transcended itself when it produced life. Though the physical and chemical processes which occur in living bodies are not fundamentally different from those found in organic nature, the patterns of these processes are different in the organic and inorganic nature. Inorganic evolution went beyond the bounds of the previous physical and chemical patternings when it gave rise to life. In the same sense, biological evolution transcended itself when it gave rise to man. There obviously exist phenomena and processes, ranging from self-awareness to the human forms of society and of history, which occur exclusively, or almost exclusively, on the human level. It seems unnecessary to labor the point that a great range of potentialities are open to man only.

The origin of life and the origin of man are, understandably, among the most challenging and also most difficult problems of evolutionary history. It would be most unwise to give a fictitious appearance of simplicity to these singularly complex issues.

I do not wish to be understood as maintaining that the origin of life [or of man] involved agents or processes that did not also earlier or later operate on earth. The flow of evolutionary events is, however, not always smooth and uniform; it also contains crises and turning points which, viewed in retrospect, may appear to be breaks of the continuity. The origin of life was one such crisis, radical enough to deserve the name of transcendence. The origin of man was another. This should not be taken to mean that the origin of life or of man was instantaneous or even very swift. A process which is very rapid in a geological (more precisely, paleontological) sense may appear to be lengthy and slow in terms of a human lifetime or a generation.

The appearance of life and of man were the two fateful transcendences which marked the beginnings of new evolutionary eras. They were, however, only extreme cases of radical innovations, other examples of which are also known. The origin of terrestrial vertebrates from fishlike ancestors opened up a new realm of adaptive radiations in the terrestrial environments, which was closed to water-dwelling creatures. The result was what Simpson has called "quantum evolution,"[7] an abrupt change in the ways of life as well as in the body structures. Domestication of fire and the invention of agriculture were among the momentous events which opened new paths for human evolution. In a still more limited compass, the highest fulfillment of an individual human life is self-transcendence.

While the origin of life on earth is an event of a very remote past, man is a relative newcomer (even though the estimates of his antiquity have been almost doubled by recent discoveries). The record is still fragmentary, but the general outlines of at least the outward aspects of man's emergence are recognizable. During the early part of the Pleistocene age (the Villafranchian time, perhaps two million to one million years ago), there lived in the east-central and in the southern parts of the African continent at least two species of *Australopithecus,* a genus of Hominidae, the family of man. One of these, larger in body size (*Australopithecus robustus,* and its race *boisei*), apparently represented an evolutionary blind alley, and eventually died out. The other species, of a more supple build (*Australopithecus africanus*), may have been one of our ancestors.

There is good evidence in the structure of their pelvic bones that both species walked erect; their teeth were, if anything, more like ours than like those of the now-living anthropoid apes. The brain-case capacity, though large in relation to the body size, was well within the ape range. Perhaps both species of *Australopithecus*, or at any rate one of them, made and used primitive stone tools. The remains of a most interesting creature have recently been found in the Villafranchian deposits in east-central Africa, and given the name *Homo habilis*. The name connotes the opinion of its discoverers that this creature had already passed from the genus *Australopithecus* to the genus *Homo*, to which we also belong. On the other hand, it is closely related to *Australopithecus africanus*, and may even have been only a race of that species. Regardless of the name by which it is classified, this is rather clearly one of the "missing links," which are no longer missing.

Later, during the mid-Pleistocene, there lived several races of the species *Homo erectus*, clearly ancestral to the modern *Homo sapiens*. Remains of *Homo erectus* have been found in Java, in China, in Africa, and probably also in Europe. Roughly 100,000 years ago, during the last, Würm-Wisconsin, glaciation, the territory extending from western Europe to Turkestan and to Iraq and Palestine was inhabited by variants of the Neanderthal race of *Homo sapiens*.

Rough stone tools have been found in association with australo-pithecine remains both in east-central and South Africa. *Homo erectus* in China is the oldest known user of fire. The Neanderthalians buried their dead. These are evidences of humanization. All animals die, but man alone knows that he will die. A burial is a sign of a death awareness, and probably of the existence of ultimate concern. The ancestors of man had begun to transcend their animality perhaps as long as 1,700,000 years ago. The process is under way in ourselves. Nobody has characterized this process more clearly than Bidney:

Man is a self-reflecting animal in that he alone has the ability to objectify himself, to stand apart from himself, as it were, and to consider the kind of being he is and what it is that he wants to do and to become. Other animals may be conscious of their affects and the objects perceived; man alone is capable of reflection, of self-consciousness, of thinking of himself as an object.[8]

And according to Hallowell:

The great novelty, then, in the behavioral evolution of the primates, was not simply the development of a cultural mode of adaptation as such. It was, rather, the psychological restructuralization that not only made this new mode of existence possible but provided the psychological basis for cultural readaptation and change.[9]

In 1871, Darwin wrote: "The difference in mind between man and the higher animals, great as it is, certainly is one of degree and not of kind."[10] This is about the state of the issue at present. The estimates of the importance of this difference have, however, varied greatly; it is belittled by those intent on proving that man is nothing but an animal, and overdrawn by those who place him outside the order of nature. To the former, there is no great difference between fish and philosopher, and nothing more remarkable about his human brain secreting thoughts than about his liver secreting bile. Leslie White, an anthropologist, believes that

whether a man—an average man, typical of his group—believes in Christ or Buddha, Genesis or Geology, Determinism or Free Will, is not a matter of his own choosing. His philosophy is merely the response of his neuro-sensory-muscular-glandular system to the streams of cultural stimuli impinging upon him from the outside.[11]

However, the same author in the same book also says,

Because *human* behavior is symbol behavior and since the behavior of infra-human species is nonsymbolic, it follows that we can learn nothing about human behavior from observations upon or experiments with the lower animals.

A. R. Wallace, the codiscoverer with Darwin of the evolutionary role of natural selection, thought that man's mind must have been implanted by a supernatural agency. Lack, one of the foremost modern evolutionary ecologists, is of the same opinion: "A Christian, agreeing to man's evolution by natural selection, has to add that man has spiritual attributes, good and evil, that are not a result of this evolution, but are of supernatural origin."[12] Brunner may be taken as representative of a type of thinking among modern theologians who concede that man is a product of biological evolution, but nevertheless insist that he has a property called the *humanum*, which is solely his attribute. The "*humanum* is characterized by something which is entirely lacking in the animal, subjectively speaking by the spirit [Geist], and objectively by the creation of culture. . . . It possesses a dimension which is lacking in biology, the law of norms, the faculty of grasping meaning, freedom, responsibility."[13]

Lists of distinctive anatomical, physiological, and psychological characteristics of the human species have often been compiled. Man is an erect-walking-primate, his brain is large in relation to the body size, his hands are fit for tool manipulation, toolmaking, and for carrying objects. He engages in play, is capable of abstract thought, laughter, formation and use of symbols, of learning and using symbolic language, of learning to distinguish between good and evil, to feel reverence and piety. At least some of these characteristics belong to the *humanum*. How did they arise in evolution? It is most certainly difficult, perhaps quite impossible, to reconstruct, even if we were given more detailed data than are now available, the exact sequence of events during the critical stages of the emergence of mankind from prehuman ancestors. There are, however, some comprehensive biological regularities which make a general understanding of the main trends of the evolutionary history of man not out of reach.

If a descended species is found to differ from its ancestor in several characteristics, the origin of each of these characteristics is not necessarily independent. The characteristics may vary together, for at least two reasons. They may be manifold effects of the same genetic factors (as are the different parts of the syndromes in many hereditary diseases and malformations dependent on single genes). Or else, they may be functionally related. There was some discussion among anthropologists as to whether the development of an erect posture and manual skill preceded or followed tool use and toolmaking. A similar issue can be raised concerning the capacity for abstract thinking, formation of symbols, symbolic language, and the beginnings of cultural transmission. To a considerable extent, if not at all points, such issues are spurious. The product grows with the instrument, and the instrument with the product. Hands freed from walking duties by an erect posture can more easily develop the dexterity needed for the manipulation of tools. An erect posture does not, however, guarantee that the anterior extremities will use tools; for example, the giant kangaroos of Australia do not utilize their front paws either for running or for handling tools. Conversely, some monkeys and apes have fairly versatile hands but do not walk upright. It is reasonably clear that an animal which begins to handle tools will derive an advantage from having a pair of its extremities become adept at such operations; vice versa, an animal which becomes bipedal may profit by using the second pair of its extremities for something else. The crux of the matter is, however, this: increasing tool use puts greater and greater selective premium on bipedalism, and vice versa.

Human languages are very different forms of behavior from the so-called animal "languages," although both serve, of course, the function of communication. This complex, and to many people puzzling, issue, has been ably clarified by Hockett[14] and Hockett and Ascher.[15] Very briefly, the differences are as follows. Animal calls or signals are mutually exclusive, meaning that the animal may respond to a situation by one or another of its repertory of calls, or may remain silent. Language is productive, i.e., man emits utterances never made before by himself or by anyone else, which are nevertheless understandable to speakers of the same language. People can speak of things that are out of sight, and of past, future, and imaginary things. This is called the property of displacement. Moreover, language has the property of a duality of patterning. It consists of units, phonemes, which have no meaning in themselves but serve to compose utterances that are meaningful. The capacity to learn a language is biologically inherited; it is a capacity to learn (at least in childhood) any one of the human languages. A chimpanzee cannot do it, although his larynx is apparently capable of producing all the necessary sounds; the incapacity is caused by the absence of certain brain mechanisms. Human languages consist of utterances the meaning of which is socially established by convention. Neologisms and changed habits of speech constantly appear.

Animal signals, calls, and the "dances" of the honey bees are admirably efficient methods of communication, but only within the narrow compass of the needs of the particular species in which they occur. There seems little point in belaboring the truism that human speech possesses the advantage of an almost infinitely greater versatility. This is because man's is a symbolic language. Cassirer,[16] Langer[17] and some others, consider the ability to form and to use symbols man's most important distinctive quality. In Cassirer's words "Signals and symbols belong to two different universes of discourse: a signal is a part of the physical world of being; a symbol is a part of the human world of meaning." And further:

The principle of symbolism with its universality, validity, and general applicability, is the magic word, the Open Sesame giving access to the specifically human world, to the world of human culture. Once man is in possession of this magic key further progress is assured.

The capacity to form and to operate with abstract ideas and symbols is correlated in evolution, if not in physiology, with the capacity to use human language. Here too, the product grows with the instrument, and vice versa. And these capacities are, in turn, correlated with toolmaking. It should be noted that tool using and tool making are performances almost as profoundly different as signs and symbols. To make a tool for a future employment one needs more than manual dexterity; what is necessary is formation of a mental picture of a situation which is expected to arise in the future but which is not yet given to the senses. In evolution, all these capacities were connected by feedback relationships. An advance in any one of them made the others adaptively more valuable, and advances in these others conferred higher selective premiums on further developments of the first. Owing to these cybernetic interrelations, the proto-human species, gradually shifting from animal to human ways of life, eventually reached a point of no return. To revert from the incipient humanity back to animality would have been genetically difficult, even if this were adaptively profitable. Extinction would become more likely than an evolutionary retreat.

At this point it is necessary to be reminded of some biological evolutionary principles which were mentioned above in another connection. Evolution arises through the action of natural selection in response to the exigencies of the environment. It is a utilitarian process; it maintains or enhances the adaptedness to the environment. But natural selection is not some sort of benevolent ghost; it is automatic, blind, and lacking foresight. It is opportunistic, in the sense that it adapts the organism to the environments existing at the time it acts, and it cannot take into account any possible changes of the conditions in the future. As pointed out above, the consequence of this opportunism and myopia may be extinction. The adaptation is liable to drive the species into a blind alley and to make retraction impossible.

Another consequence of opportunism is even more relevant in human evolution. The fitness that is selected is the overall fitness of the organism to survive and to reproduce, not the excellence of different organs, processes, and abilities taken separately. A consequence is that, especially in radical evolutionary reconstructions, the emerging product is an appalling mixture of excellence and weakness. That this is the case with man is almost a platitude. To quote Hockett and Ascher:

Language and culture, as we have seen, selected for bigger brains. Bigger brains mean bigger heads. Bigger heads mean greater difficulty in parturition. Even today, the head is the chief troublemaker in childbirth. The difficulty can be combatted to some extent by expelling the fetus relatively earlier in its development. There was therefore a selection for such earlier expulsion. But this, in turn, makes for a longer period of helpless infancy—which is, at the same time, a period of maximum plasticity, during which the child can acquire the complex extra-genetic heritage of its community. The helplessness of infants demands longer and more elaborate child care, and it becomes highly convenient for the adult males to help the mothers. Some of the skills that the young males must learn can only be learned from the adult males. All this makes for the domestication of fathers.[18]

The structure of the human psyche discovered by Freud and his followers is a bundle of contradictions and disharmonies, which at first sight seems anything but a product of adaptive evolution. It must, however, be viewed not by itself but within the total evolutionary context. So considered, man is not merely a successful biological species but the most successful one that biological evolution has ever produced. To quote Hockett and Ascher again:

As soon as the hominids had achieved upright posture, bipedal gait, the use of hands for manipulating, for carrying, and for manufacturing generalized tools, and language, they had become men. The human revolution was over.

The biological evolution has transcended itself in the human "revolution." A new level or dimension has been reached. The light of the human spirit has begun to shine. The *humanum* is born.

It remains to consider briefly some of the misgivings which arise in connection with the above account of the evolutionary transcendence giving rise to man. Those who see an unbridgeable gap between the *humanum* and the prehuman state question the presence on the animal level even of rudiments from which the *humanum* could arise. Now, the point which the believers in unbridgeable gaps miss is that the qualitative novelty of the human estate is the novelty of a pattern, not of its components. The transcendence does not mean that a new force or energy has arrived from nowhere; it does mean that a new form of unity has come into existence. At all events, no component of the *humanum* can any longer be denied to animals, although the human constellation of these components certainly can. In recent years this problem has been studied by means of many brilliantly conceived and executed experiments.[19] Birds and mammals are demonstrably able to form "abstract averbal concepts," such as those of number. In the experiments of Koehler, a bird (raven) learned to choose from among five dishes covered with cardboard disks marked with two to six spots, the dish which contained food. After training, the bird was able to pick the right dish when shown a signboard with the corresponding number of dots. Another bird (a parrot) selected the right dish when given a corresponding number of acoustic stimuli.

The "dance language" of bees is symbolic, indicating the direction of a food source by the movements of the informant bee on the honeycomb. It does not, of course, have any other of the properties of human languages discussed above. "Insight learning" has been demonstrated in chimpanzees by Koehler, Ladygina-Kots, and others. Thorpe writes that

with birds as with men, the perceptions of the experienced and mature individual are built up by a process of perceptual learning whereby the primary conceptions are combined and built into more complex gestalts; indeed, no essential differences between the principles underlying the two processes can be detected.

True play not only occurs in animals as well as in humans, but it performs the important function of training the young in preparation for adult activities. "Protoesthetic" impulses have been shown to exist in monkeys and apes, which can be made to engage in painting, with results not too dissimilar from some products of avant-garde human painters. Some precursors of cooperative, ethical, and even altruistic behavior have repeatedly been alleged to exist in animals, but in the nature of the case they are hard to prove beyond doubt (see Thorpe, *loc. cit.*, for critical reviews).

Another difficulty with the evolutionary explanation of the origin of man is of a more general sort. The occurrence of mutation and of natural selection is said to be due to "chance." Can one possibly believe that the *humanum* arose through summation of a series of accidents? De Cayeux, a French paleontologist, emphatically disagrees: "The lottery does not suffice. Mutationism is a double failure. The evolution of species cannot be due to chance alone, even if corrected by the selection." Protestations to the same effect have been made with considerable eloquence by such writers as Joseph Wood Krutch, Barzun, and others. Now, a goodly portion of all this is sheer misunderstanding. What, indeed, is meant by "chance," or "randomness," in evolution? Reference has already been made above to the fact that mutations are adaptively ambiguous. Nature has not seen fit to make mutations arise where needed, when needed, and only the kind that is needed. A geneticist would be hard put to envisage a way to accomplish this. But not even mutations are random changes, because what mutations can take place in a given gene is evidently decreed by the structure of that gene. Mutations do not, however, constitute evolution; they are only raw materials for evolution. They are manipulated by natural selection.

Natural selection is a chance process (despite the misplaced superlative "fittest" in the "survival of the fittest") only in the sense that most genotypes have not absolute but only relative advantages or disadvantages compared to others. Natural selection may act if one genotype leaves 100 surviving offspring compared to 99 left by the carriers of another genotype. Otherwise natural selection is an anti-chance agency. It makes adaptive sense out of the relative chaos of the countless combinations of mutant genes. And it does so without having a will, intention, or foresight. The classical analogy between the action of natural selection and that of a sieve is, as pointed out above, misleading. The best analogue of natural selection is a cybernetic mechanism; it transmits "information" about the state of the environment to the genotype.

The point so central that it must be pressed is that natural selection is in a very real sense creative. It brings into existence real novelties—genotypes which never existed before. Moreover, these genotypes, or at least some of them, are harmonious, internally balanced, and fit to live in some environments. Writers, poets, naturalists, have often declaimed about the wonderful, prodigal, breathtaking inventiveness of nature. They have seldom realized that they were praising natural selection. But a creative process runs the risk of failure and miscreation; this is where creation is less safe than mass production from a set pattern. Miscreation in biology is death, extinction. Paleontology abundantly shows that extinction is the most usual end of evolutionary lines. Some biologists who thought that natural selection is too soulless and mechanical believed instead in orthogenesis, a theory according to which evolution is merely an uncovering of the preformed and predetermined organic configurations. Yet, with orthogenesis evolution would create nothing really new, and extinction would be a mystery or plain nonsense.

Another aspect of the issue of creativity is the problem of determinism. A few remarks will have to suffice here. Etkin asks why only man was able to achieve the mode of adaptation by culture: "What did man have in his ancestry which no other animal had that enabled him to solve his evolutionary stress?"[20] The answer is that the human mode of adaptation is one of the many that were possibly open to our remote ancestors, and which ones of these modes were to be adopted in the different evolutionary lines of the primate order was not predetermined. It was an evolutionary invention which did not necessarily have to be made. There are, broadly speaking, two kinds of interpretations of evolution. One kind supposes that any and all evolutionary changes that ever occurred were predestined to occur. The other kind recognizes that there may be many different ways of solving the problems of adaptation to the same environment; which one, if any, of these ways is in fact adopted in evolution escapes predetermination. It is because of this lack of predestination that I am inclined to question the belief that, if life exists in different parts of the universe, it is bound to result in formation of manlike, or perhaps even of supermanlike, rational beings. Together with Simpson[21] and Blum,[22] I consider this not merely questionable but improbable to a degree which would usually mean rejection of a scientific hypothesis.

NOTES

[1] Humanism, according to Tillich, "asserts that the aim of culture is the actualization of the potentialities of man as the bearer of spirit," and "Wisdom can be distinguished from objectifying knowledge (*sapientia* from *scientia*) by its ability to manifest itself beyond the cleavage of subject and object." Cf. *Systematic Theology* (Chicago: University of Chicago Press, 1963), Vol. 3.

[2] Teilhard de Chardin, *The Future of Man* (New York: Harper, 1964).

[3] R. Dubos, *The Torch of Life* (New York: Simon & Schuster, 1962).

[4] I. I. Present, "On the Essence of Life in Connection with its Origin," in *O sushchnosti zhizni* (Moscow: Nauka, 1964).

[5] A. I. Hallowell, "The Protocultural Foundations of Human Adaptation," in *Social Life of Early Men*, ed. by S. L. Washburn (New York: Wenner-Gren Foundation, 1961), pp. 236–255.

[6] Erich Fromm, "Value, Psychology, and Human Existence," in *New Knowledge of Human Values*, ed. by A. H. Maslow (New York: Harper & Row, 1959).

[7] G. G. Simpson, *The Major Features of Evolution* (New York: Columbia University Press, 1953).

[8] David Bidney, *Theoretical Anthropology* (New York: Columbia University Press, 1953).

[9] A. I. Hallowell, "Self, Society, and Culture in Phylogenetic Perspective," in *The Evolution of Man*, ed. by Sol Tax (Chicago: The University of Chicago Press, 1960).

[10] Charles Darwin, *The Descent of Man* (New York: Modern Library Giant Ed. G27), p. 494.

[11] Leslie A. White, *The Science of Culture* (New York: Grove Press, 1949).

[12] D. Lack, *Evolutionary Theory and Christian Belief* (London: Methuen, 1957).

[13] E. Brunner, *The Christian Doctrine of Creation and Redemption* (New York: Westminster Press, 1952). A critique of Brunner's views in L. C. Birch, "Creation and the Creator," *Journal of Religion*, XXXVII (1957), 85–98; and *Nature and God* (London, SCM Press, 1965).

[14] C. F. Hockett, "Animal 'Languages' and Human Language," in *The Evolution of Man's Capacity For Culture* (Detroit: Wayne State University Press, 1959).

[15] C. F. Hockett and R. Ascher, "The Human Revolution," *Current Anthropology*, V (1964), pp. 135–168.

[16] E. Cassirer, *An Essay on Man* (New York: Anchor, 1944).

[17] S. K. Langer, *Philosophy in a New Key* (New York, Penguin, 1948).

[18] *Art. cit.*

[19] Reviews in B. Rensch, *Evolution Above the Species Level* (New York: Columbia University Press, 1960); W. H. Thorpe, *Biology and the Nature of Man* (London: Oxford University

Press, 1962); and W. H. Thorpe, *Learning and Instinct in Animals* (2nd ed.; London: Methuen, 1963).

[20] W. Etkin, "Social Behavioral Factors in the Emergence of Man," in *Culture and the Direction of Human Evolution,* ed. by S. M. Garn (Detroit: Wayne State University Press, 1964).

[21] G. G. Simpson, *This View of Life* (New York: Harcourt, Brace & World, 1964).

[22] H. F. Blum, "Dimensions and Probability of Life," *Nature* CCVI (1965), pp. 131–132.

Four stages
in the evolution of minding

Leslie A. White

CONTEXTUALIZING COMMENTS

Professor of Anthropology at the University of Michigan and one of the central and most controversial figures in the development of recent anthropological thought, Dr. White has long argued that, at least as regards man's root cultural capacity and allowing for whatever must be said concerning gradual development of the neurological bases and anatomical frame for such a capacity, the uniqueness of man is formally *sic et non*, present or wanting. For over a century, he points out,[1] two traditions of comparative psychology have developed side by side. One has declared that man does not differ from other animals in mental functionings save in degree. "The other has seen clearly that man is unique in at least one respect, that he possesses an ability that no other animal has." The question has remained open until the present day, White thinks, in consequence of the difficulty of *defining* this difference adequately. By drawing an empirically rigorous and conceptually exact distinction between *sign* behavior and *symbol* behavior, Dr. White has sought to contribute to a definitive solution of the problem of man's root evolutionary

Reprinted from *The Evolution of Man*, Volume II of *Evolution After Darwin*, the University of Chicago Centennial, edited by Sol Tax, by permission of the University of Chicago Press. Copyright © 1960 by the University of Chicago.
[1] L. A. White, *The Science of Culture* (New York: Grove Press, 1949), p. 31.

uniqueness. That Dr. White considers man's difference in kind to be superficial and not radical (in the senses already defined) will become clear as we consider in Sections II and III below the role of determinism in human life.

Four stages
in the evolution of minding

By "minding" we mean the reaction of a living organism to some thing or event in the external world.[1] It is therefore a process of interaction between an organism and a thing or event lying outside it. Minding is a function of the thing or event to which the organism reacts as well as of the organism itself: $O \times E = M$, in which O is the organism, E is the thing or event to which the organism reacts, and M is minding. Minding varies as either O or E varies.

Interaction between two bodies means a relationship between them. Minding may therefore be understood in terms of relationships between organisms, on the one hand, and events in the external world, on the other. We may deal with these relationships in terms of the *meanings* that things and events in the external world have for organisms: an organism approaches, withdraws from, or remains neutral toward some object in its vicinity, depending upon its meaning or significance to the organism. If it is beneficial (food), it may approach; if it is injurious, it may withdraw; and if it is neutral, the organism will remain indifferent. The concepts with which minding may be analyzed and interpreted are, therefore, *action, reaction, interaction, relationship,* and *meaning.* How are these relationships established? How are meanings determined?

Let us begin with inanimate bodies. According to the theory of gravitation, every particle of matter in the universe attracts every other particle, i.e., a relationship obtains between them; each has meaning for all the others. These meanings are determined by their respective masses and by the distances which intervene between them: "directly as the mass, inversely as the square of the distance." Material particles attract or repel each other in such phenomena as capillary attraction, surface tension, and electromagnetic events. But all relationships among inanimate bodies can probably be reduced to three kinds: attraction, repulsion, and indifference. And in all instances, no doubt, these relationships are determined by the inherent properties of the bodies concerned, their topological relations, and their settings (presence or absence of catalysts).

When we cross the line that divides inanimate and animate bodies and come to living organisms, we find that the simplest reactions are precisely like those of inanimate bodies. The organism's reaction is positive $(+)$, negative $(-)$, or neutral (o).[2] That is, it approaches, withdraws, or does nothing, depending upon the meaning that the object has for it.

In this simplest type of reaction, which we may for convenience call Type I (Fig. 1), the meaning that the thing or event has to the organism

$$O \longrightarrow S$$

Fig. 1.—The simple reflex: Type I behavior

is determined by the intrinsic properties of both organism and thing or event. Or, to put the matter otherwise, the relationship between organism and thing-or-event is determined by their respective intrinsic properties. (For the sake of completeness and precision, we ought, perhaps, to say: the relationship is determined by the intrinsic properties of organism and thing-or-event as conditioned by their setting and the factors—positive, or negative, or neutral catalysts—which it contains.) The organism approaches if the stimulus is positive (e.g., food), withdraws if it is injurious, and remains indifferent if it is neutral. But, whether a thing is *food* or not depends upon the intrinsic properties of the organism as well as of the thing; edibility is a function of the eater as well as of the thing eaten; what is food to one organism may be not-food to another. And so it is with injurious things or things neutral. In every instance in this simplest type of interaction the relationship between organism and thing-or-event is determined by the intrinsic properties of both.

The next stage in the evolution of minding is characterized by the conditioned reflex, and we may use the classic experiment of Pavlov with the dog and the electric bell to illustrate it. A hungry dog salivated when he smelled food; he was indifferent to the sound of an electric bell. But when stimulated by odor and bell simultaneously for a number of times, the sound of the bell alone was sufficient to excite his salivary glands and evoke the response. We may call this kind of behavior Type II, and represent it diagrammatically in Figure 2.

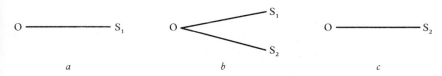

a   b   c

Fig. 2.—The conditioned response or reflex: Type II behavior

The process of condition takes place in three stages. In the first stage, *a*, in Figure 2, we have the same kind of situation that we have in Type I: the organism, O, and a significant stimulus, S_1, the odor of food, with a simple relationship between them in terms of their respective intrinsic properties. In stage b, we introduce S_2, the sound of the bell, which becomes related to S_1, the odor of food, on the one hand, and to O, the dog, on the other. Initially, S_2 is related to S_1 in time and in space, and, as a consequence of association, S_2 and S_1 become related to each other through the neurosensory-glandular system of the dog. A relationship between S_2 and the dog (O) is established at the same time and in the same way. When the relationship between S_2 and the dog has been established, S_1 may drop out. In stage c we again have a simple, direct relationship between the organism and a single stimulus.

Type II resembles Type I and grows out of it. It begins with a simple Type I reaction, and it ends with the *form* of Type I reaction. But Type II differs from Type I in a fundamental respect: Type II is characterized by a relationship between organism and stimulus which is *not* dependent upon their intrinsic properties. To be sure, the substitution of one stimulus for another could not have been effected, had not the dog been an organism capable of this kind of behavior. But the salivary-gland-meaning of the electric bell is in no sense intrinsic in the sound waves that it emits. Type II behavior ends with the *form* of Type I: the reaction takes place *as if* the relationship between dog and bell were intrinsic in them. But the response is fundamentally different in kind.

The next stage in the evolution of minding, Type III, may be illustrated by the example of a chimpanzee using a stick to knock down a banana which is suspended from the roof of his cage beyond the reach of his hand. We illustrate it in Figure 3, in which O=chimpanzee, E_1 is the banana, and E_2 is the stick.

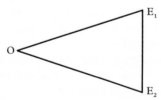

Fig. 3.—Type III behavior

Type III minding is like Type II in one respect: the organism is related simultaneously to two things-or-events in the external world (as in Fig. 2, b). But Type III differs from Type II in a number of important respects. In the first place, the two things, E_1 and E_2, are significant from start to finish in Type III, whereas they are significant in only one of the three stages of the process in Type II (Fig. 2, b). Second, the relationships established in Type III are dependent in their entirety upon their respective intrinsic properties: the chimpanzee is a banana-eating, stick-wielding animal; the banana is a knock-downable-with-a-stick thing and eatable by a chimpanzee; the stick is wieldable by a chimpanzee and can be used to knock down a banana. Third, the relationship between E_1 and E_2 in Type III is established directly and extraorganismically, whereas they are related indirectly and intraorganismically (within the neurosensory-glandular system of the dog) in Type II. And, finally, the relationships established in Type III are determined intra-organismically, i.e., by the chimpanzee himself, "of his own free will and choice," so to speak, whereas the relationships established in Type II are not determined by the organism but by its relationships to other factors—the experimenter, or circumstances, such as chance association.

We may distinguish two kinds of roles of organisms in the process of minding. Either the organism determines the configuration of behavior which it executes, or it does not; it plays either a dominant or a subordinate role. Thus, in Type I, it is not the organism alone that determines its behavior. It behaves as it does because (1) of its own intrinsic properties and (2) because of the intrinsic properties of the stimulus, E. It has neither alternatives nor choice. The flower turns its face to the sun because it must; it can do nothing else. It is something that it undergoes as well as something that it does. Its behavior is subordinate to the intrinsic properties of itself and its stimulus.

The organism plays a subordinate role in Type II, also. The dog "has nothing to say" about how he shall respond to the sound of the bell; this is determined by the experimenter (chance associations may be the determining factor in other processes of conditioning). Here, also, the organism has neither initiative nor choice.

It is different with Type III minding. Here the organism plays a dominant role. It is the chimpanzee who decides what to do and how to do it. He has initiative, alternatives, and choice. He may use the stick to reach and knock down the food, or he may, as they sometimes do, use it to pole-vault ceilingward and snatch the food when it comes within his reach. Or he may decide to build a tower of boxes from whose summit he can reach his prize. This is what we mean when we say that the pattern of action, the configuration of behavior, is determined intra-organismically: the chimpanzee solves his problem by insight and understanding, formulates a plan, then puts it into execution. He is a sublingual architect and builder.

In Type III, then, we are again dealing with relationships determined by the intrinsic properties of organisms and things, but here the organism plays a dominant, instead of a subordinate, role in the formulation and execution of patterns of behavior.

Type IV minding is well illustrated by articulate speech and may be diagrammed as in Figure 4. O is again the organism, this time a human being; E_1 is a hat; and E_2 is the word "hat." Again we have a triangular configuration as in Type III: there is a mutual and simultaneous relation-

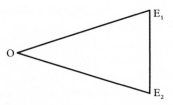

Fig. 4.—Type IV: symboling

ship between the organism and two things-or-events in the external world. And, as in Type III, the configuration is determined intra-organismically, by the organism itself, of its own will and choice; Type IV, like Type III, is characterized by alternatives and choice. But Type IV differs from Type III in that in the former the relationships are not dependent upon the intrinsic properties of the elements involved, as they are in Type III. That is to say, there is no necessary or inherent relationship between the object hat and the combination of sounds *hat*. In this respect, Type IV resembles Type II: both are independent of the intrinsic properties of the factors involved. But Type IV differs from Type II in a fundamental way: the organism plays a dominant role in Type IV, a subordinate role in Type II.

Looking back over our four types of minding, we notice similarities and differences among them. Types I and III are dependent upon the intrinsic properties of the elements involved in the configurations of behavior; Types II and IV are not so dependent. In Types I and II, the organism plays a subordinate role in the formulation and execution of patterns of behavior; in Types III and IV, it plays a dominant role. We may summarize these facts diagrammatically as follows:

	Organism plays a subordinate role	Organism plays a dominant role
Dependent upon intrinsic properties	Type I	Type III
Independent of intrinsic properties	Type II	Type IV

Fig. 5.—Comparison of four stages of minding

Another feature of our series of stages is that our types are *kinds*, not *degrees*, of minding. An organism is either capable of conditioned reflex behavior, Type II, or it is not; there are no gradations between Types I and II. Similarly, an organism is capable of Type III or Type IV behavior, or it is not. We are confronted by a series of leaps, not by an ascending continuum.

The question might be raised at this point, Does our series of stages constitute a biological evolutionary sequence or merely a logical one? It has been derived deductively, in a sense, from a consideration of a basic concept: relationship. It has been postulated that the relationship established in the process of minding is either dependent upon the intrinsic properties of organism and things-and-events in the external world, or it is not. Second, the relationship or, more specifically, the pattern of behavior in which the relationship is expressed, is determined by the organism or it is not. This gives us four categories of minding, and it might appear at first glance that our series of stages is more artificial than real.

But this is not the whole story. We did not begin with factors selected at random. Our premises and postulates were in fact derived from a careful scrutiny and analysis of the behavior of very real organisms. The fact that the series proceeds from the simple to the complex would suggest that it constitutes an evolutionary, as well as a logical, sequence. But there are other facts that make it quite clear that we do indeed have here an evolutionary sequence in a biological sense.

Let us begin, first of all, by classifying all living species with reference to our four types of minding as far as our information will permit (Fig. 6). All organisms are capable of Type I: simple reflexes or tropisms. We know from observation that some species are capable of Type I only. Type II has grown out of Type I, as we have seen; therefore, it may be assumed that organisms capable of Type II are capable of Type I also; this assumption is validated by observation. But, obviously, there are fewer species capable of Type II than are capable of Type I. We know, also from observation, that there are organisms capable of Type II that are incapable of Type III, but all organisms capable of Type III are capable of Types II and I also. Only one species, man, is capable of Type IV, and this species is capable of Types III, II, and I, also. Hence the following generaliza-

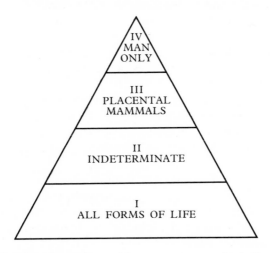

Fig. 6.—Classification of organisms with respect to types of minding

tions: (1) our series of stages is incremental and cumulative, a new stage being added to the one, or ones, that preceded it; (2) the number of species capable of a given type of behavior diminishes as we proceed from Type I to Type IV; and (3) organisms classified according to our series of types of minding are thereby arranged in a biological evolutionary series, the lowest and simplest organisms being at the bottom, the higher and more complex at the top.

Specifically, we cannot assign each and every species to one or another stage in our series of types of minding for the simple reason that we do not have the requisite information. But we have every reason to believe that we could do this if we possessed full information as to their behavior and that no species would fail to be accommodated by our series.

We do not know at what point in the evolutionary scale organisms become capable of Type II behavior. Snails, it appears, are capable of conditioned reflex behavior, which puts this ability fairly low in the scale of biological evolution, but what other species belong here is a question that we cannot answer.

We know that apes are capable of Type III minding, and there appears to be much evidence that dogs and elephants also possess this ability. It seems reasonable to suppose that all placental mammals fall within this class, but whether marsupials and monotremes belong here or not is a question to which comparative psychology provides us with no answers.

Incidentally, although we have illustrated Type III minding with an instance of tool-using (Fig. 3), it must be made clear that the ability to wield tools is not essential to this kind of behavior. The characteristic of Type III minding is the ability to formulate intra-organismically—i.e., by means of insight and comprehension and in terms of alternatives and choice—a configuration of behavior in which the organism is related simultaneously to two or more things or events in the external world in such a way that all elements of the configuration are related to one another in terms of their intrinsic properties, as in the case of the chimpanzee, the stick, and the banana.

But an organism need not wield tools to exhibit this kind of behavior; it may move itself, rather than a thing (tool), with reference to two (or more) things, all of which are related to one another in a configuration of behavior in terms of their intrinsic properties. Let us illustrate with an example.

The dog (D) in Figure 7 has formulated a plan of action, intra-organismically, in which he has related himself, the food, and the open gate to one another in terms of their respective intrinsic properties, and he then executes this plan in overt behavior. He has done, in effect, what the chimpanzee did in reaching the banana with the stick, except that the dog moved his body with reference to the two other factors instead of moving one of those external factors. Thus tool-wielding may be the most characteristic form of expression of Type III minding, but it is not essential to it. In this example (Fig. 7) we are reporting the results of some experiments in which it was demonstrated that a dog, but not a chicken, was capable of solving this problem by insight and comprehension. The chicken, placed in this situation, simply ran back and forth aimlessly along the fence at point *P*.

Our series of types of minding constitute a progressive series of advantages for the life-process, for living organisms: they emancipate organisms from limitations imposed upon them by their environments, on the one hand, and confer positive control over the environment, on the

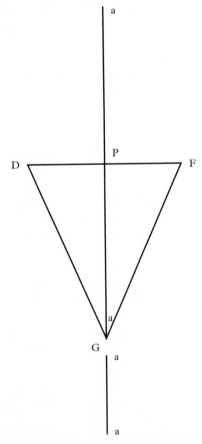

Fig. 7.—D is a dog; a, a is a fence; F is dogfood; G is an open gate. The dog can see and smell the food. But he cannot approach and obtain it directly because of the intervening fence which he cannot jump over or crawl through or under at P. But he knows, either by previous experience or by observation, that there is that open gate, G, down the line. He therefore promptly runs to the gate, passes through it, and on to the food.

other. This is another significant indication of the biological evolutionary character of our sequence of stages.

Every living organism exists in a setting which imposes limitations upon its behavior in many ways: gravitation, temperature, atmospheric pressure, and humidity are universal factors that have to be reckoned with and which circumscribe the behavior of the organism. Food, enemies, and incidental obstacles further condition behavior. As we have seen, organisms capable of only Type I minding are to a great extent subordinate in their behavior to factors of their environment: the petals of a flower close at night to reopen after sunrise; the paramecium approaches or withdraws from a stimulus; and so on. The organism has neither alternatives nor choice; it must do what its own inherent properties and those of its environment dictate. All living organisms are dynamic (thermodynamic) systems, to be sure; but the behavior of organisms capable of only Type I is something that they *undergo* as well as something that they do.

The advent of the conditioned reflex brought about a revolution in minding. At a single stroke it emancipated organisms from many limitations imposed upon them by their environment. To be sure, the organism can live only with boundaries circumscribed by such factors as gravity and temperature, but it need no longer limit its behavior to responses to stimuli as determined by their intrinsic properties. A stimulus (a thing or event) can have only one meaning in Type I minding; it may mean any one of many things in Type II. Thus the sound of the electric bell has only auditory meaning in Type I; it may mean food, danger, sex, or something else in Type II. The conditioned reflex was, so to speak, an emancipation proclamation for evolving life-forms: it emancipated organisms from limitations imposed upon them by the natural properties of things in their environment; it multiplied the number of kinds of responses that could be made to a given thing as a stimulus.

But, under Type II, the organism was still subordinate to outside things and circumstances. With a conditioned reflex, the organism can acquire new meanings for things or events (Pavlov's dog acquired a new meaning for the sound of the bell), *but it cannot determine what this new meaning shall be;* this is done by an experimenter or by other circumstances. The organism still plays a passive role with reference to his environment. The advent of Type III gives him positive control over it. The extent of this control is not, of course, complete or absolute, *but it is control,* and it was this that was lacking in Types I and II. In Type III the organism rises above his environment. It now has alternatives, and it can make choices. It is the ape himself who decides how he is to reach the banana and with what means. In Type III the organism plays a dominant role in his interaction with his environment.

In Type IV we have the emancipation from the limitations of intrinsic properties of external things or events which was won in Type II, combined with the ascendance to control over environment that was achieved in Type III. In Type IV, it is the organism, man, who determines what meaning the sound of the electric bell shall have, and he can give it any meaning he chooses. Emancipation *and* control are thus united in Type IV.

The career of life is a struggle between thermodynamic processes: a building-up process coping with a running-down, breaking-down process. Life is simply the name that we give to a thermodynamic process that moves in a direction opposite to that specified by the Second Law of Thermodynamics for the universe as a whole. Life sustains itself by the capture and utilization of free energy, obtained in the form of food. Life is enabled to extend itself by the multiplication of numbers and to evolve new and higher forms because of its ability to capture and utilize free energy in increasing amounts.[3] To do this, it must overcome the obstacles of its natural environment. The emancipation from limitations imposed by the intrinsic properties of things, on the one hand, and the assertion of positive control over environment, on the other, that are won by Types II and III, respectively, and which are combined to produce Type IV have been the ways and means by which biological evolution has been achieved.

SYMBOLING

Type IV behavior, or minding, is characterized by freely and arbitrarily bestowing meaning upon a thing or an event and in grasping meanings thus bestowed. These meanings cannot be comprehended by the senses alone. Holy water is not the same kind of thing, from the standpoint of human experience, as mere H_2O; it has a distinctive quality, or attribute, in addition to hydrogen and oxygen in molecular organization. This new quality, or meaning, that holy water has, which distinguishes it from ordinary water, was bestowed upon it by human beings, and this meaning can be grasped and appreciated by other human beings. It is this ability to originate and bestow, on the one hand, and to grasp, on the other, meanings that cannot be comprehended with the senses that we have termed "symboling."[4] Symboling is a kind of behavior that is characterized by traffic in non-sensory meanings.

To be sure, a symbol must have a physical basis or form, otherwise it could not enter our experience. But the *meaning* of the symbol and its physical basis or form are two quite different things, and there is no inherently necessary relationship between the two. The combination of sounds *s-ee* may be the vehicle of an indefinite number of meanings—to use the eyes, seat of power of a bishop, yes, or anything else that we please. Articulate speech is the most important and characteristic form of symboling. A symbol may have any kind of physical basis: an object (keepsake or fetish), an act (tipping one's hat), a color (royal purple), or a sound (a spoken word).

Things that we call "symbols" have often been confused with things that we call "signs." Some psychologists who have worked with rats or apes have described their behavior in terms of symbols. The red triangles that mean food, the green circles that mean an electric shock, the blue or yellow poker chips that are used in the "chimp-o-mat," etc., have been called symbols because their meaning or significance is not inherent in them; their meanings have been assigned to them. Therefore, they conclude, they are just like our symbols—holy water, words, crucifixes, fetishes, etc. This reasoning is unsound because it has allowed a similarity to obscure a fundamental difference between these two kinds of situations.

The red circles and the blue poker chips have indeed acquired a meaning which is not inherent in their physical structure and composition, just as holy water or the combination of sounds *see* has acquired a meaning. This is the similarity. But there is a fundamental difference also. It is not the rats and apes that have determined the meanings which the red circles or the blue poker chips have acquired, and, what is more, *they are incapable of doing this*. Only man is capable of freely originating such kinds of meanings and of bestowing them upon things or events. The apes and rats can *acquire* meanings for things, but they cannot originate or determine them. They are acquired by the mechanism of the conditioned reflex, on the level of Type II; the ability to originate and to bestow meanings is found only on the level of Type IV. The behavior of the rats responding to red circles or the poker-chip-using apes is not, therefore, symbol behavior but *sign* behavior.

A sign is a thing that indicates something else. Its meaning may be either inherent in it and its context (dark clouds a sign of rain) or extrinsic to it (yellow quarantine flag). In either case, the meaning of the sign has become identified with its physical form and context through experience and the mechanism of the conditioned reflex (or, in the case of man, by means of observation and reason, good or bad). The meaning of the sign, having become identified with its physical form, may be grasped and appreciated with the senses: we distinguish blue from red poker chips with our eyes.

But a thing or event which is significant in a context of symboling may be translated and become significant in a context of sign behavior. Thus we create a symbol: *boko*. We make it mean "hop on your left foot." The processes of originating and bestowing this meaning and the act of grasping it by non-sensory means constitute, as we have said before, an instance of Type IV behavior, of symboling, a kind of behavior peculiar to the human species. But, after we have used this word as a command several times, its meaning becomes identified with its physical (phonetic) form, becomes a sign, and we grasp and comprehend its meaning with our senses: we distinguish it from *loko*, "hop on your right foot," with our ears.

THE PRINCIPLE OF REGRESSION

The fact that a meaningful thing or event may have its origin on the level of Type IV but subsequently descends to the level of Type II brings us to an important principle in behavior, namely, the tendency for behavior to regress from higher to lower levels. We have just analyzed an instance of the translation of a thing from a symbol context (Type IV) to a sign context (Type II). But it may go even lower and assume the level of Type I. Let us illustrate with an example.

A word is a physical thing or event, a sound or a combination of sounds, or a visible mark on paper or some other substance. A word may be significant in a variety of contexts. It may be significant and have meaning on all levels of minding that we have distinguished. Words and their meanings originate only on the level of Type IV, where they function as symbols. But they may function in Type III contexts also. The chimpanzee reaches his banana with a stick, but a human being can use the word to obtain the same result: he need merely say "bananas" in the store, and the clerk brings him some. When the word descends from the symbol level to the sign level, it has descended to the level of Type II, as we have just seen. But it may go even lower and become significant as a thing or event in terms of its intrinsic properties. We use words because of their phonetic qualities in poetry: "charms" rhymes with "arms," not "legs." Some names are thought to be pretty (Sylvia); others ugly (Bridget). Primitive peoples use words in some situations as if their properties were intrinsic in their phonetic structure (spells, incantations). Members of a certain tribe sought prescriptions from a European physician, but, instead of taking the prescription to the pharmacist and exchanging it for medicine, they soaked it in water until the ink was dissolved and then drank the water; they availed themselves of the therapeutic values of the words by drinking them. And even in our culture we treat certain four-letter words of Anglo-Saxon derivation as if their meanings were inherent in their phonetic structure (their Latin cognates may, however, be used with propriety). Words are born as symbols, but they may descend to the level of things in themselves.

One reason for regression of behavior to lower forms is that an economy of effort or time is effected thereby. If one had to stop and think, when driving an automobile, whether the red light symbol means stop or go, he would probably violate a traffic rule or possibly kill someone. Traffic lights originated as symbols. Meanings were freely and arbitrarily bestowed upon colored lights (any colors would do). But these meanings quickly become identified with their physical structures (wave lengths), descended to the level of Type II, and became signs. But we react to the red light *as if* its meaning were inherent in it. Thus we end up with the *form* of Type I minding: a simple, straight-line reaction of stop or go (Fig. 1). This fact is recognized and expressed sometimes by saying "I instinctively did so and so."

Each stage in the evolution of minding has given life a new dimension: Type II, the conditioned reflex, emancipated living organisms from limitations imposed upon them by the intrinsic properties of the environment; Type III gave organisms positive control over things and events in the external world. But Type IV, symboling, has been especially significant, at least as far as human beings are concerned; indeed, it was symboling that made human beings of certain primates.

Both human behavior and culture are expressions and products of symboling. Human behavior consists of acts and things, dependent upon symboling, considered in terms of their relationship to the human organism. Tipping one's hat is an example of human behavior. And the hat, too, may be considered as a form, or product, of human behavior; it is human behavior locked up in a form and a fabric. But all these things and events that are dependent upon symboling may be considered in another context, also: an extra-somatic context. That is, instead of regarding them in relationship to human organisms, we can consider them in their relationships to one another and without reference to the human organism. Thus tipping one's hat may be thought of as a ritual and in terms of its relationship to other rituals, to customs of kinship, to social or class structure, and so on. In short, things and events dependent upon symboling may be thought of as constituting a continuum, a flow of culture traits from one individual and generation to another. We may think of words as items of human behavior, as acts of the human organism, as things and events in terms of their relationship to the human organism. But we may think of words as constituting a class of things and events *sui generis*, and we may study them without reference to the human organism. This is the science of linguistics and is concerned with such things as phonemics, grammar, syntax, word order, etc., whereas the scientific study of words in their relationship to the human organism is the science of the psychology of speech and is concerned with such things as habit formation, imagination, conception, attitude, and so on.

Now, just as we may think of words in a somatic context (i.e., in terms of their relationship to the human organism) or as a self-inclosed continuum, as a process *sui generis*, so we may consider all things and events dependent upon symboling in terms of their relationship to the human organism, in which case they are human behavior; or we may consider them as a self-inclosed flow, a process *sui generis*, in which case they are *culture*. "Culture" is the name of a flow of things and events dependent upon symboling considered in an extra-somatic context.[5]

Symboling has brought a certain kind of things and events into existence. They constitute a continuum, a flow of tools, customs, and beliefs, down through the ages. Into this flow, this extra-somatic continuum called "culture," every human individual and group is born. And the behavior of these human beings is a function of this extra-somatic continuum: an organism born into Tibetan culture behaves in one way (as a Tibetan); an individual born into Scandinavian culture behaves in another way. Thus the determinants of human behavior, insofar as the individuals may be considered as typical or average, are no longer the properties of the biological organism; the determinants are to be found in the extra-somatic tradition (culture). It is not the nature of the lips, palate, teeth, tongue, etc., that determine whether the human organism will speak Tibetan or Swedish; it is the linguistic tradition that determines this. Therefore, in contrast with all other kinds of living organisms, if we wish to learn why a typical individual—a typical Crow Indian or a typical Englishman—behaves as he does, we must concern ourselves not with their bodies, their neuromuscular-sensory-glandular systems, but with the cultures into which they have been born and to which they respond. Similarly, if we wish to learn why the Japanese behave differently in 1960 than in 1860, we must concern ourselves with the changes that have taken place in their culture.

In the scientific study of behavior, or minding, therefore, we must concern ourselves with organisms when we are dealing with Types I, II, and III. But when we come to Type IV and man, a new world has been created, an extra-somatic, cultural environment, and it is this which determines the behavior of peoples[6] living within it and not their bodily structures. This means that we must have a new science: a science of culture rather than a science of psychology if we are to understand the determinants of *human* behavior. But this is not the place to go into that, and, besides, we have touched upon it elsewhere.[7]

NOTES

[1] Cf. Leslie A. White, "Mind is Minding," *Scient. Monthly*, XLVIII (1939), 169–171; reprinted in *ETC., a Review of General Semantics*, I, No. 2 (1943–44), 86–91, and in White, *The Science of Culture*, pp. 49–54.

[2] If rest is a form of motion in which the velocity is zero, as Alfred North Whitehead says (*Introduction to Mathematics* [1st American ed.; New York: Oxford Univ. Press, 1948], p. 29), then we may say that indifference or neutrality of an organism toward a body is a response in which the action is zero.

[3] See L. A. White, *The Evolution of Culture* (New York: McGraw-Hill, 1959), pp. 33–38, for fuller discussion of this subject.

[4] We have discussed this subject at some length in "The Symbol: The Origin and Basis of Human Behavior," *Philosophy of Science*, VII (1940), 451–463. This essay was reprinted, in slightly revised form, in Leslie A. White, *The Science of Culture* (New York: Farrar, Straus &

Co., 1949; New York: Grove Press, 1958). It has also been reprinted in *ETC., a Review of General Semantics*, I (1954), 229–237; *Language, Meaning, and Maturity*, ed. S. I. Hayakawa (New York: Harper, 1954); *Readings in Anthropology*, ed. E. Adamson Hoebel *et al.* (New York: McGraw-Hill, 1955); *Readings in Introductory Anthropology*, ed. Elman R. Service (Ann Arbor, Mich.: Univ. of Michigan Press, 1956); *Sociological Theory*, ed. Lewis A. Coser and Bernard Rosenberg (New York: Macmillan, 1957); and *Readings in the Ways of Mankind*, ed. Walter Goldschmidt (Los Angeles: 1957).

[5] See Leslie A. White, "The Concept of Culture," *Amer. Anthropologist*, LXI (1959), 227–251, for a fuller discussion of this point.

[6] The behavior of *individuals* is, of course, determined, or conditioned, by their biological make-up as well as by cultural factors. But we know of no biological differences among *peoples*—tribes, races, etc.—that would produce corresponding differences in human cultural behavior. *Peoples*, therefore, may be considered as a constant biologically, as an independent variable; it is the exosomatic cultural factor that varies.

[7] "The Science of Culture" and other essays, in White, *The Science of Culture* (1949).

The emergence of man:
An inquiry into the operation of
natural selection in the making of man

John N. Deely

CONTEXTUALIZING COMMENTS

This third reading attempts to do no more than bring traditional insights of philosophical anthropology to bear within the context of man's evolutionary origins. It attempts an analytical projection of the structural uniqueness of human being to the point of man's temporal origin accomplished on the basis of the evolutionary data alone. The argument as developed in this article accords fundamental interpretive significance to the fact that man is an animal characterized by a capacity for modes of strictly non-genetic behavior, and argues in terms of this capacity to a radical difference in kind between hominoid and hominid, animal and man.

To grasp the structure of the argument here brought forward, therefore, it is necessary to realize that even though "evolution does not happen in individuals, but in populations," it does not inevitably follow from this that, as some authors assert, "there is no known mechanism by which the human species might have arisen by a single step in one or two individuals only."[1] This illation depends not on the known processes of

Reprinted with slight alteration from *The New Scholasticism*, XL, 2 (April 1966), pp. 141–176. By permission of the Editor, *The New Scholasticism*.
[1] Francisco José Ayala, "Man in Evolution: A Scientific Statement and Some Theological and Ethical Implications," *The Thomist*, XXXI (January, 1967), 15.

genetic transmission, but on the definition of man which one formulates. That the concepts of genetics in themselves do not necessitate the illation in question is clear from this simple observation: "even though whole populations are involved in the total process of species change," so that it is truly species and not individuals which evolve, still, *"species evolve through individual generations;"* [2] and just as it is impossible to remove this ultimate discontinuity of individual reproductions of organisms as a primary *datum* of evolutionary science, so also is it invalid to dismiss in advance the possibility of a logically rigorous presentation of evidence for radical human uniqueness referenced by this same datum.

Keeping these ground-rules in mind, we may proceed in our discussion to a consideration of how far and in what sense the same causes are at work in the evolutionary origins of mankind as are at work in the origins of all other animal species.

The emergence of man:
An inquiry into the operation of natural selection
in the making of man

The more deeply science plumbs the past of our humanity, the more clearly does it see that humanity, *as a species*, conforms to the rhythm and the rules that marked each new offshoot on the tree of life before the advent of mankind.

Teilhard de Chardin

By way of preface, I wish only to recall to mind that every line of thought has its starting point. The thought in this essay begins with two premises. The first is agreement with Theodosius Dobzhansky's contention that "no one who takes the trouble to become familiar with the pertinent evidence has at present a valid reason to disbelieve that the living world, including man, is a product of evolutionary development," that, in short, "man is a biological species which has evolved from ancestors who were not men." [1]

And the second premise is acceptance of the claim of Sir Julian Huxley, that "the discovery of the principle of natural selection made evolution comprehensible; together with the discoveries of modern

[2] R. J. Nogar, *The Wisdom of Evolution* (New York: Doubleday, 1963), p. 290. For an extended critique of this fact and of its implications in evolutionary explanations, see J. N. Deely, "The Philosophical Dimensions of the Origin of Species, Parts I & II," *The Thomist*, XXXIII (January and April, 1969), esp. pp. 108–111 in Part I, and pp. 305–316, 320–321 in Part II.

genetics, it has rendered all other explanations of evolution untenable."[2]

In a word, because the so-called synthetic theory of evolution so adequately embraces all the known facts in the history of life and is judged virtually irrefutable by a consensus of the experts, in writing this paper I take the point of view that any attempts at formulating philosophical affirmations about human origins which either leave this theory out of account, or oppose its legitimate claims, must remain highly tentative and indeed suspect.[3]

It is necessary that philosophy examine what this theory says about the operation of evolution, and direct its inquiry into human origins in a manner that harmonizes with the patterns of development which the facts themselves disclose. Since, however, in research "real progress comes not so much from collecting results and storing them away in 'manuals' as from inquiring into the ways in which each particular area is basically constituted,"[4] throughout our study of human emergence "we must equally resist the temptation to regard man either as something completely unlike any animal or as something devoid of all novelty."[5]

Such are the point of view and method which define what follows.

I. THEMATIC REMARKS

We know that the organisms on earth today have issued from one or a few simplest forms, and that the entire development has required something more than two billion years. All around us, we see a world which has built itself up, so to speak, from scratch; and the biological community traces its descent from very different beings which lived in the past. In this respect, the history of the world and the history of life correspond, point by point.

This knowledge, however, is an acquisition of quite recent times. Before the discovery and verification of biological evolution, for example, the various kinds of living things had always seemed to have a fundamental permanence which remained unaffected by the passage of time. Structurally, the universe seemed to be something "given" once and for all. Gilson—to give one illustration—describes "the eternal and uncreated cosmos of Aristotle, peopled with species immutably fixed under their present appearances," as "completely alien to history both in its origin and its duration."[6]

Considered in such a cross-sectional perspective, the interval separating man from the other animals was obvious. He could be defined with Porphyrian precision as "Animal rationale," and the definition would be readily appreciated in terms of the ontological gap it denotes.

But the re-setting of the world in general and mankind in particular within a perspective of temporal development and a dimension of emergent phenomena has altered this cross-sectional viewpoint considerably. A. Irving Hallowell, professor of anthropology at the University of Pennsylvania and in the Psychiatry Division of the School of Medicine, summarizes the new point of view as it bears on man succinctly:

The advent of Darwinism helped to define and shape the problems of modern psychology as it did those of anthropology. An evolution of the 'mind' within the natural world of living organisms was envisaged. Now a bridge could be built to span the deep and mysterious chasm that separated man from other animals and which, according to Descartian tradition, must remain forever unbridged. Darwin himself explicitly set processes of reasoning, long considered an exclusively human possession, in an evolutionary perspective. . . . He argued that mental differences in the animal series present gradations that are quantitative rather than qualitative in nature.[7]

It is a fact that much of contemporary research centers on questions concerning the emergence of psychic structures in an effort to span the deep and mysterious abyss that has come to separate man psychologically from the animal. As Sir Julian Huxley explains, the evolutionary scientist "wants to understand something of the way in which the dual-aspect system of mind and behavior evolves; of how mental organization is specialized and improved during evolution."[8]

Once it was established that man represents a terminal product in a development spanning millions of years and in no wise human in origins, the questions concerning him must be raised in new terms and directed in the first place toward his origins. Above all, we can no longer think primarily in terms of an *ontological gap* between man and the so-called brute animal. That there is such a gap today is undeniable. But in an evolving world, the first question is, how did it get there?

St. Thomas, in analyzing the ontological difference between man and animal cross-sectionally, claimed to show that there is in man a strictly spiritual principle which in some way transcends the conditions of material existence. And while it is true that every development is more accurately understood at its term than in its inception, yet contemporary thinkers assuredly cannot confine themselves to the horizon of an outmoded problematic. It is necessary to come to grips with the very data of evolution, and see whether or not they provide an intelligible ground for determining an *ontogenic* or *epigenic break* at the threshold of human emergence, the onset of hominization.

II. LIMNING THE FACT OF HUMAN EMERGENCE

Once evolutionary science entered its phase of synthesis in the first half of our own century, the attention of students of human development "became focussed more upon the mysterious evolutionary changes which are believed to have taken place between the behavioral systems of the highest primates and those of the earliest men,"[9] all of which changes are subsumed under the expression "man's capacity for culture." In this connection, a crucial point on which all students of evolutionary science agree, regardless of their philosophical stance, is the fact that with the appearance of man in the biological community "the basic novelty was, however, the development of unprecedented intellectual abilities, which made possible the control of environment by culture."[10] It is this more restricted sense of a root capacity that will primarily guide our inquiry.

Let us begin by placing this "basic novelty" in its historical perspective. We know that the history of modern man, *Homo sapiens*, spans little more than a quarter of a million years. He is preceded by another group, *Homo erectus*, anatomically quite different but, by the classical definition "Animal rationale," unquestionably human. This stage of human anatomical development is itself preceded by the controversial *Australopithecine* fossils. These fossils include the first known tool-makers and place them at the astonishing depth in time of more than one and three-quarter million years.[11] If we maintain, as perhaps we must, that *Homo faber* is an operational definition coextensive with the metaphysical *Animal rationale*, then the age of true man approaches two million years.

That only one truly hominid species existed at this time level is a view seriously challenged on the basis of Louis S. B. Leakey's very recent and most important finds in what may prove to be the cradle of civilization, the Olduvai Gorge in Tanzania, East Africa. The April 2–4, 1965 international "Origins of Man" symposium, sponsored by the Wenner-Grenn Foundation for Anthropological Research and called by Professor Sol Tax of the University of Chicago, focussed the controversy over the interpretation to be placed on Leakey's finds, but failed to disclose a consensus of expert opinion.

We may say in any case that so far as mankind today is concerned, only one species of *Australopithecus* could have been our ancestor. Species contemporaneous with that ancestor became extinct without issue. "This is so because there is only one hominid species now living, and it cannot be derived from two or more reproductively isolated species."[12] A considered and consistently expressed opinion of Dobzhansky, one of the country's leading geneticists and an outstanding authority on evolution, is this:

The evidence now available is compatible with the assumption that, at least above the australopithecine level, there always existed only a single prehuman and, later, a human species (which evolved with time from *Homo erectus* to *Homo sapiens*). Mankind was and is a single inclusive Mendelian population, endowed with a single corporate geno-type, a single gene pool.[13]

Commenting on this very statement in the light of the Origins of Man symposium, Professor Sol Tax writes:

I think it would be difficult to improve upon Dr. Dobzhansky's cautious manner of stating this matter, for the main disputes do occur at the Australopithecine level and below. Many palaeontologists now take serious consideration of the possibility that the hominid "bush" contains some branches that died out some time ago, as urged by Dr. Leakey and by analogy to the history of other mammals. However, Dr. Robinson contends that there was only one such extinct branch, and Dr. Simpson seems not yet to be convinced that this branching occurred at all.[14]

Leaving the dialectic of controversy to the specialists concerned and keeping to the question which guides our own inquiry, we may in line with current thinking in population genetics envisage at the onset of human emergence a speciation process operating through a group of reproductively and proximally related individuals spread out over a limited but sufficiently broad "surface of evolution." By assuming at this stage in our problematic such a wide and complex cross-section at the base of the human stem, we make allowance for marginal and divergent forms as well as for the truly ancestral forms, thus taking account of the bush-like (or at least "bushy") structure evidenced in the various fossil stages of human development.[15] (Against this backdrop, Homo habilis might be perhaps best interpreted as a transitional form marking the general line of advance toward Pithecanthropus.)

We may say that the structural evolution of man has progressed through three main stages or groups: the Australopithecine group, the Homo erectus group, and our own group, Homo sapiens. Together with the development of upright posture, the unmistakable, constant, and main trend throughout has been the development of a larger and more neurologically complex brain.

"Drs. Leakey and Robinson agree that the absolute brain size of Australopithecus has probably been overestimated in the past. Dr. Robinson has just completed a rather thorough restudy of this question and came up with a mean capacity of 430 c.c. for Australopithecus."[16] (The brain size of Homo habilis may have been somewhat larger than that of the Australopithecines of South Africa.) Richard Carrington, though assigning apparently too large a brain to the Australopithecines as a group, makes this valuable observation:

The average brain size of the australopithecines . . . does not at first sight suggest that the creatures were very intelligent, but when we remember that they were comparatively small and lightly built, and that intelligence does not depend on absolute brain capacity but on the ratio between the size of brain and body, we can believe that their mental powers were considerable. In fact, in relation to their body size they had a larger capacity than any animal that had previously appeared.[17]

Passing on to the more clearly "human" phases of pre-history, the meagre cranial capacity of the Australopithecines gives way to the considerable 900–1100 c.c. of the *Homo erectus* group, and ends up in the very considerable 1200–1500 c.c. of *Homo sapiens*. Moreover, this expansion of the brain has not been uniform but differential, occurring primarily in the associative areas which relate to "tool use, speech, and to increased memory and planning."[18] "An important adaptation for culture," explains James N. Spuhler, a physical anthropologist and human geneticist from the University of Michigan, was "the change from built-in nervous pathways to neural connections over associative areas (where learning and symboling can be involved) in the physiological control of activities like sleep, play, and sex."[19] S. L. Washburn, a physical anthropologist from the University of California, summarizes this key development in the human evolutionary line concisely. From the immediate point of view, he remarks, the association-type brain makes culture possible. "But from the long-term, evolutionary point of view, it is culture which creates the human brain."[20]

In an important article entitled "The Human Revolution" which appeared in the June, 1964, issue of *Current Anthropology*, Charles Hockett, professor of linguistics and anthropology at Cornell University, and Robert Ascher, associate professor of anthropology at the same university, expressed the present consensus on this point:

We must therefore assume that if a species has actually developed a bigger and more convoluted brain, there was survival value in the change. For our ancestors of a million years ago the survival value was obvious if and only if they had *already* achieved the essence of language and culture.[21]

This is admittedly a rather crude way to follow the movement of human evolution. As von Bonin remarks in his close study of the expansive development of the central nervous system defining an important distance between *Australopithecus* and modern man:

How the cortex does things is the only element in the operation of the brain that is laid down at birth; what the cortex does and what man thinks about are elements that develop only during the life of the individual. To disentangle these two strands and to write an evolution of the mind from the point of view of the brain is not yet feasible.[22]

Nonetheless, it is a sobering reflection to realize that much of what we think of as normal human nature does not date back to anywhere near the point of human emergence; and that man's brain, the immediate instrument of consciousness, has approximately trebled in size since man first appeared. Spuhler observes with a healthy wonder that "there has been an unusually rapid rate of hominoid evolution during the past twelve million years, and especially in the past million. . . . Something has speeded up hominoid evolution." And he is right in guessing "that selection (perhaps within-species or inter-group selection) for a new type of environment—a cultural environment—has a lot to do with it."[23]

The way of life of pre-*Homo sapiens* was very different from anything known to us. Not only has "much of what we think of as human evolved long after the use of tools," but in general, it is more correct to think of much of our structure as the result of culture than it is to think of men anatomically like ourselves slowly discovering culture.[24] "The inter-relationships between biology and culture," in short, "especially since the development of civilization, have influenced the evolutionary patterns of the human species so decisively that human biology is incomprehensible apart from the human frame of reference."[25]

It is of course not certain what limits their anatomical configuration placed on those earliest men as regards intelligence, capacity for language, art, and social organization. Von Bonin marks off the limits of his special-ized research in similar terms: "The results of our inquiries into the brains of fossil men are somewhat meagre: we cannot deduce any details about their mental life—whether they believed in God, whether they could speak or not, or how they felt about the world around them."[26]

What is certain is that these substantial anatomical differences which a longitudinal view of human history discloses are not distinct in fact from the specific essence of man;[27] so that we must say that the very root determinants of the uniquely human properties of personality, creativity, and freedom have undergone an expansion and development in time. Cultural anthropology attests in no uncertain terms to the spiritual transcendence of human intelligence. "Yet man's 'substance' is not spirit as a synthesis of soul and body; it is rather *existence*."[28] That is to say, the being of man in the world is manifestly a structure which is primordially and constantly *whole*. This truth gives our history an ontological content, a contingent *substantial* dimension. Man in his totality is an evolutionary phenomenon.

In summary we can say that *the fact of the emergence of man* consists of, or is constituted by, seven elements:

(1) Toward the end of the Tertiary era and on into early Pleistocene times, that is, about twelve million years ago, over a well-defined but immense area extending from South Africa to Southern China and Malaya, the anthropoid apes were far more numerous than they are today.[29]

(2) Besides the gorilla, chimpanzee, and orangutan, there was a whole popula-tion of other large primates, some of which, like the African Australopithecines, were much closer to man, more hominoid, than any alive today.

(3) All living descendants from this generalized stage of primate development have been forced into their own lines of specialization. In other words, the terrestrial

conditions which governed the initial appearance of man, the onset of hominisation, have long since vanished.[30]

(4) One of these specialized lines, the hominid line, terminated (as we can know by introspection) in *Homo sapiens*. However, in the view of the prehistorian, "no one can tell just when the human phase of evolution started or how long its different stages took. The critical step from animal to man might have taken some hundreds of thousands of years,"[31] since all that appears with certainty is that man does not stand at the end of a phyletic line which includes the living primates. We can say that although man is today an almost isolated figure in nature, in his beginnings he was much less so, and indeed would not have been readily discernible at all. From this standpoint, the "humanness" of man arose gradually rather than overnight.

(5) This line of descent which led to man did not simply specialize in the biological sense, but developed or emerged with an entirely unique mode of adaptation which is extraorganic or cultural—that is, not transmissible by any biological mechanism.

(6) Nevertheless, as we have seen, it cannot be inferred from the historical evidences that the cultural capacity of the human species appeared suddenly in some remote ancestor of ours.[32] On the contrary, this capacity seems to have developed gradually in every aspect—an interpretation reinforced *prima facie* by the fact that we find in the present day "evidence for culture of a rather thin sort among the hominoid apes,"[33] the late Tertiary lines of specialization nearest our own.

(7) What is to be affirmed definitely is that there are two altogether fundamental aspects in the emergence of man: *immanence*—that is, man underwent all the conditions of a progressive emergence in psychological expression for the simple reason that the somatic determinants of his human capacities underwent an expansive evolution; *transcendence*—that is, there was a factor operative in human evolution from its earliest definitely discernible beginnings in the depths of prehistory, namely, a radical capacity underlying cultural development, the precise content of which is by definition not physiologically given. Even from the isolated vantage of the prehistorian, this duality constitutes "the first, basic, fundamental fact about human evolution."[34]

III. THE LIMITS OF HISTORICAL INQUIRY INTO THE FACT OF HUMAN EMERGENCE

Once we have described and temporally circumscribed the fact of the emergence of man, however, the paleo-sciences are of little further help to us. For while they enable us to approximate the locale and moment of man's emergence from the late Tertiary anthropoid population, they leave the nature of that emergence almost entirely veiled. These sciences uncover the manifestations of man's capacity for culture, but what about the capacity in itself? There lies the meaning of human emergence. Heidegger is right to assert that "if this beginning is inexplicable, it is not because of any deficiency in our knowledge of history. On the contrary, the authenticity and greatness of historical knowledge reside in an understanding of the mysterious character of this beginning."[35]

Unless we can somehow effect conceptually a radicalization of the human capacity for culture we shall never quite touch on the meaning of man, never quite establish an evolutionary uniqueness for the inception of human emergence as contrasted with all other evolutionary beginnings. This truth has received forceful and definitive formulation in Ernst Mayr's magistral study of *Animal Species and Evolution*:

The gradualness of man's becoming man must be stressed in opposition to continuing attempts to present the origin of man as a single-step phenomenon. What stage in this continuum could one arbitrarily pick out and designate as the "real" origin of man? Would it be the branching off from the pongids, or the first manufacturing of tools, or the use of fire, or a speech development indicated by a brain size of 1000 cm[3], or the first attainment of 1500 cm[3]? There is not one "missing link" but a whole series of grades of "missing links" in hominid history.[36]

Frank B. Livingstone, an anthropologist from Ann Arbor, Michigan, summarizes our dilemma this way:

. . . as much as we might wish it, we will never discover enough facts to reveal to us the total way of life of our transitional ancestors. Thus, the development of an adequate explanation of the origin of the hominids and their peculiar capacities will result as much from the sifting of theoretical questions as from the sifting of new facts.[37]

I want to propose in these pages, tentatively, what may prove a way out of this dilemma. But to do this, we must take up our inquiry on another ground. We must look to the very source of life's movement. We must analyze the very causal mechanisms of evolution and see if on this basis we can establish an objective uniqueness at the point of human origins, a structural principle unique to the mode of being human and conferring on it a significant measure of biological advantage. Such a principle would accordingly have to constitute at least one critical factor operative from the earliest moments of human emergence which did not enter into the development of life at any other point, i.e., an operative factor upon which could be based "an adequate explanation of the origin of the hominids and their peculiar capacities" as they have unfolded and developed in time.

In undertaking such an inquiry we must keep our reasoning precise and our data clearly in view. This exactness is all the more necessary since we have at this point determined our problematic in a manner which, in the considered judgment of many of the clearest thinking students of evolutionary science, will not admit of a resolution and is in fact fallacious.

In any event, whatever the outcome in terms of our problematic, we want to know *in evolutionary terms* what it means to be human. This is the intentional content of our inquiry.

IV. THE CAUSATIVE FACTORS IN THE EMERGENCE OF MAN

What then, fundamentally, is evolution? Dobzhansky answers this question as directly and clearly as one could wish. "Evolution," he says, "is change in the heredity, in the genetic endowment of succeeding generations." "No understanding of evolution is possible," he warns, "except on the basis of a knowledge of heredity." [38]

Therefore if we are to get at the meaning and inner moment of the emergence of man we must try to define human origins in terms of the established facts of inheritance. Fortunately, while these facts are almost incredibly complex on the experimental plane, the concepts underlying them are easy to grasp.

The biological inheritance of every creature consists of genes received from its parents. The totality of genes constitutes the *genotype* or genetic endowment of an organism. This is contrasted with the so-called *phenotype*, which is not transmissible biologically and comprises "the total of everything that can be observed or inferred about an individual, excepting only his genes." [39]

A basic operation of the genes is self-reproduction. Consequently, "every genotype exerts a 'pressure' on its environment; it tends to organize and transform into its replicas all the materials available in the environment which it can use." This means that "any phenotype that may be formed is necessarily a response of the environment to the activity of a genotype."

"The total range of phenotypes which a given genotype can engender in all possible environments constitutes the *norm of reaction* of the genotype." Therefore "whatever change is induced in the phenotype is *of necessity* within the norm of reaction circumscribed by the genotype," [40] i.e., fixed in the zygote at fertilization.

Since during the process of hereditary transmission the genetic units segregate, rearranging more or less randomly in the establishment of a new genotype, what a given individual inherits "is the result [both] of the self-reproduction of genes determining modes of reaction to the environment and of the transmission of copies of parental genes" in some one of the (in higher organisms almost astronomical number of) possible combinations. [41] This random rearrangement of parental genes is designated *recombination* by geneticists, and its evolutionary significance lies in the fact that the effects of a gene on the organism's development "depend not only on the structure of the gene itself but on its position among the neighboring genes as well." [42]

These so-called *position effects* are complemented, so to speak, by the phenomenon called *pleiotropism*, the manifold manifestation of a gene: "Every gene may affect many visible traits; most traits are influenced by several or by many genes."[43]

The concept of a pleiotropic action of genes . . . leads us to the idea of the genotypic milieu which acts from the inside on the manifestation of every gene in its character. An individual is indivisible not only in its soma but also in the manifestation of every gene it has.[44]

"Sexual reproduction is thus a creative process; it originates biological novelty."[45] In consequence of sexual recombination, an evolving population becomes a veritable sea of latent genetic (that is, developmental) potential. Since at this level "every individual is biologically unique and non-recurrent,"[46] sexual reproduction "is by far the most important source of genetic variation": "Through recombination a population can generate ample genotypic variability for many generations without any genetic input (by mutation or gene flow) whatsoever."[47]

To get at the ultimate source, however, of the sequence of genetic changes which constitute the movement of evolution, we must return to a consideration of the self-reproducing function of the genes. The extreme atomic complexity of the genetic unit coupled with a high sensitivity to forces external to itself makes it inevitable that the self-copying process will sometimes be imprecise. An imprecise gene *which retains the capacity for replication* is called a mutant. Usually a mutant gene is detrimental to the organism; but occasionally, depending on the environmental situation, it confers a measure of biological advantage— that is, it has a survival value.

In any case, the inaccuracies in the mutant gene "are then faithfully reproduced by the self-copying process, so that the original mutation becomes a strain of mutant genes."[48] In this way, what in the parent organism was "essentially a dislocation taking place in the delicate self-reproducing mechanism of the gene"[49] becomes for succeeding generations "the origin of an hereditary trait which did not exist at all in the parents of the mutant."[50]

So far as is known, mutations occur in all genes, and they occur with a relative constancy, so that there is a radical variability operative within the inheritance factors themselves. Since "individuals which carry a mutant gene possess a new norm of reaction to the environment,"[51] traditional scholastic thought was seriously mistaken in regarding all deviations from the parental pattern as strictly accidental, in the sense of not entering into the fundamental process of the organism's development. However, it is of foundational importance to realize that the great majority of mutational changes are so slight that they are hard, sometimes impossible, to detect; and that it is hardly credible that the development of the living world ever involved major changes produced by single mutational steps. The dynamic balance of the genotypic milieu is simply too delicate to accommodate itself viably to macromutations. Indeed, within the genetic environment the chances that a mutation will long survive stand in an inverse ratio to its size. As has always been known, *agens facit simile sibi.*

Of course, what kinds of mutation are or are not possible in a given individual's sex cells is determined by the historically established composition of its genotype—i.e., the mutability of a gene is controlled (as in the case of position effects) not only by its own structure but as well by the complex of other genes among which it finds itself. *Agere sequitur esse.*

To sum up:

Mutation causes changes in the genes and variants of the gene structure; these are the raw materials of evolution. In those organisms which reproduce sexually, these variants are combined and recombined to form countless different genotypes.[52]

Such are the deceptively simple facts which provide the basis for biological evolution. The "creative evolution" which so fascinated Bergson has turned out to be but the necessary eventuation of mutation and recombination in encounter with varying environmental conditions. The process, mediated at its upper levels by recombination,[53] whereby the occurrence of favorable mutations increases the adaptive fitness of an organism to meet the requirements of its environment, and subsequently to better survive and reproduce, is the process of *evolutionary* or *natural selection.* "The environment is in a state of flux, and its changes, whether slow or rapid, make the genotypes of the bygone generations no longer fit for survival. The ensuing contradictions can be resolved either through extinction of the species, or through reorganization of its genotype."[54] This latter is the way of evolutionary selection, in the course of which "the mutant genes replace the old genes that were 'normal' . . . and become the new adaptive norm."[55]

Thus "selection occurs in any environment in which the carriers of different genotypes do not transmit their genes to their progeny at equal rates."[56] "Therefore, modern evolutionists include in natural selection any factor that contributes to differential reproduction."[57]

It is clear then that in the effort to understand the inception of evolutionary changes, "instead of considering the genotypes of the parents who mate, it is more convenient to consider the sex cells which they produce."[58] Nonetheless, in stressing the fact that "infirmity or well being, survival or death of an individual or a population in a given environment are determined in the last analysis by the genes which they carry," we must not lose sight of the equally important fact that "it is the phenotype which is adaptive in some environments and unfit in other environments." These two factors are readily correlated if we keep in mind that "the success or failure of a genotype in evolution is determined by its reaction norm, by the adaptedness of the modifications evolved in response to recurring environmental influences." In an integral perspective, "what changes in evolution is the norm of reaction to the environment."[59]

There is accordingly a kind of two-way causality continually operative on the origin and extinction of species: "mutations are limited by the structure of the gene which mutates and this structure is determined by the evolutionary forces . . . active in the history of the gene,"[60] especially the force of selection—so subtle that "any variation observable by us is under favorable circumstances sufficient for its operation,"[61] so complex because it is "a process intricately woven into the whole life of the group, equally present in the life and death of the individuals, in the associative relationships of the population, and in their extra-specific adaptations."[62] "Selection selects not only a gene which determines the character under selection, but it affects the whole genotype (the genotypic milieu), leads to an intensification of the trait selected, and in this participates actively in the evolutionary process."[63]

In this way, careful analysis reveals that while in nature today individuals differ in many genes, races in more, and species and genera in still more again, all these differences arose through the sorting out by evolutionary selection of the genetic raw materials provided by mutations in the proximate, close, remote, and very remote ancestors of the biological community. "The immediate causes of the origin of all genotypes are mutation and recombination," but "in the long run, selection is the directing agent because it determines which genotypes are available for new mutations to occur in."[64] Thus "the reaction norms of the genotypes which occur frequently in populations are molded in the evolutionary history controlled by natural selection. . . . These reaction norms are so adjusted that environmental agencies which the species commonly meets evoke adaptively valuable modifications,"[65] i.e., viable and "fit" phenotypes.

Dobzhansky pictures the evolution of life this way:

Species arise gradually by the accumulation of many mutational steps which may have taken place in different countries at different times. And species arise not as single individuals, but as diverging populations, breeding communities, and races which do not reside at geometric points but occupy more or less extensive territories.

Since evolution is a continuous process, we would not find gaps anywhere, if we could know all the links between the common ancestor and the now living species.[66]

Let us take it then, as the life sciences assure us, that in every generation "the occurrence of mutation adds a variety of genes to the gene pool;"[67] and that evolutionary selection fails to perpetuate harmful mutants (though these are the great majority), stores up a rich potential variation by perpetuating neutral mutants in proportion to their frequency, and multiplies the useful ones (though these are few)—that, in fine, evolution is a gradual and continuous process resulting primarily from the recombinatorial summation of many discontinuous changes, mutations, the vast majority of which are small.[68]

How does man emerge from this general pattern of biological development? In terms of its genetic basis, what must be said of human evolution?

V. HUMAN EMERGENCE IN GENETIC PERSPECTIVE

In genetic reference, the fact of the emergence of man can again be stated in seven propositions.

(1) "The evolution of man was not predetermined by a few conditions in a population of Miocene apes. Mutations are the fundamental genetic events in the historical acquisition of the capacity for culture."[69] In short, "natural selection found in the ancestors of our species the raw materials from which the present genetic endowment of mankind was compounded."[70]

(2) The process was vastly slow and complex, "probably thousands of genes undergoing changes, many genes going through several consecutive changes." Moreover: "The changes in the genes of our ancestors, or at any rate a majority of these changes, happened because they enabled their possessors to outbreed the unchanged forms in the environments in which our ancestors lived at a given time."[71] "No modern geneticist thinks that there existed in our ancestors or that there appeared by mutation some special genes 'for culture.'" As we have already seen, "there is no gene 'for' . . . anything but causing the organism to develop in co-operation with all the other genes it has." "The transformation of the prehuman species into the 'political animal' involved [ultimately] mutational changes in most or all gene loci. It is the *whole genetic system* which makes us human."[72]

(3) "Through the higher animals, and most strikingly in man, there has been a trend toward replacing rigidly genetically determined behavior patterns by behavior that is subject to learning and conditioning. The 'closed' program of genetic information is increasingly replaced in the course of this evolution by an 'open' program, a program which is so set up that it can incorporate new information. In other words, the behavior phenotype is no longer absolutely determined genetically, but to a greater or lesser extent is the result of learning and education."[73] From the standpoint of *phenomenal continuity*,[74] the transition from the adaptive zone of a pre-human primate to the human adaptive zone was brought about by the development of the biological basis for . . . educability."[75]

(4) "The newcomer, the human species, proved fit when tested in the crucible of natural selection; this high fitness is a product of the genetic equipment which made culture possible." Concomitantly: "It was the development of culture that prevented the human species from breaking into several species. . . ." In this way "the adaptive function of the genetic variability has been altered in the human species by cultural development."[76]

(5) "The biological and cultural components of human evolution . . . serve the same basic function—adaptation to and control of man's environments."[77] Functionally, "it is a demonstrable fact that human biology and human culture are part of a single system, unique and unprecedented in the history of life,"[78] which means that "no theory of human evolution that ignores its pragmatic aspect can be valid."[79] In this regard, remarks Dobzhansky, "a super-organic culture proved to be the most powerful method of adaptation to the environment ever developed by any species," so that "judged by any reasonable criteria, man represents the highest, most progressive, and most successful product of organic evolution."[80]

(6) As we have already witnessed then, "man's capacity for culture did not appear all at once, complete and finished." "Nor is this capacity a constant."[81] Rather, "Culture arose and developed hand in hand with the genetic basis which made it possible," and which "varies from time to time and from individual to individual."[82]

(7) Correlatively, "the physical and genetic endowments of the human species now living have evolved as a result of and in hand with the development of culture. That is to say biological and cultural evolution are interdependent,"[83] mutually reinforcing, "connected by what is known as a circular feedback relationship" in that "human genes stimulate the development of culture, and the development of culture stimulates genetic changes which facilitate further developments of culture."[84] Thus: "The only known processes which could have transformed the genotype of the prehuman apelike animal into the present human genetic endowment are mutations, sexual reproduction, genetic drift, geographic isolation, and social regulation of marriage."[85]

Now we are in a position to define in evolutionary terms the meaning of being human: To be human is to possess a genetic endowment which creates the setting for cultural traits without exercising power of compulsion over any particular ones—"Genes do not transmit and do not determine specific components of our cultural heredity."[86]

As Dobzhansky plainly expresses it: "To be a man one has to have a human genotype, and an ape genotype will not do regardless of any amount of training and of any known environmental influences."[87]

VI. MAN'S CAPACITY FOR CULTURE "RADICALITER SUMPTA"

The fact that a non-human organism is unable to become a carrier of human culture means that there is a capacity within the norm of reaction materially circumscribed by the human genetic endowment which is not present in the case of any other genetic structure. To get at the root of the uniqueness of man's cultural capacity we must therefore bypass all secondary manifestations of this capacity. Without losing sight of the fact that "the genetic basis of man's capacity to acquire, develop or modify, and transmit culture emerged because of the adaptive advantages which this capacity conferred on its possessors," we must concentrate not on those adaptive advantages, which are secondary phenomena (cf. sec. V, no. 3); rather we must look through them and isolate the primary or root phenomenon, that "in producing the genetic basis of culture, biological evolution has transcended itself—it has produced the superorganic." [88]

In terms of the adaptive phenotype, the radical basis of human cultural capacity means this: "The human organism has a constitutional capacity to react to objects . . . without the specific content or form of the reaction being in any way physiologically given," and on the basis of this capacity "the human attains levels of organization beyond those open to animals." [89] This is the genetic and hence evolutionary essence of the phenomenon of man so far as biology can define it.

The point is precisely this. Since the phenotype "is the necessary outcome of the development brought about by a certain genotype in a certain succession of environments," [90] while "the *only possible* way in which genes could influence the development of an organism is through physiological, and ultimately chemical, processes in the living body," [91] there remains at every moment in human existence a latent reaction potential which can never be exhausted by the "sequence" of purely biological environments. Dobzhansky puts it this way. Since "there is no 'culture' planted by nature in the genes," "the genotype of the human species is a necessary but not a sufficient condition for cultural development." [92] One must be careful to understand this rightly, for it means that there is a cognitive capacity in man which in a certain aspect stands "outside" the frontiers of strictly genetic possibilities—i.e., the human norm of reaction embraces a manner of knowing on the basis of which certain modalities of man's phenotypic response to environment exhibit no specific physiological valence. In man, in the human mode of being,

knowledge means . . . the initial and persistent looking out beyond what is given at any time. In different ways, by different channels, and in different realms, this transcendence (Hinaussein) effects (setzt ins Werk) what first gives the datum its relative justification, its potential determinateness, and hence its limit. Knowledge is the ability to put into work the being of any particular [entity]. . . . The passion of knowledge is inquiry. [93]

In evolutionary perspective, man is that being capable of a superior, realizing opening and keeping open that derives from an ability to perceive things as having an existence independent of their proximate affective reference; and so, seeing beyond each immediate need of physiological valence, able to carve out for himself, so to speak, a transcendent environmental niche always a little wider and more supple than the biologically given. For that individual existent which (who) stands in the mode of being human, "understanding signifies one's projecting oneself upon one's current possibility of Being-in-the-world; that is to say, it signifies existing as this possibility."[94]

The radical power underlying man's cultural capacities is the very capacity from which conjecture springs, and so hypothetical problem-solving—the ability of man "to form mental images of things and situations which do not yet exist but which may be found, brought about, or constructed by his efforts."[95] To this root capacity the term "intellectus" is traditionally applied.[96]

Its high survival value is evident, so that *once given*,[97] intellect inevitably generated an intense selection pressure toward the improvement of its genetic basis. "If natural selection has not developed genes for philosophy," muses Dobzhansky, "it has favored genetic endowments which enable their carriers to become, among other things, philosophers."[98]

VII. THE SPIRIT OF MAN: TRANSCENDENT EPIGENESIS

Once we have focused this fact, namely, that the human evolutionary uniqueness is constituted by a norm of reaction which admits of certain non-physiological phenotypic modalities, and that such a reaction range is established in consequence of a kind of awareness which (in escaping the direct influence of genic, or physiological, processes) transcends the biologically given, we have the evolutionary basis for the very potency-act analysis by which, at another observational level and in a non-evolutionary perspective, St. Thomas pointed out the spirit in man (cf. sec. II, fn. 28) that is not strictly educible from the potentiality of matter—even though it appears in time exactly as though by way of biological causation.

To keep to the terms of our own analysis: At the interior moment of man's initial emergence, we encounter an epigenic break in the unfolding evolutionary pattern. (The term "break" is carefully chosen to avoid any orthogenic connotation of an inner dynamism overarching the movement of life as a whole. I deliberately set aside such terms as "leap," "jump," "saltation," etc.) For perhaps the only time in the history of biological development, a specific discontinuity arose and could only have arisen *between two individuals*: i.e., *either* the entire development of the anthropoid organism was circumscribed by the limitations imposed through the biological mechanism of gene transfer—in which case it did not possess a human genotype; *or* there was a reaction capacity in the organism which transcended the limitations of biological heredity—in which case it was endowed with a human genetic structure.[99]

I take such a capacity as defining the human norm of reaction independently of all other variables. The kind of mental awareness from which it flows establishes then a centrality in the historical emergence of man, a reference point itself undergoing advance and growth, yet in relation to which any developmental ambivalence within the human group as a whole must have been poised.[100] Viewed from this vantage, the conclusion that *the first true man entered the movement of biological evolution with a single step* becomes virtually inescapable.

I realize full well the difficulties such a judgment entails. It is by no means clear how such a statement can be squared operationally with the established principles of population statics and dynamics which structure the scientific understanding of evolution. "In modern evolutionary studies concern has shifted from the organism as a describable object to the more sophisticated view of it as a morphological expression of the genetic and environmental status of an evolving population,"[101] for the very good reason that "in sexual organisms, Mendelian populations, rather than individuals, have become the units of the adaptively most decisive forms of natural selection."[102]

The crucial difficulty comes down to appreciating how a developmental beginning which admits of such a rigidly circumscribed conceptualization could have eventuated in a surface of evolution adequate for interplay of the forces of directional selection. It may be some time before the precise nature of this difficulty can be adequately characterized.[103] What is clear even at the present stage of research is that even at the level of the individual organism "as a describable object" "the pre-human animal was not transformed one fine day into man by a single lucky mutation."[104] The context of the present analysis could hardly render more inadmissible such an impoverished, naive and simplistic interpretation.

Yet I do not see how the conclusion can be denied or avoided: *at whatever point in space-time the phenomena of genetic mutation and recombination brought into play the factors of position effects and pleiotropism in the establishment of a genotypic milieu defining (and here the scholastic distinction between material and formal establishment proves decisive) a norm of reaction that enclosed a capacity (however limited) for phenotypic response to environment in nowise physiologically given as regards its specific content or form—precisely at that point in the gradated series of hominid forms must the real origin of man be located* (cf. sec. III, fn. 36).

Since this interval defines an "either—or" phenomenon, we must postulate a difference in kind *strictly understood* rather than one sharp simply in degree or based on degree as separating the human from the pre-human realm of awareness. The hypothetical conception of animal intelligence leading by insensible gradations to the faculty of intellect in man can therefore no longer be maintained.[105] In fact, the emergence of man is radicated in a new way of looking out on the world. To neglect this is to overlook what even so thoroughgoing an evolutionary philosopher as Bergson was able to indicate as "the one point on which everyone is agreed, to wit, that the young child understands immediately things that the animal will never understand, and that in this sense intelligence, like instinct, is an inherited function, *therefore an innate one.*" [106] *Rooted* in a new way of seeing, *caused* by a new mode of adaptation: such is the fact of human emergence, the unique development gradual and substantial and immanent, yet in a radical cognitive modality transcendent.

Understanding "culture" in terms of this radicalization which we have now effected, we may capsulize our analysis by appropriating another of Dobzhansky's customarily insightful formulations: "*Culture* is man's most potent means of adaptation to environment; genetically conditioned *educability* [cf. sec. V, no. 3] is his most potent biological adaptation to culture." [107] Thus were set up the whole series of selection pressures, both somatic and social, of a kind which did not exist on the prehuman level.

In this way, we can take into account the established principle that "mosaic evolution is the characteristic form of evolution of all types that shift into a new adaptive zone," as well as the established fact that "the evolution of the hominids is an almost classic demonstration of mosaic evolution." [108] Similarly, our interpretation gives concrete specification to Hockett's and Ascher's cryptic reference to the "essence" of language and culture (sec. II, fn. 21). The initial human emergence was defined morphologically simply by an animal, of the super-family hominidae (cf. fn. 66); yet functionally, it was characterized by a radical advance over every other form. The evolutionary principle of human being, upon which the constitutional whole as such is ontologically supported, becomes accessible to us, therefore, only if and when we look all the way *through* this whole to a single *primordially* constitutive phenomenon which is already in this whole in such a way that it provides the ontological foundation for each structural item in its structural possibility. At the focus of human emergence we must envisage accordingly something much like what Gardner Murphy describes as

a '*first human nature*'—that human nature produced by the gradual development of a raw distinctive humanness differing from the nature of all other creatures and possessing sharper wits, greater capacity to learn and, above all, keener exploratory functions, the capacity to discover and use new relationships. It is from this raw— or "original"—human nature that the more complex cultural processes have developed, the human nature that we see today.

The first human nature is essentially the product of the evolutionary process, and of the particular stocks from which man descended; the emotional and impulsive equipment of man is essentially produced by his place in the evolutionary scheme. One of the fundamental features, however, of such a broad picture of man is the conception of sensitivity, modifiability, capacity to learn, adapt, adjust, remake the world in the image of his needs. It will therefore be not a fixed biological human nature that is involved but a human nature precariously balanced, always ready to be thrown into a new equilibrium. The first human nature will be not a frozen but a constantly changing model upon which ever changing forces work. Thus there is a cultural molding of this biological framework. . . .[109]

Causal analysis succeeds, then, precisely where the paleosciences proved inadequate. It enables us to penetrate the very inner moment of hominization and define its fundamental condition in experimental terms, however invisible that moment may remain in the eyes of physical and cultural anthropology. Père Teilhard de Chardin was right to assert that "the access to thought represents a threshold which had to be crossed at a single stride;" but he was mistaken in regarding this threshold as "a 'transexperimental' interval about which scientifically we can say nothing."[110] An observation of Huxley's has particular relevance in this connection: "The paleontologist, confronted with his continuous and long range trends, is prone to misunderstand the implications of a discontinuous theory of change such as mutation, and to invoke orthogenesis or Lamarckism as explanatory agencies."[111]

However the tension may be resolved between this conclusion (that the appearance of the first human being in the biological community must have represented a single-stride transition) and the exigencies of population thinking, a study of human evolution in terms of its genetic basis discloses to us exactly what Père Sertillanges projected might be disclosed when, in 1945, he wrote his influential *L'Idée de création*:

The appearance of spirit in no wise interrupts the course of the physiobiological phenomena. We have here a supercreation, or as Leibniz puts it, a transcreation, but in the strictest phenomenal continuity. It is a crowning, not autogenous, but conjoined. It is a blossoming *in materia*, in the matter itself, although it is not *vi materiae*, that is, not in virtue of the matter alone.[112]

Regarded historically, we summarize, the fact of the emergence of man is before all else an evolutionary phenomenon—just the continuation by one avant-garde line of a general developmental movement in the community of living things. But when analyzed causally, though it reveals itself as no less a developmental unfolding of the very being of man, yet a certain transcendent, expansive energy discloses itself at the center of human emergence, a radical capacity unique to man and constituting the human phenomenon precisely as such; so that if we want to capture in some one expression the fact and meaning of human emergence, then perhaps we should say with Bergson that "Everything happens as though the grip of intelligence on matter were, in its main intention, to *let something pass* that matter is holding back,"[113] namely, the spirit of man.

NOTES

[1] *The Biological Basis of Human Freedom* (New York: Columbia Univ. Press, 1956), pp. 6, 9. This means in negative terms that from the vantage afforded by the neo- and paleo-sciences, there is simply no coherent basis for regarding man as unrelated genetically *omni ex parte* to the biological community. Watson's and Crick's breakthrough with the DNA model in 1953, together with the critical hominid fossil finds of L. Leakey in 1959 and 1964 heavily underscore this remark of Dobzhansky.

[2] *Evolution in Action* (New York: Harper, 1953), p. 35.

[3] The broad basis for this point of view is set forth in my two-part article, "Evolution: Concept and Content," *Listening* (Autumn, 1965), 27–50, and (Winter, 1966), 38–66.

[4] M. Heidegger, *Being and Time*, trans. J. Macquarrie and E. Robinson (New York: Harper, 1962), p. 29. Cf. pp. 28–30, 350, 371, 413–414, and 490 note x.

[5] T. Dobzhansky, *Mankind Evolving* (New Haven: Yale Univ. Press, 1962), p. 203.

[6] E. Gilson, *The Philosopher and Theology*, trans. C. Gilson (New York: Random House, 1962), p. 129.

[7] "Self, Society, and Culture in Phylogenetic Perspective," in *Evolution after Darwin*, ed. S. Tax (Chicago: Univ. of Chicago Press, 1960), II, 309.

[8] *Op. cit.*, p. 81.

[9] M. Critchley, "The Evolution of Man's Capacity for Language," in *Evolution after Darwin*, II, 294–295.

[10] Dobzhansky, *The Biological Basis of Human Freedom*, p. 108.

[11] "Dr. Leakey believes, and many other palaeontologists and archaeologists seem to agree with him, that he has found stone and bone that was either simply utilized or deliberately shaped at the bottom of Bed I in Olduvai Gorge, which has been dated at between 1.9 and 1.75 million years ago. He has also found a circle of stones, perhaps a hut circle, at that level."— Personal letter from Professor Sol Tax of the University of Chicago Department of Anthropology, dated May 18, 1965. Cf. Dobzhansky, *Mankind Evolving*, pp. 174–176; R. L. Fleischer, P. B. Price, R. M. Walker, of the General Electric Research Laboratory, Schenectady, New York, and L. S. B. Leakey of Coryndon Museum Centre for Prehistory and Palaeontology, Nairobi, "Fission-Track Dating of Bed I, Olduvai Gorge," in *Science*, CXLVIII (1965), 72–74; "The Problem of Man's Emergence," in *Scientific American*, CCXII (1965), 51–52.

[12] Dobzhansky, *Mankind Evolving*, p. 188: "But we cannot lightly dismiss the possibility that the now-living human species carries in its gene pool genetic elements derived from most or from all the fossil races of *Homo erectus* and *Homo sapiens*, though in very unequal proportions. On the other hand, we may be descended from only one ancient race, all others having petered out." Cf. also pp. 183, 192.

[13] *Ibid.*, pp. 220–221. Cf. pp. 188, 220, 221, 239, 269; also Dobzhansky's *Genetics and the Origin of Species* (3rd ed., rev.; New York: Columbia Univ. Press, 1951), pp. 305–306.

[14] Personal letter of May 18, 1965.

[15] Cf. Pierre Teilhard de Chardin, "The Idea of Fossil Man," in *Anthropology Today*, ed. by S. Tax (Chicago: Univ. of Chicago Press, 1962), pp. 35–36. See *infra*, ad fn. 66.

[16] Tax, Letter of May 18, 1965.

[17] *A Million Years of Man* (Cleveland: World Pub. Co., 1964), pp. 78–79. Tax assures me that "most people would still agree with Carrington's statement about relative brain size, called 'encephalization' by Dr. J. N. Spuhler. . . . I might add that, while everyone seemed to agree that the concepts of encephalization and excess neurones are more meaningful than absolute brain size, Dr. S. L. Washburn is convinced that the greatest stress should be placed on the details of the structure and organization of the brain."—Letter of May 18, 1965.

[18] S. Washburn, "Speculations on the Interrelations of the History of Tools and Biological Evolution," in *The Evolution of Man's Capacity for Culture*, essays arranged by J. Spuhler (Detroit: Wayne State Univ. Press, 1959), p. 29.

[19] "Somatic Paths to Culture," *ibid.*, p. 7.

[20] *Loc. cit.*, p. 29. Dobzhansky explicates this interaction nicely in *The Biological Basis of Human Freedom*, p. 105.

[21] V, 145, author's emphasis.

[22] G. von Bonin, *The Evolution of the Human Brain* (Chicago: Univ. of Chicago Press, 1963), p. 80.

[23] *Loc. cit.*, p. 10.

[24] Washburn, *loc. cit.*, p. 21. Cf. S. Washburn and F. Howell, "Human Evolution and Culture," in *Evolution after Darwin*, II, 49: Washburn and Howell are right to see in this conclusion "the most important result of the recent fossil hominid discoveries," one which "carries far-reaching implications for the interpretation of human behavior and its origins."

[25] Dobzhansky, *Genetics and the Origin of Species*, ed. cit., pp. 304–305.

[26] *Op. cit.*, p. 76.

[27] Cf. J. Gredt, *Elementa Philosophiae Aristotelico-Thomisticae*, 13th ed. rev. by E. Zenzen (Freiburg i Br.: Herder, 1961), I, 450, n. 531, cor. 7 & 8. On this crucial point, and in connection with the reference to Heidegger given in fn. 28, cf. J. Deely, "Evolution: Concept and Content, Part II," *Listening*, Vol. I, No. 1 (Winter, 1966), 38–66, esp. the section headed "The error of univocally ontologized kind-essences," pp. 42–48.

[28] Heidegger, *op. cit.*, p. 153. A radical appropriation of Heidegger's formulary at this point brings out in a striking manner a Thomistic insight which will be foundational in the effort to construct integrally a philosophical anthropology—namely, the realization that the intelligibility of the matter-form analogy tends to disintegrate when focussed reflexively on the being of man unless it is reinforced by the essence-existence analogy. For St. Thomas, only an understanding of man achieved in terms of the unity and uniqueness of *ipsum esse humanum* can, in confrontation with the *datum* of the transcendence of human understanding, secure itself conceptually against oscillation between the counterposed interpretations of materialism and idealism. Cf. *Q. D. de an.*, I ad 1; *Summa Theol.*, I, 76, 1 ad 5, and 75, 6; *in II de an.*, I, n. 215. The plight of Cajetan in this area is evidence clear enough—cf. A. Maurer, *Medieval Philosophy* (New York: Random House, 1962), pp. 350–351, and 179–182, 338ff.

[29] Pierre Teilhard de Chardin, *The Phenomenon of Man*, trans. B. Wall (New York: Harper, 1960), p. 184.

[30] Cf. J. Huxley, *Evolution: The Modern Synthesis* (New York: G. Allen & Unwin, 1964), p. 571.

[31] Dobzhansky, *The Biological Basis of Human Freedom*, p. 106.

[32] Cf. R. Nogar, *The Wisdom of Evolution* (Garden City, N.Y.,: Doubleday, 1963), pp. 194–195, p. 214 note 16, pp. 200–201.

[33] C. Hockett, "Animal 'Languages' and Human Language," in *The Evolution of Man's Capacity for Culture*, p. 36. On the question of language and its degree of primordiality for the mode of being human, cf. Heidegger, *Being and Time*, pp. 203 ff.

[34] T. Dobzhansky, *Heredity and the Nature of Man* (New York: Harcourt, Brace & World, 1964), p. 140. Cf. Nogar, *op. cit.*, pp. 151–155.

[35] M. Heidegger, *An Introduction to Metaphysics*, trans. R. Manheim (Garden City, N.Y.,: Doubleday, 1961), p. 131. Cf. *Being and Time*, p. 195; also p. 76: "But since the positive sciences neither 'can' nor should wait for the ontological labours of philosophy to be done, the further course of research will not take the form of an 'advance' but will be accomplished by *recapitulating* what has already been ontically discovered, and by purifying it in a way which is ontologically more transparent." (Cf. *supra*, ad fn. 4.)

[36] (Cambridge, Mass.: Harvard Univ. Press, 1963), p. 637.

[37] "Comments" on Hockett's and Ascher's "The Human Revolution," *loc. cit.*, p. 150. Cf. Heidegger, *Being and Time*, p. 275.

[38] Dobzhansky, *The Biological Basis of Human Freedom*, pp. 10 and 11.

[39] Dobzhansky, *Mankind Evolving*, pp. 40–42.

[40] Dobzhansky, *Genetics and the Origin of Species*, pp. 20, 21: emphasis supplied. These facts define the *principle of the reaction range*, a principle altogether central in our interpretation of human origins, as will become clear. Problematically, the principle means this: "The sensible question about any characteristic is not 'Is it hereditary or environmental—due to nature or to nurture?' but 'What is the reaction range of its genotype, and what are the environmental factors correlated with this particular position in the range?'"—cf. G. Simpson, C. Pittendrigh, and L. Tiffany, *Life: an Introduction to Biology* (New York: Harcourt, Brace & World, 1957), p. 334.

[41] L. Dunn, *Heredity and Evolution in Human Populations* (New York: Atheneum, 1965), p.11.

[42] Dobzhansky, *Genetics and the Origin of Species*, p. 36.

[43] Dobzhansky, *Evolution, Genetics, and Man* (New York: Science Editions, 1963), p. 37.

[44] S. Cheterikov, "On Certain Aspects of the Evolutionary Process from the Standpoint of Modern Genetics," cited by E. Mayr, *op. cit.*, p. 265. Cf. trans. of M. Barker, ed. I. Lerner, in *Proceedings of the American Philosophical Society*, CV (1961), 192, n. 18.

[45] Dobzhansky, *The Biological Basis of Human Freedom*, p. 57.

[46] Dobzhansky, *Mankind Evolving*, p. 219.

[47] Mayr, *op. cit.*, p. 179.

[48] Huxley, *Evolution in Action*, pp. 33–34.

[49] Dobzhansky, *Evolution, Genetics, and Man*, p. 81.

[50] L. Dunn and T. Dobzhansky, *Heredity, Race, and Society*, rev. and enlarged ed. (New York: New American Library, 1952), p. 107.

[51] Dobzhansky, *Evolution, Genetics, and Man*, p. 81.

[52] Dobzhansky, *The Biological Basis of Human Freedom*, p. 56.

[53] Cf. Mayr, *op. cit.*, pp. 180–181: "There seems to be an inverse correlation between the number of generations per unit of time and the relative importance of recombination (as compared to mutation) as a source of new genotypes. In an organism with a slow sequence of generations, a new mutation can be tested on a different genetic background only at intervals of many years; in a large tree in a climax forest, a new mutation may be tested only once every 100-200 years. Mutation pressure as a determinant of evolutionary change is of negligible importance in such organisms. Such types rely for evolutionary plasticity on the storage of genetic variation and the production of a great diversity of genotypes."

[54] Dobzhansky, *Genetics and the Origin of Species*, pp. 73–74.

[55] Dobzhansky, *Evolution, Genetics, and Man*, p. 106.

[56] Dobzhansky, *The Biological Basis of Human Freedom*, p. 81.

[57] Mayr, *op. cit.*, p. 184.

[58] Dobzhansky, *Evolution, Genetics, and Man*, p. 117.

[59] Dobzhansky, *Genetics and the Origin of Species*, pp. 21–22: *passim.*

[60] Spuhler, *loc. cit.*, p. 4.

[61] G. Simpson, *The Life of the Past* (New Haven: Yale Univ. Press, 1953), p. 145.

[62] G. Simpson, *The Meaning of Evolution* (New Haven: Yale Univ. Press, 1949), p. 223.

[63] Cheterikov, as cited by Dobzhansky in *Mankind Evolving*, p. 136. Cf. *loc. cit.* (in fn. 44), pp. 192–193, n. 20. Also "Correlated Responses to Selection," in Dobzhansky's *Genetics and the Origin of Species*, pp. 99–101.

[64] Dobzhansky, *Evolution, Genetics, and Man*, p. 132.

[65] Dobzhansky, *Genetics and the Origin of Species*, p. 82.

[66] *Mankind Evolving*, pp. 181, 186 (see text *supra* ad fn. 15). This conception has fundamental interpretive significance for the analysis of human origins, for in terms of the appearance of man on earth, at the very least it means that we cannot be startled in realizing that the crossing of the human threshold is in its datability concealed probably forever under a veil of historically gradated morphological structures or "types." This is a point Mayr forcefully illustrates in the key passage cited above (sec. III, fn. 36). W. Boyd, in reviewing "The Contributions of Genetics to Anthropology" (in *Anthropology Today*, p. 66), observes that this hermeneutical circumstance is practically guaranteed by the structural stability of the individual organism which underpins evolutionary fluctuations at the level of the population; and indeed the analysis of "Existence and Causality" (cf. *The Thomist* XXVIII [1964], 89–90) carried out by W. Kane would seem to afford Boyd's guarantee a metaphysical seal of approval. Cf. also P. Medawar, *The Uniqueness of the Individual* (New York: Basic Books, 1957), p. 14.

[67] Dunn and Dobzhansky, *op. cit.*, p. 79.

[68] Cf. *inter alia*, Dobzhansky, *Evolution, Genetics, and Man*, pp. 82–83; *Mankind Evolving* p. 139.

[69] Spuhler, *loc. cit.*, p. 4.

[70] Dobzhansky, *Evolution, Genetics, and Man*, p. 340.

[71] Dobzhansky, *Heredity and the Nature of Man*, p. 146.

[72] Dobzhansky, *Mankind Evolving*, pp. 322, 280, 322, respectively, emphasis supplied. Cf. also pp. 199 and 203; and Spuhler, *loc. cit.*

[73] Mayr, *op. cit.*, p. 636.

[74] As a phenomenal horizon this continuity is of course in some way preparatory for what we

can call with Heidegger phenomenological disclosure of the *a priori* defining the area of this transition—cf. *Being and Time*, pp. 359; 71 and 490, I, 1, note x; also 249, 270–271. The point is, however, that in this instance the preparation is not proximate. In this connection, cf. A. Sertillanges, *L'Idée de création et ses retentissements en philosophie* (Paris: Aubier, 1945), pp. 147–149. Also see *infra*, sec. VII fn. 106. And St. Thomas, *In I Met.*, 1, esp. n. 20.

[75] Dobzhansky, *The Biological Basis of Human Freedom*, p. 121. Cf. also pp. 52, 96–97, 99, 105–106, 118, 130; also "Selection for Educability," in *Mankind Evolving*, pp. 250–252; and *Heredity and the Nature of Man*, pp. 108, 111, 144–145.

[76] Dobzhansky, *Mankind Evolving*, pp. 287, 224, 223, respectively.

[77] *Ibid.*, p. 20.

[78] Dobzhansky, *Genetics and the Origin of Species*, p. 304.

[79] Dobzhansky, *Mankind Evolving*, p. 213. The metaphysical dimensions of this empiriological formulation can be glimpsed in Heidegger's "Analysis of Environmentality and Worldhood in General" (*Being and Time*, pp. 95–122), of "The Aroundness of the Environment and Dasein's Spatiality" (*ibid.*, pp. 134–148), and of "Understanding and Interpretation" (*ibid.*, pp. 188 ff.).—Cf. esp. pp. 95, 101, 116–117, 120–122, 137, and 187. Cf. also Carrington, *op. cit.*, pp. 164–165.

[80] Dobzhansky, *The Biological Basis of Human Freedom*, pp. 121 and 86. What Bergson remarked as early as 1907 continues to hold true: "The controversy bears less on the main lines of the movement [of evolution] than on matters of detail."—*Creative Evolution*, trans. A. Mitchell (New York: The Modern Library, 1944), p. 117. Among the leading experts in evolutionary science, agreement with the cited formulation of Dobzhansky is substantial and practically unanimous.

[81] Dobzhansky, *Mankind Evolving*, p. 19, and *Heredity and the Nature of Man*, p. 143, respectively.

[82] Dobzhansky, *Mankind Evolving*, p. 75, and *Heredity and the Nature of Man*, p. 143, respectively.

[83] Dobzhansky, *Mankind Evolving*, p. 193. Cf. also p. 319.

[84] Dobzhansky, *Heredity and the Nature of Man*, p. 14.

[85] Dobzhansky, *Genetics and the Origin of Species*, p. 304. In the authoritative view of Mayr (*op. cit.*, pp. 204–214), the term "genetic drift" is used ambiguously in evolutionary discussion. For a clarification of Dobzhansky's usage in specific reference to the earliest times of human emergence, see *Evolution, Genetics, and Man*, pp. 29–30.

[86] Dobzhansky, *The Biological Basis of Human Freedom*, p. 132. Cf. *Mankind Evolving*, p. 322.

[87] *Ibid.*, p. 25. (Anima enim humana accurate respondet subjecto cui infunditur tamquam educatur.)

[88] Dobzhansky, *Mankind Evolving*, p. 20.

[89] T. Parsons and E. Shils, eds. *Toward a General Theory of Action* (Cambridge: Harvard Univ. Press, 1951), pp. 10 and 17. St. Thomas touches on this point obliquely in the *Summa*, II-II, 57, 3.

[90] Dobzhansky, *Evolution, Genetics, and Man*, p. 77. That the strict understanding of this statement applies to man as well as other organisms is evident at the level of full existential individuality. Less evident, but no less true, is the fact that such an application does not contradict human freedom. This can be brought out clearly by developing in a fuller way the analysis sketched incisively by Bergson in his essay on *Time and Free Will* (trans. F. Pogson [New York: Harper, 1960], pp. 165–167). What is involved in such an approach is of course not human freedom radically considered, but rather the suffusion of spontaneity with intellectuality consequent upon the "esse" of human "essentia." See A.-D. Sertillanges' analysis of Bergson on this point, "Le libre arbitre," *Le Christianisme et les philosophies* (Paris: Aubier, 1941), II, 396–400. The proposed expansion of Bergson's oblique mention is carried out in effect by G. Murphy, "Freedom of Choice," *Human Potentialities*, pp. 277–283. Cf. also *Being and Time*, p. 239.

[91] Dobzhansky, *Genetics and the Origin of Species*, p. 13: emphasis supplied. Cf. also p. 30.

[92] Dobzhansky, *The Biological Basis of Human Freedom*, pp. 28, 42. Cf. pp. 119–120.

[93] Heidegger, *An Introduction to Metaphysics*, pp. 133–134: *passim*. Cf. also pp. 70–71. *Being and Time*, p. 27: "This entity which each of us himself is and which includes inquiring as one of the possibilities of its Being, we shall denote by the term Dasein." Also pp. 88–89, 215. St. Thomas, *In II de an.*, 1, 13, nn. 396–398.

[94] Heidegger, *Being and Time*, p. 439. Cf. also pp. 100, 106, esp. 119, 188–195, esp. 189 and 194.

[95] Dobzhansky, *Mankind Evolving*, p. 338. This is indeed only the Aristotelian notion of "suppositional" or "hypothetical" necessity—"the necessity that the requisite antecedents be there, if the final end is to be reached" (*De partibus animalium*, Bk. I, 642a33–4)—considered in the context of its potential survival value, its biological advantage. A clear and interesting illustration of this is provided in the *Metaphysics*, Bk. VII, ch. 7, 1032b6–12. The important point for our context is that such "reality" or causal thinking presupposes a virtual cognitive awareness of individual existents precisely as such—that is, as having altogether *proper* possibilities of being (and therefore, limits *qua* limits) in their structure. As Heidegger pointedly remarks: "Transcendence does not *consist* in Objectifying, but is *presupposed* by it." (*Being and Time*, p. 415, emphasis supplied.)

[96] Cf. St. Thomas, *In I de an.*, 1. 2, nn. 19–20; A. Carrel, *Man the Unknown* (New York: Harper, 1939), p. 81.

[97] Selection pressure, after all, is not creative: it can only work on what already exists to force a direction. The somatic determinants of the human *propria* would never have expanded beyond those of the late Tertiary peak hominid psychic development if the radically non-physiological cognitive capacity in man had not, by reason of its phenotypic consequences, provided evolutionary selection with a privileged hold on the human genetic structure.

[98] Dobzhansky, *The Biological Basis of Human Freedom*, p. 133. In this connection, it is worth remarking Loren Eiseley's conjecture: "It is not beyond the range of possibility that . . . reduction of instincts in man in some manner forced a precipitous brain growth as a compensation—something that had to be hurried for survival purposes. Man's competition, it would thus appear, may have been much less with his own kind than with the dire necessity of building about him a world of ideas to replace his lost animal environment." *The Immense Journey* (London: Gollancz, 1958), pp. 92–93.

[99] The intrinsic proportionality at this stage of our problematic to the act/potency categories of metaphysical analysis is unmistakable: an act can never exceed its limiting potency. Obviously, this consideration is prior to and independent of an accounting for whatever environmental factors will enter into the organism's actual course of development. To keep to an empiriological formulation: what is fixed in the zygote at fertilization is its genotype, the norm of reaction of the organism to the environment, and not necessarily any characters or functions. One may distinguish in principle the genotype from its manifestation, the phenotype. The proposed dichotomous argument, therefore, depends in its thrust on a primary datum of observation, namely, that *the absolute limitations on SUBSTANTIAL developmental variation potential are established once and for all at the moment of that organism's genetic origin* (cf. Dobzhansky's explication of the dynamic nature of genotypic continuity, with its degree of tolerance for accidental developmental fluctuations: *Heredity and the Nature of Man*, p. 50; *Genetics and the Origin of Species*, pp. 20–21; *Mankind Evolving*, pp. 40–42).

[100] E.g., cf. Dobzhansky, *The Biological Basis of Human Freedom*, p. 84 in juxtaposition with p. 106. Heidegger, *Being and Time*, pp. 119 and 183.

[101] G. Jepsen, "Foreword" to *Genetics, Paleontology and Evolution*, ed. G. Jepsen, G. Simpson, and E. Mayr (New York: Atheneum, 1963), p. ix.

[102] Dobzhansky, *Genetics and the Origin of Species*, p. 260. Cf. Medawar, *op. cit.*, p. 14.

[103] An important step in this direction may have been achieved, however, with the appearance of George C. Williams' study, *Adaptation and Natural Selection* (Princeton, New Jersey: Princeton University Press, 1966).

[104] Dobzhansky, *Heredity and the Nature of Man*, pp. 145–146. Cf. *Mankind Evolving*, pp. 47, 203; G. Simpson, *This View of Life* (New York: Harcourt, Brace & World, 1964), p. 272.

[105] Aristotle, *Metaphysics*, Bk. X, ch. 7, 1057a21: "For we call those things intermediates, into which that which changes must change first."

[106] *Creative Evolution*, ed. cit., p. 162, emphasis supplied. Cf. *Being and Time*, p. 385.

[107] Dobzhansky, *Mankind Evolving*, p. 251; emphasis supplied.

[108] Mayr, *op. cit.*, p. 637. Cf. Spuhler, *loc. cit.*, p. 5.

[109] *Human Potentialities* (New York: Basic Books, 1958), pp. 15–16, 16–17: ". . .—and there are further biological considerations in the question of how the biological stuff itself is changing today as a result of the cultural conditions that influence new health techniques, new types of family living, new types of blending human stocks."

[110] *The Phenomenon of Man*, p. 172.

[111] *Evolution: The Modern Synthesis*, p. 31. This critical application of Huxley's words cannot however be made without reservation. Indeed, in the framework of a different problematic,

they might have to be withdrawn entirely. Cf. the remarks made in this connection by Dobzhansky in *Heredity and the Nature of Man*, pp. 151–152 and 171.

[112] *L'Idée de création*, p. 152. Cf. also pp. 159–169 in juxtaposition with Pierre Teilhard de Chardin, *Hymn of the Universe*, trans. S. Bartholomew (New York: Harper, 1965), pp. 100–101 and *The Phenomenon of Man*, pp. 171–172.

[113] *Creative Evolution*, p. 201, author's emphasis.

2

The humanness of man

Rationale of this section

Man emerges as biologically unique in that he evolves a capacity for developing culture. The fruit of this root uniqueness is nothing other than the cultural world itself, that environment thanks to which man becomes overtly human. In truth, from the evolutionary point of view, the most remarkable feature of our planet's recent past is the extraordinary and in a basic sense unparalleled upsurge of life which swept man from his genetic, biological origins in the early Pleistocene to the complexities and splendor of civilization. In something like two million years—an astonishingly short span for evolutionary transformations of such scope—the strictly hominid lines underwent the transformation into self-conscious man. In a fraction of that time the primitive hunter and food-gatherer ("upper Paleolithic") learned to grow crops and domesticate animals, to build towns and villages and construct simple machines, to create vessels of copper and bronze, and to establish gigantic monuments and elaborate rituals for the worship of his gods.

Yet this rapid, spectacular transformation was, from a phenomenological point of view, as much the result of natural processes as the gradual evolution of the whole of life from the primeval self-replicating units. Even so, to see the emergence of recognizably human man as a simple and direct extension of biological trends would be almost totally misleading.

The ant we discover imbedded in a block of Baltic amber, which the geologist dates 20,000,000 years ago, may be trusted to reproduce its typical ant behavior wherever it can survive. . . . The repetitious pattern in which countless generations of a single species repeat a pattern more complicated than the dreams of a technocratic Utopian is protected by these two circumstances: behaviour imbedded in physical structure and an inability to communicate new learnings. But man does not even carry the simplest forms of his behaviour in such a way that a human child without other human beings to teach it can be relied upon to produce spontaneously a single cultural item.[1]

Thus, "the clear distinction between the biological and the psycho-social worlds is the greatest advance in the general theory of evolution since Darwin."[2] And it must be said that "our humanity rests upon a series of learned behaviors woven together into patterns that are infinitely fragile and never directly inherited."[3] The humanness of man, in short, is entirely bound up with the traditional wisdom—or lack of it—of (any given) society. It is for that reason and to that extent something fragile, something that can be torn and changed or lost. Doubtless this is why social groups have always guarded "the way one does things" with the preservatives of offering, sacrifice, totem, taboo, and discipline. Humanness is never derived from individual instinct as such, but rests rather on an order of value and meaning of which psychology can deal with the experience, but remains powerless to explain. Still less can the nature of man's humanness be gleaned from the study of chemistry applied to our

---

[1] Margaret Mead, *Male and Female* (New York: Mentor Books, 1955), p. 143.

[2] Benjamin Farrington, *What Darwin Really Said* (London: Macdonald, 1966), p. 106. "When one attempts to consider the problem of human evolution," Waddington astutely comments (*The Ethical Animal*, p. 103), "the type of phenomena which should rise to one's mind as presenting the problem to be discussed are all the most crucial changes which have occurred between, say, the late Stone Age and the present. If one compares the Paleolithic population of scattered nomadic hunters with modern highly populous and complex societies, it is not the comparatively slight changes in bodily structure which differentiate us from Cromagnon man that makes the greatest impression. Human evolution has been in the first place a cultural evolution. Its achievements have been the bringing into being of societies in which contributions deriving from such sources as Magna Carta, Confucius, Newton and Shakespeare can be both perpetuated and utilized.

"It is important not to overlook this first impression of what human evolution is all about. It is clear that for an understanding of how the human race has come into the possession of those characteristics which we now think most valuable in human life, a theory is needed which is primarily one of cultural evolution. In man, we have, in addition to the biological evolutionary system, a second one in which the mechanism of social transmission fills the role which in the biological realm falls to genetics, that of passing information from one generation to the next." Just as the biological mode of hereditary transmission primarily and directly transmits the genetic materials which control variation in our bodily structure, so the socio-cultural mode of hereditary transmission primarily and directly transmits the mental materials which control variation in our attitudes and outlook. And just as physical variation can have an impact on attitudes and outlook, so also variations in outlook can have an impact on the physical environment. But this circular feedback does not alter the fact that the two heredities differ in primary formal effect, and therefore in essential definition.

[3] Mead, *op. cit.*, p. 143.

flesh and blood—the radical error behind all racist ideologies. In fact, the paradox of the human zoological group lies in the fact that the *humanitas* of *homo sapiens* cannot be in any adequate sense correlated with his physical, biological frame—any more than the meaning of a book can be gathered from a physico-chemical analysis of print and page.

It is this irreducibly immaterial context, thanks to which *homo sapiens* (*homo biologicus*) becomes *homo humanus*, on which we now wish to focus attention.

Cultural evolution

Julian H. Steward

CONTEXTUALIZING COMMENTS

The distinction between biological and cultural evolution is one of the fundamental distinctions structuring contemporary anthropological thought. The distinction is expressed in various ways—sometimes such terms as "psychosocial," "psychocultural," "sociopsychological," or simply "human" are exchanged for the term "cultural"; but in every case what is intended is the demarcation of man's social heredity (which consists in the cumulative transmission of acquired characters) with respect to his organismic or biological heredity (which is entirely governed by the mechanisms of gene transfer and which precludes direct transmission of acquired characters). In a very wide sense, sometimes truly analogical, sometimes simply metaphorical, there are similarities between biological development and cultural development; but the differences are great enough to demand a different name when speaking strictly and formally, and this is agreed to by those who argue that the root of man's humanness is only relatively qualitative (superficially different in kind) as well as by those who argue that this difference is simply qualitative (radically different in kind). In the following selection, Dr. Julian H. Steward, Research Professor of Anthropology and member of the Center for Advanced Study at the University of Illinois, explains how the distinction between biological and cultural development came to be established within evolutionary thought, and sketches some of the implications of this distinction.

Cultural evolution

It is almost 100 years since evolution became a powerful word in science. The concept of evolution, which Charles Darwin set forth so clearly and convincingly in his *Origin of Species* in 1859, came like a burst of light that seemed to illuminate all of nature—not only the development of the myriad forms of life but also the history of the planet earth, of the universe and of man and his civilization. It offered a scheme which made it possible to explain, rather than merely describe, man's world.

In biology the theory of evolution today is more powerfully established than ever. In cosmology it has become the primary generator of men's thinking about the universe. But the idea of evolution in the cultural history of mankind itself has had a frustrating career of ups and downs. It was warmly embraced in Darwin's time, left for dead at the turn of our century and is just now coming back to life and vigor. Today a completely new approach to the question has once more given us hope of achieving an understanding of the development of human cultures in evolutionary terms.

Before considering these new attempts to explain the evolutionary processes operating in human affairs, we need to review the attempts that failed. By the latter part of the 19th century Darwin's theory of biological evolution had profoundly changed scientists' views of human history. Once it was conceded that all forms of life, including man, had evolved from lower forms, it necessarily followed that at some point in evolution man's ancestors had been completely without culture. Human culture must therefore have started from simple beginnings and grown more complex. The 19th-century school of cultural evolutionists—mainly British—reasoned that man had progressed from a condition of simple, amoral savagery to a civilized state whose ultimate achievement was the Victorian Englishman, living in an industrial society and political democracy, believing in the Empire and belonging to the Church of England. The evolutionists assumed that the universe was designed to produce man and civilization, that cultural evolution everywhere must be governed by the same principles and follow the same line, and that all mankind would progress toward a civilization like that of Europe.

Among the leading proponents of this theory were Edward B. Tylor, the Englishman who has been called the father of anthropology; Lewis

H. Morgan, an American banker and lawyer who devoted many years to studying the Iroquois Indians; Edward Westermarck, a Finnish philosopher famed for his studies of the family; John Ferguson McLennan, a Scottish lawyer who concerned himself with the development of social organization, and James Frazer, the Scottish anthropologist, historian of religion and author of *The Golden Bough*. Their general point of view was developed by Morgan in his book, *Ancient Society*, in which he declared: "It can now be asserted upon convincing evidence that savagery preceded barbarism in all the tribes of mankind, as barbarism is known to have preceded civilization." Morgan divided man's cultural development into stages of "savagery," "barbarism" and "civilization"—each of which was ushered in by a single invention.

These 19th-century scholars were highly competent men, and some of their insights were extraordinarily acute. But their scheme was erected on such flimsy theoretical foundations and such faulty observation that the entire structure collapsed as soon as it was seriously tested. Their principal undoing was, of course, the notion that progress (i.e., toward the goal of European civilization) was the guiding principle in human development. In this they were following the thought of the biological evolutionists, who traced a progression from the simplest forms of life to *Homo sapiens*. Few students of evolution today, however, would argue that the universe has any design making progress inevitable, either in the biological or the cultural realm. Certainly there is nothing in the evolutionary process which preordained the particular developments that have occurred on our planet. From the principles operating in biological evolution—heredity, mutation, natural selection and so on—an observer who visited the earth some half a billion years ago, when the algae represented the highest existing form of life, could not possibly have predicted the evolution of fishes, let alone man. Likewise, no known principle of cultural development could ever have predicted specific inventions such as the bow, iron smelting, writing, tribal clans, states or cities.

THE FACTS

When, at the turn of the century, anthropologists began to study primitive cultures in detail, they found that the cultural evolutionists' information had been as wrong as their theoretical assumptions. Morgan had lumped together in the stage of middle barbarism the Pueblo Indians, who were simple farmers, and the peoples of Mexico, who had cities, empires, monumental architecture, metallurgy, astronomy, mathematics, phonetic writing and other accomplishments unknown to the Pueblo. Field research rapidly disclosed that one tribe after another had quite the wrong cultural characteristics to fit the evolutionary niche assigned it by Morgan. Eventually the general scheme of evolution postulated by the 19th-century theorists fell apart completely. They had believed, for example, that society first developed around the maternal line, the father being transient, and that marriage and the family as we know it did not evolve until men began to practice herding and agriculture. But field research showed that some of the most primitive hunting and gathering societies, such as the Bushmen of South Africa and the aboriginal Australians, were organized into patrilineal descent groups, while much more advanced horticultural peoples, including some of the groups in the Inca Empire of South America, had matrilineal kin groups. The Western Shoshonis of the Great Basin, who by every criterion had one of the simplest cultures, were organized in families which were not based on matrilineality. Still another blow to the evolutionists' theory was the discovery that customs had spread or diffused from one group to another over the world: that is to say, each society owed much of its culture to borrowing from its neighbors, so it could not be said that societies had evolved independently along a single inevitable line.

The collapse of the theory that cultural evolution had followed the same line everywhere (what we may call the "unilinear" scheme) began with the researches of the late Franz Boas, and the *coup de grâce* was dealt by Robert H. Lowie in his comprehensive and convincing analysis, *Primitive Society*, published in 1920. When the evolutionary hypothesis was demolished, however, no alternative hypothesis appeared. The 20th-century anthropologists threw out the evolutionists' insights along with their schemes. Studies of culture lost a unifying theory and lapsed into a methodology of "shreds and patches." Anthropology became fervently devoted to collecting facts. But it had to give some order to its data, and it fell back on classification—a phase in science which F. S. C. Northrop has called the "natural history stage."

The "culture elements" used as the classification criteria included such items as the bow and arrow, the domesticated dog, techniques and forms of basketry, the spear and spear thrower, head-hunting, polyandrous marriage, feather headgear, the penis sheath, initiation ceremonies for boys, tie-dyeing techniques for coloring textiles, the blowgun, use of a stick to scratch the head during periods of religious taboo, irrigation agriculture, shamanistic use of a sweat bath, transportation of the head of state on a litter, proving one's fortitude by submitting to ant bites, speaking to one's mother-in-law through a third party, making an arrowhead with side notches, marrying one's mother's brother's daughter. Students of the development of culture sought to learn the origin of such customs, their distribution and how they were combined in the "culture content" of each society.

Eventually this approach led to an attempt to find an over-all pattern in each society's way of life—a view which is well expressed in Ruth Benedict's *Patterns of Culture*. She contrasted, for example, the placid, smoothly functioning, nonaggressive behavior of the Pueblo Indians with the somewhat frenzied, warlike behavior of certain Plains Indians, aptly drawing on Greek mythology to designate the first as an Apollonian pattern and the second as Dionysian. The implication is that the pattern is formed by the ethos, value system or world-view. During the past decade and a half it has become popular to translate pattern into more psychological terms. But description of a culture in terms either of elements, ethos or personality type does not explain how it originated. Those who seek to understand how cultures evolved must look for longer-range causes and explanations.

MULTILINEAR EVOLUTION

One must keep in mind Herbert Spencer's distinction between man as a biological organism and his functioning on the superorganic or cultural level, which also has distinctive qualities. We must distinguish man's needs and capacity for culture—his superior brain and ability to speak and use tools—from the particular cultures he has evolved. A specific invention is not explained by saying that man is creative. Cultural activities meet various biological needs, but the existence of the latter does not explain the character of the former. While all men must eat, the choice of particular foods and of how they are obtained and prepared can be explained only on a superorganic level. Thanks to his jaw and tongue structure and to the speech and auditory centers of his brain, man is capable of speech, but these facts do not explain the origin of a single one of the thousands of languages that have developed in the world. The family is a basic human institution, but families in different cultures differ profoundly in the nature of their food-getting activities, in the division of labor between the sexes and in the socialization of the children.

The failure to distinguish the biological basis of all cultural development from the explanation of particular forms of culture accounts for a good deal of the controversy and confusion about "free will" and "determinism" in human behavior. The biological evolutionist George Gaylord Simpson considers that, because man has purposes and makes plans, he may exercise conscious control over cultural evolution. On the other hand, the cultural evolutionist Leslie A. White takes the deterministic position that culture develops according to its own laws. Simpson is correct in making a biological statement, that is, in describing man's capacity. White is correct in making a cultural statement, that is, in describing the origin of any particular culture.

All men, it is true, have the biological basis for making rational solutions, and specific features of culture may develop from the application of reason. But since circumstances differ (*e.g..*, in the conditions for hunting), solutions take many forms. Moreover, much culture develops gradually and imperceptibly without deliberate thought. The growth of settlements, kinship groups, beliefs in shamanism and magic, types of warfare and the like are not planned.

This does not mean that there is no rhyme or reason in the development of culture, or that history is random and haphazard. It is possible to trace causes and order in the seeming chaos. In the early irrigation civilizations of the Middle East, Asia and America the inventions were remarkably similar and ran extraordinarily parallel courses through several thousand years. There was clearly a close connection between large-scale irrigation agriculture, population increase, the growth of permanent communities and cities, the rise of specialists supported by agricultural workers, the appearance of unprecedented skills in technology, the need for a managerial class or bureaucracy and the rise of states.

There have been other patterns in the development of man's institutions, each adapted at different times and places to the specific circumstances of a specific society. The facts now accumulated indicate that human culture evolved along a number of different lines; we must think of cultural evolution not as unilinear but as multilinear. This is the new basis upon which evolutionists today are seeking to build an understanding of the development of human cultures. It is an ecological approach—an attempt to learn how the factors in each given type of situation shaped the development of a particular type of society.

Multilinear evolution is not merely a way of explaining the past. It is applicable to changes occurring today as well. In the department of sociology and anthropology of the University of Illinois my colleagues and I are studying current changes in the ways of rural populations in underdeveloped areas of the world: it is called "The Project to Study Cross-Cultural Regularities." During the past three years my colleagues—Eric Wolf, Robert F. Murphy, F. K. Lehman, Ben Zimmerman, Charles Erasmus, Louis Faron—and I have constructed research models to be tested by investigations in the field. These models consist of several types of populations—peasants, small farmers, wage workers on plantations and in mines and factories, primitive tribes. The objective is to learn how the several types of societies evolved and how their customs are being changed by economic or political factors introduced from the modern industrial world. Such studies should obviously have practical value in guiding programs of technical aid for these peoples.

HUNTERS, TRAPPERS, FARMERS

To illustrate the ecological approach let us consider very briefly several different types of societies, using the ways in which they made their living as the frame of reference. The first example is the form of society consisting of a patrilineal band of hunters. This type of organization was found among many primitive tribes all over the world, including the Bushmen of the deserts in South Africa, the Negritos of the tropical rain forest in the Congo, the aborigines of the steppes and deserts in Australia, the now extinct aboriginal islanders in Tasmania, the Indians of the cold pampas on the islands of Tierra del Fuego and Shoshoni Indians of the mountains in Southern California. Although their climates and environments differed greatly, all of these tribes had one important thing in common: they hunted cooperatively for sparsely scattered, nonmigratory game. In each case the cooperating band usually consisted of about 50 or 60 persons who occupied an area of some 400 square miles and claimed exclusive hunting rights to it. Since men could hunt more efficiently in familiar terrain, they remained throughout life in the territory of their birth. The band consequently consisted of persons related through the male line of descent, and it was required that wives be taken from other bands. In sum, the cultural effects of this line of evolution were band localization, descent in the male line, marriage outside the group, residence of the wife with the husband's band and control by the band of the food resources within its territory.

Another line of evolution is exemplified by rubber-farming Mundurucú Indians in the Amazon Valley and fur-trapping Algonquian Indians in eastern Canada, of whom Murphy and I recently made a comparative study. The common feature in these two groups is that both were transformed by contact with an outside economy from simple farmers or hunters to barterers for manufactured goods. Although the aboriginal Mundurucu villagers and the Algonquian bands had had very different forms of social organization, both converged to the same form after they began to pursue similar ways of making a living. As the Indians came to depend on manufactured goods, such as steel axes and metal utensils, obtained from traders, they gradually gave up their independent means of subsistence and spent all their time tapping rubber trees and trapping beaver, respectively, eventually depending upon the trader for clothing and food as well as for hardware. Since tapping and trapping are occupations best carried out by small groups on separate territories, the Indians' villages and bands broke down into individual families which lived in isolation on fairly small, delimited areas. The family became part of the larger Canadian or Brazilian national society, to which it was linked through the trader. Its only relations with other families were the loose social contacts created by dealing with the same trader.

IRRIGATION CIVILIZATIONS

Irrigation farming is the major organizing factor of another line of evolution, which covered a considerable span of the early prehistory and history of China, Mesopotamia, Egypt, the north coast of Peru, probably the Indus Valley and possibly the Valley of Mexico. This line had three stages. In the first period primitive groups apparently began to cultivate food plants along the moist banks of the rivers or in the higher terrain where rainfall was sufficient for crops. They occupied small but permanent villages. The second stage started when the people learned to divert the river waters by means of canals to irrigate large tracts of land. Irrigation farming made possible a larger population and freed the farmers from the need to spend all their time on basic food production. Part of the new-found time was put into enlarging the system of canals and ditches and part into developing crafts. This period brought the invention of loom weaving, metallurgy, the wheel, mathematics, the calendar, writing, monumental and religious architecture, and extremely fine art products.

When the irrigation works expanded so that the canals served many communities, a coordinating and managerial control became necessary. This need was met by a ruling class or a bureaucracy whose authority had mainly religious sanctions, for men looked to the gods for the rainfall on which their agriculture depended. Centralization of authority over a large territory marked the emergence of a state.

That a state developed in these irrigation centers by no means signifies that all states originated in this way. Many different lines of cultural evolution could have led from kinship groups up to multi-community states. For example, feudal Europe and Japan developed small states very different from the theocratic irrigation states.

The irrigation state reached its florescence in Mesopotamia between 3000 and 4000 B.C., in Egypt a little later, in China about 1500 or 2000 B.C., in northern Peru between 500 B.C. and 500 A.D., in the Valley of Mexico a little later than in Peru. Then, in each case, a third stage of expansion followed. When the theocratic states had reached the limits of available water and production had leveled off, they began to raid and conquer their neighbors to exact tribute. The states grew into empires. The empire was not only larger than the state but differed qualitatively in the ways it regimented and controlled its large and diversified population. Laws were codified; a bureaucracy was developed; a powerful military establishment, rather than the priesthood, was made the basis of authority.

The militaristic empires began with the Sumerian Dynasty in Meso-potamia, the pyramid-building Early Dynasty in Egypt, the Chou periods in China, the Toltec and Aztec periods in Mexico and the Tiahuanacan period in the Andes.

Since the wealth of these empires was based on forced tribute rather than on increased production, they contained the seeds of their own undoing. Excessive taxation, regimentation of civil life and imposition of the imperial religious cult over the local ones led the subject peoples eventually to rebel. The great empires were destroyed; the irrigation works were neglected; production declined; the population decreased. A "dark age" ensued. But in each center the process of empire building later began anew, and the cycle was repeated. Cyclical conquests suc-ceeded one another in Mesopotamia, Egypt and China for nearly 2,000 years. Peru had gone through at least two cycles and was at the peak of the Inca Empire when the Spaniards came. Mexico also probably had ex-perienced two cycles prior to the Spanish Conquest.

Our final example of a specific line of evolution is taken from more recent times. When the colonists in America pre-empted the Indians' lands, some of the Indian clans formed a new type of organization. The Ute, Western Shoshoni and Northern Paiute Indians, who had lived by hunting and gathering in small groups of wandering families, united in aggressive bands. With horses stolen from the white settlers, they raided the colonists' livestock and occasionally their settlements.

Similar predatory bands developed among some of the mounted Apaches, who had formerly lived in semipermanent encampments con-sisting of extended kinship groups. Many of these bands were the scourge of the Southwest for years. Some of the Apaches, on the other hand, yielded to the blandishments of the U.S. Government and settled peace-fully on reservations; as a result, there were Apache peace factions who rallied around chiefs such as Cochise, and predatory factions that followed belligerent leaders such as Geronimo.

The predatory bands of North America were broken up by the U.S. Army within a few years. But this type of evolution, although transitory, was not unique. In the pampas of South America similar raiding bands arose after the Indians obtained horses. On an infinitely larger scale and making a far greater impression on history were the Mongol hordes of Asia. The armies of Genghis Khan and his successors were essentially huge mounted bands that raided entire continents.

BIOLOGY AND CULTURE

Human evolution, then, is not merely a matter of biology but of the inter-action of man's physical and cultural characteristics, each influencing the other. Man is capable of devising rational solutions to life, especially in the realm of technical problems, and also of transmitting learned solutions to his offspring and other members of his society. His capacity for speech gives him the ability to package vastly complicated ideas into sound symbols and to pass on most of what he has learned. This human potential resulted in the accumulation and social transmission of an incalculable number of learned modes of behavior. It meant the perpetuation of established patterns, often when they were inappropriate in a changed situation.

The biological requirements for cultural evolution were an erect posture, specialized hands, a mouth structure permitting speech, stereoscopic vision, and areas in the brain for the functions of speech and association. Since culture speeded the development of these requirements, it would be difficult to say which came first.

The first step toward human culture may have come when manlike animals began to substitute tools for body parts. It has been suggested, for example, that there may have been an intimate relation between the development of a flint weapon held in the hand and the receding of the apelike jaw and protruding canine teeth. An ape, somewhat like a dog, deals with objects by means of its mouth. When the hands, assisted by tools, took over this task, the prognathous jaw began to recede. There were other consequences of this development. The brain centers that register the experiences of the hands grew larger, and this in turn gave the hands greater sensitivity and skill. The reduction of the jaw, especially the elimination of the "simian shelf," gave the tongue freer movement and thus helped create the potentiality for speech.

Darwin called attention to the fact that man is in effect a domesticated animal; as such he depends upon culture and cannot well survive in a state of nature. Man's self-domestication furthered his biological evolution in those characteristics that make culture possible. Until perhaps 25,000 years ago he steadily developed a progressively larger brain, a more erect posture, a more vertical face and better developed speech, auditory and associational centers in the brain. His physical evolution is unquestionably still going on, but there is no clear evidence that recent changes have increased his inherent potential for cultural activities. However, the rate of his cultural development became independent of his biological evolution. In addition to devising tools as substitutes for body parts in the struggle for survival, he evolved wholly new kinds of tools which served other purposes: stone scrapers for preparing skin clothing, baskets for gathering wild foods, axes for building houses and canoes. As cultural experience accumulated, the innovations multiplied, and old inventions were used in new ways. During the last 25,000 years the rate of culture change has accelerated.

The many kinds of human culture today are understandable only as particular lines of evolution. Even if men of the future develop an I.Q. that is incredibly high by modern standards, their specific behavior will nonetheless be determined not by their reason or psychological characteristics but by their special line of cultural evolution, that is, by the fundamental processes that shape cultures in particular ways.

Cultural determinants of mind

Leslie A. White

CONTEXTUALIZING COMMENTS

The development of the human mind has plainly led to an awareness of certain features of the universe which are not inextricably bound up with the physico-chemical environment which offers the main challenges for response among non-human organisms. The human being must adapt itself not only to the physical challenge of climate and competing life-forms, but must cope as well with his own sense of isolation in the immensity of space-time, feelings of inadequacy or despair at moments of emotional or social crisis, dependency needs, kinship ties, etc.,—and at bottom with the threat of his own dissolution. Thus the human *prise de conscience* has given rise to unique problems, based on less tangible but no less real "selection pressures" than those operative in the non-human world. Such immaterial influences have become in truth the substance of the environment of a self-conscious being, and the adaptations which man has attempted to make to an environment become human are expressed in the characteristic phenomena of art, religion and philosophy— themselves continuations (in one or another way) of the "material" culture expressed in technology. We have already introduced Dr. White in the preceding section. In this article he considers the unique "nature" that must be recognized in man if we are to make sense out of the endless diversity that ethnography and prehistory reveal in human behavior. Although biological heredity and social determinants of behavior ("inter-action") mutually involve each other, guaranteeing that human behavior will contain always a residue of response to basic organismic needs, the cultural components of biocultural evolution must be seen as the dominant determinant, inasmuch as they are peculiar to man, exhibit his essential distinction from other life forms, and preclude an a-priori determination of the range of human potentialities.

A point that is of particular interest in our dialectical presentation of polar views is the extent to which L. White pushes his argument on the *sui-generis* character of cultural reality.

Cultural determinants of mind

"When I fulfil my obligations as brother, husband, or citizen, when I execute my contracts, I perform duties which are defined, externally to myself and my acts, in law and in custom. Even if they conform to my own sentiments and I feel their reality subjectively, such reality is still objective, for I did not create them; I merely inherited them through my education. . . . Similarly, the church-member finds the beliefs and practices of his religious life ready-made at birth; their existence prior to his own implies their existence outside of himself. . . . Here, then, are ways of acting, thinking, and feeling that present the noteworthy property of existing outside the individual con-sciousness.

"These types of conduct or thought are not only external to the individual but are, moreover, endowed with coercive power, by virtue of which they impose themselves upon him, independent of his individual will . . ."

Emile Durkheim, *The Rules of Sociological Method*

Human behavior is, as we have just seen, a compound of two separate and distinct kinds of elements: psychosomatic and cultural. On the one hand we have a certain type of primate organism, man; on the other, a traditional organization of tools, ideas, beliefs, customs, attitudes, etc., that we call culture. The behavior of man as a human being—as distin-guished from his non-symbolic, primate behavior—is an expression of the interaction of the human organism and the extra-somatic cultural tradition. Human behavior is, therefore, a function of culture as well as of a biological organism. In the preceding chapter we examined the relationship between man and culture at some length. We endeavored to show that psychological interpretations of cultures—of institutions, customs, attitudes, etc.—which have been, and still are, so popular, are unsound; that cultures cannot be explained psychologically but only culturologically. In the present chapter we shall continue our inquiry into the relationship between man and culture, but this time our focus will be upon the human organism rather than upon the external cultural tradition. If we cannot explain cultures psychologically, and, if human behavior is a product of culture as well as of nerves, glands, muscles, sense organs, etc., perhaps some of the phenomena commonly regarded as psychological are actually culturally determined. If, on the one hand, there has been a widespread tendency to regard cultures as psychologically determined, perhaps there has been a corresponding failure to recognize cultural determinants of mind. The point of view and habit of thought that sees in a custom or institution merely the expression of an innate desire, need or ability, is likely also to think of the "mind" of man as something innate in his organism, biologically determined. Just as culture is naively thought to be a simple and direct expression of "human nature," so is the "human mind" thought to be a simple and direct expression of the neuro-sensory-glandular-etcetera organization of man.

This view is, however, an illusion. Just as scientific analysis discovers a non-anthropomorphic, culturological determination of culture, and demonstrates the irrelevance of psychological explanations of cultures, so does it find that many of the elements or attributes of "the human mind" are not to be explained in terms of the action of nerves, brains, glands, sense organs, etc., but in terms of culture. This does not mean that the reactions of the human organism to cultural elements in the external world are not "psychological" or "mental;" they are. It simply means that in the *minding* of man as a human being there are non-psychosomatic, i.e., extra-somatic cultural, determinants. The "human mind" is the *reacting* of the *human organism* to external stimuli; mind is minding here as elsewhere. But this reacting, this minding, varies. The Hottentot mind, or minding, is not the same as Eskimo, or English, minding. The "human mind"—human minding—is obviously a variable. And its variations are functions of variations of the cultural factor rather than of the psychosomatic factor, which may be regarded as a constant. The whole concept of "the human mind" is thus thrown into a new light and perspective.

In other animal species, the "mind" is a function of the bodily structure, of a particular organization of nerves, glands, sense organs, muscles, etc. Thus the mind of the gorilla differs from that of the chimpanzee; the mind of a bear differs from that of a cat or squirrel. In each case, the minds are functions of their respective bodily structures, differences of mind are correlated with differences of bodily structure. In the case of the human species, however, this is not the case. The mind (minding) of the Chinese is not like the mind of the Sicilians or the Hopi Indians. But here the differences of mind are not due to differences of bodily structure for, from the standpoint of the human behavior of races or other groups, this may be considered as a constant. Differences of mind among different ethnic groups of human beings are due to differences of cultural tradition. Thus we have a radical and fundamental difference between the determination of mental variation among sub-human species and mental variation within the human species. For the sub-human species the formula is: $V_m = f(V_b)$—variations of mind are functions of variations of bodily structure. For the human species the formula is: $V_m = f(V_c)$—variations of human minding are functions of the extra-somatic tradition called culture.

In the realm of human behavior we are concerned of course with organisms; organizations of bones, muscles, glands, nerves, sense organs, and so on. And these organisms react to external stimuli, cultural as well as otherwise. The human mind is still the reacting of the human organism. But we now see that the specific content of the human mind in any particular expression—speaking here of peoples rather than of individuals—is determined by the extra-somatic factor of culture rather than by the neurologic, sensory, glandular, muscular, etc., constitution of the human organism. In other words, the Chinese mind, the French, Zulu, or Comanche mind, as a particular organization of human behavior, is explainable in cultural terms, not biological.

In the category "the human mind," therefore, in the minding of human beings, we discover cultural determinants as well as psycho-somatic factors. And, furthermore, we learn that in an explanation of differences among types of the human mind, such as Eskimo, Zulu, or English, it is the cultural determinant that is significant, not the psycho-somatic. A comparative, ethnographic survey of the human mind leads to a realization that many of its attributes are not due to an inborn "human nature" at all, as was formerly supposed, but to differences of external cultural stimulation.

One of the most popular formulas of interpretation of human behavior is that of "human nature." People behave as they do, have the institutions, beliefs, attitudes, games, etc., that surround them, because "it is human nature." And, incidentally, most people—however much they may be willing to admit their ignorance in other respects—usually feel that they "understand human nature." The human mind and organism are so constituted, according to this view, as to make certain kinds of response simply and directly forthcoming. One has only to know human nature to understand society and culture and to predict their course of development. The fallacy or illusion here is, of course, that what one takes for "human nature" is not *natural* at all but cultural. The tendencies, emphases, and content that one sees in the overt behavior of human beings are often not due to innate biological determination—though such determinations do of course exist—but to the stimulation of external cultural elements. Much of what is commonly called "human nature" is merely *culture* thrown against a screen of nerves, glands, sense organs, muscles, etc. We have a particularly fine example of this illusion, this mistaking of culture for nature, in a passage from Thomas Wolfe's *You Can't Go Home Again*[1]:

For what is man?

First, a child, unable to support itself on its rubbery legs, befouled with its excrement, that howls and laughs by turns, cries for the moon but hushes when it gets its mother's teat; a sleeper, eater, guzzler, howler, laugher, idiot, and a chewer of its toe; a little tender thing all blubbered with its spit, a reacher into fires, a beloved fool.

After that, a boy, hoarse and loud before his companions, but afraid of the dark; will beat the weaker and avoid the stronger; worships strength and savagery, loves tales of war and murder, and violence done to others; joins gangs and hates to be alone; makes heroes out of soldiers, sailors, prize fighters, football players, cowboys, gunmen, and detectives; would rather die than not out-try and out-dare his companions, wants to beat them and always to win, shows his muscle and demands that it be felt, boasts of his victories and will never own defeat.

Then the youth: goes after girls, is foul behind their backs among the drugstore boys, hints at a hundred seductions, but gets pimples on his face; begins to think about his clothes, becomes a fop, greases his hair, smokes cigarettes with a dissipated air, reads novels, and writes poetry on the sly. He sees the world now as a pair of legs and breasts; he knows hate, love, and jealousy; he is cowardly and foolish, he cannot endure to be alone; he lives in a crowd, thinks with the crowd, is afraid to be marked off from his fellows by an eccentricity. He joins clubs and is afraid of ridicule; he is bored and unhappy and wretched most of the time. There is a great cavity in him, he is dull.

Then the man: he is busy, he is full of plans and reasons, he has work. He gets children, buys and sells small packets of everlasting earth, intrigues against his rivals, is exultant when he cheats them. He wastes his little three score years and ten in spendthrift and inglorious living; from his cradle to his grave he scarcely sees the sun or moon or stars; he is unconscious of the immortal sea and earth; he talks of the future and he wastes it as it comes. If he is lucky, he saves money. At the end his fat purse buys him flunkeys to carry him where his shanks no longer can; he consumes rich food and golden wine that his wretched stomach has no hunger for; his weary and lifeless eyes look out upon the scenery of strange lands for which in youth his heart was panting. Then the slow death, prolonged by costly doctors, and finally the graduate undertakers, the perfumed carrion, the suave ushers with palms outspread to leftwards, the fast motor hearses, and the earth again.

To many, no doubt, Wolfe's characterization of man is both true and apt. This is what man really is, they feel. Others, perhaps, would disagree and say, "No, man is not as Wolfe depicts him; he is *this* sort of being." Each view may seem plausible; each can be supported with evidence. And, however much Wolfe's characterization of man may differ from that of another, both may agree that the *method of interpretation* is sound. You place man before you; you study him, analyze him, and then report your findings. Plausible and reasonable as this may seem, it is but an illusion. The Wolfes are not describing Man at all, but Culture.

This is not quibbling in any way. The distinction is real, profound, and important. What Wolfe describes as Man is merely the way the human organism responds to a certain set of cultural stimuli. In another kind of culture the organism would respond quite differently. His characterization of man would certainly not be applicable to the Zuni Pueblo Indians nor to the Pygmies of the Congo, the aborigines of Australia, or the peasant folk of Mexico. And, as a matter of fact, he all but says that it is not man's "real nature" that he is describing. Does he not suggest at least that man *is* a being who *could* "see the sun, moon and stars and be conscious of the immortal sea and earth" were it not for the culture which holds him in its grip and compels him to waste his precious life selling real estate, cheating rivals? Wolfe is describing a *culture* in terms of its effects upon the human organism.

But what difference does it make, one might ask, whether "human nature" or "culture" is the cause so long as man actually performs the acts and must suffer their consequences? What difference does it make whether a gangster murders a cashier and robs a bank because he was born and reared in a certain type of culture or because he was "by nature" murderous, vicious, and rapacious? The cashier is dead in either case, the money gone, and the police are hot on the gangster's trail. True enough; things are what they are. But it makes all the difference in the world whether the man did the killing and the robbing because it is human nature to do so, or whether his behavior was determined by the type of culture, the kind of social system, he happened to be living in. All the difference, that is, to the scientist who wishes to provide an adequate *explanation* of the behavior. And a great deal of difference to the layman, too, because of the implications inherent in the two alternatives: cultures may change—they are constantly changing in fact; but human nature, biologically defined, is virtually constant—it has undergone no appreciable change in the last 30,000 years at least.

Wolfe's description of man is a philosophy of behavior, an explanatory device. It is based on certain premises. It may be supported by much evidence, but the premises are wrong for all that, and much confusion and error flow inevitably from them.

Let us consider a few areas of behavior. Take food habits for example. Man is one but his tastes vary enormously. A food loathed by one people may be a delicacy to another. Many Chinese cannot bear the thought of eating cheese, whereas most Europeans are very fond of it, and the choicest cheeses are often those with an odor of putrefaction or ordure. Neither do the Chinese like milk—even Grade A. Some tribes will not eat chicken or eggs. Others will eat eggs but prefer rotten eggs to fresh ones. The choicest porterhouse steak has no charms for the Hindu, nor baked ham or pork chops for the Jew. We have an aversion for worms and insects as food but many peoples eat them as delicacies. The Navajos will not eat fish. We will not eat dogs. The eating of human flesh is regarded with extreme revulsion by some peoples; to others it is the feast supreme. It would be hard indeed to name an edible substance that is regarded everywhere as food. The aversions and loathings likewise vary. What then can we attribute to "human nature?" Virtually nothing. What a people likes or loathes is not determined by the innate attractions and repulsions of the human organism. On the contrary, the preferences and aversions are produced within the human organism by a culture acting upon it from the outside. Why cultures vary in this respect is another matter; we shall turn to it later on.

Is it human nature to kiss a loved one? If it were, then the practice would be universal. But it is not. There are peoples who do not kiss at all. Some rub noses. Others merely sniff the back of the neck of children. And in some societies a parent or elder relative will spit in the face of a child; saliva is here regarded as a magical substance and this act is therefore a sort of blessing. Among some peoples adult males kiss each other. I once witnessed greetings between men in one of the isolated valleys of the Caucasus mountains. They kissed each other fervently, pushing aside a thick growth of whiskers to reach the lips. Other peoples regard kissing among adult males as unmanly. Where does human nature enter this picture? It does not enter at all. The attitude toward kissing as well as its practice is not determined by innate desires of the human organism. If this were so, kissing behavior would be uniform throughout the world as the organism is uniform. But this is not the case. Behavior varies because cultures differ. You will do, or taboo, what your culture calls for.

Human behavior varies widely at other points. Sexual jealousy is so powerful and so poignant in some societies that to doubt that it is a simple and direct expression of human nature might seem almost absurd. It is "just natural" for a lover to be jealous of a rival. If a man kills the "seducer" of his wife, a jury of his peers may let him go scot free; it was only natural that he should do this, they observe. Yet, we find societies, like the Eskimo, where wives are loaned to guests as a part of hospitality. And Dr. Margaret Mead reports that the Samoans simply cannot understand jealousy among lovers, and find our sentiments in this respect incredible or preposterous.

In some groups premarital sexual intercourse is not only permitted to girls but the practice forms an integral part of the routine of courtship. Out of these intimacies come an acquaintance, a sympathy, and an understanding that make for an enduring marriage. In other groups, brides may be subjected to chastity tests and killed if they fail to pass them. The unmarried mother is stigmatized in some societies, taken for granted in others. Attitude toward homosexuality varies likewise; in some groups it is a mark of shame and degradation, in others it is recognized and accepted. Some societies recognize and give status to a third, or intermediate, sex—the *berdache*, transvestite—in addition to man and woman. A man must avoid his mother-in-law assiduously in some societies; he must not speak to her or allow himself in her presence. In other tribes, a man must have no social intercourse with his sister. Some peoples regard polygamy with aversion, even horror. To marry one's deceased wife's unmarried sister is a crime in some societies, a sacred obligation in others.[2] In none of these instances can we explain custom or institution in terms of the innate desires, sentiments, and aversions of the people concerned. It is not one set of sentiments and desires that produces monogamy here, another set polygamy there. It is the other way around; it is the institution that determines the sentiments and behavior. If you are born into a polygamous culture you will think, feel and behave polygamously. If, however, you are born into a Puritan New England culture you will look upon polygamy with marked disapproval.

There are still other aspects or expressions of the human mind that were once thought to be determined by innate psychobiological factors but which we now recognize as being largely determined by culture. Take the Oedipus complex for example. It was once thought that a boy's hostility toward his father and his love for his mother were simply expressions of his biological nature. But, as Malinowski and others have shown, these attitudes vary with type of family organization. In some societies the husband is not the head of the family, the disciplinarian. It is the mother's brother who takes this role, and the father is merely the kindly, indulgent friend and companion. The attitude of boys toward father and mother are not the same here as in the patriarchal household known to Freud. Polygynous and polyandrous households produce other orientations of attitude. In some cultures it is the sister rather than the mother who becomes the primary object of incestuous desire. The definition of incest, and consequently one's attitude toward sexual union with cross or parallel, first or second, cousins, varies with the culture as we shall see later on.

One's conscience is often thought to be the most intimate personal and private characteristic of one's ego. Here if anywhere one ought to find something that is wholly one's own, a private and unique possession. To an ordinary individual the conscience seems to be a mechanism, an inborn ability, to distinguish between right and wrong, just as he possesses a mechanism for distinguishing up from down, the vertical from the horizontal. Except, perhaps, that conscience seems deeper within one, a more intimate part of one's make-up, than semi-circular canals. After all, these canals are merely a mechanical device, whereas a conscience is an integral part of one's self, one's ego. Yet, for all the conviction that immediate experience carries, we can still be tricked by illusion. And this is exactly what has happened in the present instance. Our sense of balance, our distinction between up and down, is indeed a private faculty; it is built into our psychosomatic structure and has no origin or significance apart from it. But our conscience has a sociocultural origin; it is the operation of supraindividual cultural forces upon the individual organism. Conscience is merely our experience and our awareness of the operation of certain sociocultural forces upon us. Right and wrong are matters of sociocultural genesis; they are originated by social systems, not by individual biological organisms. Behavior that is injurious, or thought to be harmful, to the general welfare is *wrong*; behavior that promotes the general welfare is *good*. The desires inherent in an individual organism are exercised to serve its own interests. Society, in order to protect itself from the demands of the individual as well as to serve its own interests, must influence or control the behavior of its component members. It must encourage *good* behavior and discourage the *bad*. It does this by first defining the *good* and the *bad* specifically, and secondly, by identifying each *good* or *bad* with a powerful emotion, positive or negative, so that the individual is motivated to perform good deeds and to refrain from committing bad ones. So effective is this socio-psychologic mechanism that society not only succeeds in enlisting individuals in the cause of general welfare but actually causes them to work against their own interests—even to the point of sacrificing their own lives for others or for the general welfare. A part of the effectiveness of this social mechanism consists in the illusion that surrounds it: the individual is made to feel that it is *he* who is making the decision and taking the proper action, and, moreover, that he is perfectly "free" in making his decisions and in choosing courses of action. Actually, of course, this still small voice of conscience is but the voice of the tribe or group speaking to him from within. "What is called conscience," says Radcliffe-Brown, "is . . . the reflex in the individual of the sanctions of the society." [3] The human organism lives and moves within an ethical magnetic field, so to speak. Certain social forces, culturally defined, impinge upon the organism and move it this way and that, toward the good, away from the bad. The organism experiences these forces though he may mistake their source. He calls this experience conscience. His behavior is analogous to a pilotless aircraft controlled by radio. The plane is directed this way and that by impulses external to it. These impulses are received by a mechanism and are then transmitted to motors, rudders, etc. This receiving and behavior-controlling mechanism is analogous to conscience.

That conscience is a cultural variable rather than a psychosomatic constant is made apparent of course by a consideration of the great variation of definition of *rights* and *wrongs* among the various cultures of the world. What is right in one culture may be wrong in another. This follows from the fact that an act that will promote the general welfare in one set of circumstances may injure it in another. Thus we find great variety of ethical definition and conduct in the face of a common and uniform human organism, and must conclude therefore that the determination of right and wrong is social and cultural rather than individual and psychological. But the interpretation of *conscience*, rather than custom and mores, in terms of social and cultural forces serves to demonstrate once more that the individual is what his culture makes him. He is the utensil; the culture supplies the contents. Conscience is the instrument, the vehicle, of ethical conduct, not the cause. It is well, here as elsewhere, to distinguish cart from horse.

The *unconscious* also is a concept that may be defined culturologically as well as psychologically. Considered from a psychological point of view, "the unconscious" is the name given to a class of determinants of behavior inherent in the organism, or at least, having their locus in the organism as a consequence of the experiences it has undergone, of which the person is not aware or whose significance he does not appreciate. But there is also another class of determinants of human behavior of which the ordinary individual may be—and usually is—unaware, or at least has little or no appreciation of their significance. These are extra-somatic cultural determinants. In a general and broad sense, the whole realm of culture constitutes "an unconscious" for most laymen and for many social scientists as well. The concept of culture and an appreciation of its significance in the life of man lie beyond the ken of all but the most scientifically sophisticated. To those who believe that man makes his culture and controls its course of change, the field of cultural forces and determinants may be said to constitute an unconscious—an extra-somatic unconscious.

The unconscious character of the operation of culture in the lives of men can be demonstrated in many particular instances as well as in a general way. A moment ago we distinguished the unconscious factor in ethical behavior. The determinants of ethical behavior—why, for example, one should not play cards on Sunday—lie in the external cultural tradition. The individual, however, unaware of either the source or the purpose of the taboo, locates it in his inner self: his conscience is but the screen upon which the unconscious factors of society and culture project themselves.

Incest is defined and prohibited in order to effect exogamous unions so that mutual aid may be fostered and, consequently, life made more secure for the members of society. But of the existence and significance of these cultural factors all but a few are unconscious. To the individual, incest is simply a sin or crime that is inherently and absolutely wrong.

Or, take the rules of etiquette: A man in a certain society is not permitted to wear earrings or to use lipstick. The purpose of these restrictions is to define classes of individuals within society: a man, woman, priest, etc., is an individual who behaves positively in a certain manner and who must refrain from certain kinds of acts. By means of these definitions, prescriptions, and prohibitions, each individual is made to conform to his class and the classes are thereby kept intact. Thus, order is achieved in society, order both structurally and functionally. And, to conduct its life effectively a society must have order. But the individual seldom has any appreciation of the source and purpose of these rules; he is apt to regard them, if he thinks about them at all, as natural and right, or as capricious and irrational. Another example of the cultural unconscious.

The church is an organ of social control; it is a mechanism of integration and regulation. In this respect it has political functions just as does the State. It operates to preserve the integrity of society against disintegration from within and against aggression from without. It is thus an important factor in a nation's war machine; it mobilizes the citizenry to fight against foreign foes. It must also strive to harmonize conflicting class interests at home. This it does frequently by telling the poor and the oppressed to be patient, to be satisfied with their lot, not to resort to violence, etc.[4] In these ways the Church like the State exercises political functions that are essential to the life of the society. Yet how many members of a congregation or of the clergy have any awareness of this aspect of the rituals, paraphernalia, theology, and dogma that occupy them?

The determinants of our form of the family lie so deep within our cultural unconscious that even social science has yet no adequate answer to the question why we prohibit polygamy. The Chinese, according to Kroeber, were long unaware that their language had tones. "This apparently simple and fundamental discovery," he says, "was not made until two thousand years after they possessed writing, and a thousand after they had scholars."[5] And they might not have made it even then had not "the learning of Sanskrit for religious purposes . . . made them phonetically self-conscious." Like the rustic who had been talking prose all his life without realizing it, the peoples of the Western world, too, have long been unconscious of much of the structure and processes of Indo-European languages.

Thus, in addition to the determinants of behavior that lie deep within the tissues of our own organisms, below the level of awareness, there is another class of determinants of which we are equally unconscious: forces and factors within the extra-somatic cultural tradition. The science of culture is endeavoring to discover, define and explain these unconscious cultural factors as psychoanalysis has undertaken to explore and make known the intra-organismal unconscious.[6] We may illustrate these two realms of the unconscious in the following diagram:

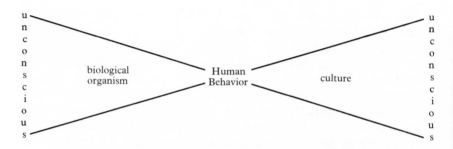

Human behavior is a function of the biological organism on the one hand, and of the extra-somatic cultural tradition or process, on the other. The individual is more or less aware of some of the determinants of his behavior in each category, the cultural and the biological. But of others he is quite unaware, or has no adequate appreciation of the role they play as determinants of his behavior. These are the realms of the unconscious: the biological and the cultural.

The nature of the relationship between the mind of the individual human organism on the one hand and the external cultural tradition on the other may be illuminated by a critical examination of a certain thesis widely held in recent and current anthropological circles in the United States. Briefly stated, this thesis asserts that man has created culture, that culture is the accumulated product of the creative acts of countless individuals, that the individual is the *fons et origo* of all cultural elements, and, finally, that the culture process is to be explained in terms of the individual.

Thus Ralph Linton writes: ". . . the individual . . . *lies at the foundation of all social and cultural phenomena.* Societies are organized groups of individuals, and *cultures* are, in the last analysis, *nothing more* than the organized and repetitive *responses* of a society's members. For this reason the *individual* is *the logical starting point* for *any investigation* of the *larger configuration*"[7] (emphasis ours). "If we had the knowledge and the patience to analyze a culture retrospectively," says Goldenweiser, "*every element* of it would be found to have had its beginning in the creative act of *an individual mind*. There is, of course, *no other source* for culture to come from. . . . An *analysis of culture*, if fully carried out, *leads back to the individual mind*"[8] (emphasis ours). Edward Sapir asserts that the "currency [of 'any cultural element'] in a single community is . . . an instance of diffusion that has radiated out, at last analysis, from a single individual."[9] Ruth Benedict declares that "no civilization has in it any element which in the last analysis is not the contribution of an individual. Where else could any trait come from except from the behavior of a man, woman or a child?"[10] Clark Wissler said that "the inventive process resides in individual organisms; so far as we know, it is a function of the individual organism."[11] Linton asserts that "it is the individual *who is responsible* in the last analysis, for *all additions to culture*"[12] (emphasis ours). Hallowell finds the conception of *cultural* influence unrealistic; "In the last analysis," he says, "it is *individuals* who respond to and influence one another."[13] Both Goldenweiser and Malinowski place the individual "at the beginning and the end" of the sociocultural process.[14] And, finally, we cite Sapir's categorical dictum: "It is *always* the *individual* that *really thinks* and *acts* and *dreams* and *revolts*."[15]

The import of the foregoing is clear. It is the individual who "is responsible" for culture change; it is the individual who *really* does things; every cultural element has its beginning in the creative act of an individual mind, etc., etc. It would appear from our quotations that their authors feel that they are expressing a fundamental proposition and point of view. Nearly all of them use the phrase "in the last analysis" in setting forth their position. Their premises seem to appear to them so simple and so realistic as to be virtually axiomatic: "Every cultural element originates in the mind of an individual—of a man, woman, or child. Where else *could* it come from?" Culture is pictured as a great structure built by countless individuals, much as a coral reef is produced by myriads of marine organisms during the course of time. And, as the coral reef is explained in terms of the activities of marine organisms, so culture may be explained by citing the "creative acts of the individual human mind."

This view seems plausible enough: as a matter of fact, it appears to be virtually self-evident. Anyone can see for himself that it is man, human individuals, who chop down trees, build houses, pass laws, write sonnets, worship gods, etc. But we have become a bit wary of the self-evident and the obvious: anyone can see for himself that it is the sun, not the earth, that moves. But, thanks to Copernicus, we now know better.

Obvious and self-evident though the proposition that culture is made by individuals may appear to be, we must reject it as a means of explaining cultural processes or traditions. As a matter of fact, we regard it as an expression of the primitive and prescientific philosophy of anthropomorphism. Man has been explaining the world he lives in by attributing its existence and nature to the action of some mind, his own or a god's, for ages on end. William James accounted for machines, instruments, and institutions by asserting that they "were flashes of genius in an individual head, of which the outer environment showed no sign."[16] To Newton "this most beautiful cosmos could only proceed from the counsel and dominion of an intelligent and powerful Being."[17] To Plato, the material world was but the expression of "ideas in the mind of God." "Let there be light," said Yahweh, "and there was light." In the mythology of ancient Egypt, everything came from the thinking and willing of the great artificer deity, Ptah.[18] Among our preliterate Keresan Pueblo Indians, Tsityostinako, or Thought-Woman, brought things to pass by acts of thought and will.[19] And today, in line with this ancient and primitive philosophic tradition, we are told that culture has issued from the mind of man—of men, women, and children—and therefore if we are to understand culture and explain its content and course of change, we must do so in terms of the individual.

It is obvious, of course, that culture has emanated from the organisms of human beings: without the human species there would be no culture. We recognize also that a *generic* relationship obtains between culture as a whole and the human species in its, or their, entirety; the general character of culture is an expression of the biological properties of the human species. But, when it comes to an explanation of any particular culture—and all the cultures of the world are particular, specific cultures—or to an explanation of the process of culture change in general, a consideration of the human organism, either in its collective or individual aspects, is irrelevant. The culture process is not explainable in terms of races, physical types, or of individual minds. It is to be explained in terms of culture itself. In short, the culture process is to be explained culturologically rather than biologically or psychologically.[20]

Thus we do not account for differences between Chinese and Swedish culture by appeal to the physical, somatological, and innate psychological differences between the Chinese and the Swedish peoples. We know of no differences between cultural traditions, no specific feature of the culture process, that can be explained in terms of innate biological properties, physical or mental. On the other hand, we can explain the human behavior of Chinese and Swedish peoples as biological organisms in terms of their respective cultures.

The proposition just enunciated is generally accepted in the social sciences today. We no longer subscribe to racial explanations of culture. But the thesis that the sociocultural process is explainable in terms of individuals rests upon the same premise, namely, that biological factors are relevant to interpretations of the culture process. Thus, it is admitted that the biological factor is extraneous to an interpretation of the culture process when taken in its collective (i.e., racial) aspect, but, many scholars contend, it is not only relevant but fundamental when taken in its individual aspect. We regard this reasoning as unsound; a single individual organism is as irrelevant to an interpretation of the culture process as a group of individuals.

It might be well at this point to draw a distinction between two fundamentally different propositions. The individual himself is not irrelevant to the actual culture process. On the contrary, he is an integral and in one sense a fundamental part of it. Individuals do indeed enamel their fingernails, vote, and believe in capitalism as Lynd has observed. But *the individual is irrelevant to an explanation of the culture process*. We cannot explain the culture trait or process of enameling nails in terms of innate desire, will, or caprice. We can however explain the behavior of the individual in terms of the culture that embraces him. *The* individual, the average, typical individual of a group, may be regarded as a constant so far as *human*, symbolic behavior is concerned. The typical Crow Indian organism may be regarded as biologically equivalent to the typical English, Zulu, or Eskimo organism so far as his capacities and inclinations for human behavior are concerned. The alternative to this proposition is acceptance of a racial determinant of human behavior and culture. In the process of interaction between the human organism on the one hand and the extra-somatic cultural tradition on the other, the cultural factor is the variable, the biological factor the constant; it is the cultural factor that determines the variations in the resulting behavior. The human behavior of the individual organism is therefore a function of his culture. The individual becomes then the locus of the culture process and the vehicle of its expression. Thus we arrive at a culturological conception of individuality to add to those of anatomy, physiology, and psychology.

NOTES

[1] Thomas Wolfe, *You Can't Go Home Again* (New York: Sun Dial Press, 1942), pp. 432–433.

[2] "... in modern England ... marriage with a deceased wife's sister became equivalent to incest and the thought of such marriage was defined as 'psychic incest.' ... Around the year 1850, when Lord Russell's bill for the repeal of the law against such marriages was being debated, countless sermons were preached and thousands of pamphlets and letters were printed protesting against repeal:

" 'It would be difficult (says Lecky) to overstate the extravagance of language employed. ... One gentleman (Lord Hatherley), who had been Lord Chancellor of England, more than once declared that if marriage with a deceased wife's sister ever became legal "the decadence of England was inevitable," and that for his part he would rather see 300,000 Frenchmen landed on the British coasts.' " (Wm. I. Thomas, *Primitive Behavior* [New York: McGraw-Hill, 1937], pp. 196–197.)

Contrast this with the command, in *Deuteronomy* (XXV: 5–12) that a man shall marry his deceased brother's wife. Should he refuse, the woman shall disgrace him publicly, taking off his sandal "in the presence of the elders ... and spit in his face." Note, also, that Onan was killed by the Lord for avoiding his duty to his deceased brother's widow (*Genesis* XXXVIII: 6–11).

[3] "Sanction, Social," *Encyclopedia of the Social Sciences*, Edwin R. A. Seligman, ed. (New York: Macmillan), XIII, 531. See also Emile Durkheim, *The Rules of Sociological Method*, trans. by S. A. Solvay and J. H. Mueller (Chicago: Univ. of Chicago Press, 1938), pp. 1–2.

[4] "Religion teaches the laboring man and the artisan to carry out honestly and fairly all equitable agreements freely entered into; never to injure the property, nor to outrage the person, of an employer; never to resort to violence in defending their own cause, nor to engage in riot or disorder ..." (Pope Leo XII's Encyclical on Condition of Labor, May 15, 1891, *The Official Catholic Year Book Anno Domini* [New York: P. J. Kenedy & Sons, 1928]), p. 540.

[5] *Configurations of Culture Growth* (Berkeley: Univ. of California Press, 1944), p. 224.

[6] Kroeber has a fine appreciation of the unconscious character of cultural determinants of human behavior as the section "Unconscious Factors in Language and Culture" in his *Anthropology* (New York: Harcourt, Brace, 1923), pp. 125–131, makes clear. But despite certain examples which he cites and which show quite clearly that the locus of the unconscious

is in the culture process, he locates it "in the mind." Thus he says: "It is difficult to say where the creative and imitative impulses of fashion come from; which, inasmuch as the impulses obviously reside somewhere in human minds, means that they spring from the unconscious portion of the mind" (p. 127).

[7] R. Linton, *The Cultural Background of Personality* (New York: Appleton-Century, 1945), p. 5.

[8] Goldenweiser, *History, Psychology, and Culture* (New York: A. A. Knopf, 1933), p. 59: White's emphasis.

[9] "Time Perspective in Aboriginal American Culture" (Ottawa: Canada Department of Mines, Memoir 90, 1916), p. 43.

[10] R. Benedict, *Patterns of Culture* (New York: Houghton Mifflin, 1934), p. 253.

[11] "Recent Developments in Anthropology," in *Recent Developments in the Social Sciences*, ed. by E. C. Hayes (Philadelphia, 1927), p. 87.

[12] "The Present Status of Anthropology," *Science*, LXXXVII (1938), p. 248.

[13] Hallowell, "Sociopsychological Aspects of Acculturation," in *The Science of Man in the World Crisis*, ed. by Ralph Linton (New York: Columbia Univ. Press, 1945), p. 174.

[14] A. Goldenweiser, "Why I am not a Marxist," *Modern Monthly*, IX (1935), p. 75, and B. Malinowski, "The Group and Individual in Functional Analysis," *American Journal of Sociology*, XLII (1939), p. 964.

[15] "Do We Need a Superorganic?" *American Anthropologist*, XIX (1917), p. 442. We recall at this point Georg Simmel's emphatic assertion: "It is certain that *in the last analysis* only individuals exist" (emphasis ours)—"The Persistence of Social Groups," *American Journal of Sociology*, III (1898), p. 665.

[16] "Great Men, Great Thoughts and the Environment," *Atlantic Monthly*, XLVI (1880), pp. 441–459.

[17] Newton, *Principia Mathematica*, ed. by F. Cajori (Berkeley: Univ. of California Press, 1934), p. 544.

[18] J. H. Breasted, *A History of Egypt* (rev. ed.; New York: Scribner's 1909), p. 357.

[19] L. A. White, "History, Evolutionism and Functionalism: Three Types of Interpretation of Culture," *Southwestern Journal of Anthropology*, I (1945), p. 82.

[20] Ed. Note. As the author explains in other places throughout his work, "to explain culture in terms of culture is merely to say that cultural elements act and react upon one another, form new syntheses, eliminate some elements as obsolete, and so on. Thus a change in the process of forming plurals is a *linguistic* phenomenon, not a psychological or sociological one. Matrilineal organization is a combination of certain cultural elements; patrilineal organization, a synthesis of other elements. We discover the relationship between the manufacture of automobiles and the use of buggies directly, and so on. In short, culture may be interpreted culturologically rather than sociologically or psychologically. More than that, there are many problems that can be solved *only* by culturological techniques, psychological and sociological interpretations being illusory or irrelevant." (p. 201.)

Society and culture: The fallacy of reductionism and the theory of causality

David Bidney

CONTEXTUALIZING COMMENTS

David Bidney's work on *Theoretical Anthropology* constitutes one of the landmarks in contemporary anthropological thought. A measure of the importance of Dr. Bidney's researches is the fact that it was his criticisms which led to a modification in the views of the great American anthropologist, Alfred Louis Kroeber (1876–1960). From a view of culture as a superpsychic reality *sui generis* much like that of White, Kroeber came to consider—owing "to Bidney and his bringing the four Aristotelian kinds of 'causes' to bear on the problem"—that "the efficient causes of cultural phenomena unquestionably are men."[1] But for Bidney, it is not licit even as a purely pragmatic, methodological device to approach the study of culture "as if" it were a closed, self-determined system or level or reality, inasmuch as such an attitude makes it impossible to face what Bidney considers a central issue, namely, the problem of the *locus* and reality of culture. For Bidney, the priority of man in society as the author of culture must be acknowledged, and *per consequens* the ontological priority of social man to culture; and moreover this humanist interpretation of

Reprinted with permission of Columbia University Press from *Theoretical Anthropology* by David Bidney. Copyright © 1953 by Columbia University Press.
[1] Alfred Louis Kroeber, "White's View of Culture," *American Anthropologist*, L (1948), 410.

culture history is the antithesis of the superorganic theory of the autonomy of the cultural reality. The free human person, subject to intelligence and to all those natural conditions which affect human power of action, is, according to Bidney, if not at the center of the cultural process, at least a necessary postulate for an understanding of human life and history.

Society and culture

I. THE FALLACY OF REDUCTIONISM AND THE THEORY OF CAUSALITY

It seems to me that modern social scientists, in their anxiety to avoid the fallacy of reductionism, have presupposed it all along by reducing effect to cause. A cultural effect, they argue, is to be explained only through a cultural cause; the cultural present is to be explained through the cultural past; if culture were to be explained through some noncultural process or phenomenon, that would mean the reduction of cultural to noncultural phenomena. The whole argument presupposes the classic notion so clearly formulated in Spinoza's *Ethics*, namely, that a knowledge of the effect involves a knowledge of the cause, so that an effect may be deduced from its cause. This basic assumption leads straight to monistic, causal determinism and involves ultimately the reduction of all given phenomena to some one underlying causal principle or substance.

If, instead, one were to adopt an emergent theory of causality, according to which a cause is but the limiting condition of an effect, then one avoids the danger and fallacy of reductionism entirely. On this basis, an effect is a newly emergent novelty requiring certain necessary and sufficient conditions for its existence, but the effect cannot be deduced from the cause. There is no a priori logical necessity, as Hume pointed out long ago, for a given cause to produce a given effect; the effect is known only a posteriori, as given in experience. Thus, water, for example, is the effect of a synthesis of hydrogen and oxygen in proper proportions, but the phenotype "water" cannot be deduced from either of these elements taken separately. Similarly, given a human organism and a viable environment, the human being will proceed in time to invent a system of artifacts to minister to his needs, for example, fire, utensils, and so forth. The given artifacts may be explained as the effects of natural causes—man in society working upon the materials of his environment—but these artifacts may not be deduced from these antecedent conditions. Similarly, culture as a whole may be understood as the emergent effect of certain noncultural, causal conditions, but culture would not thereby be reduced to human nature, society, or the environment, since it cannot be deduced from them.

The paradox of modern culture theory lies in the fact that modern ethnologists began by postulating the emergent character of cultural phenomena and insisting upon their autonomy, as may be seen from Kroeber's classic paper on "The Superorganic," but then employed a reductionistic logic or methodology in order to justify their thesis. On this basis, a cultural effect could be explained only through a cultural cause, since to introduce noncultural explanatory factors would be to reduce cultural to noncultural phenomena. Thus, culture became a transcendental entity without roots in the nature of man and society, and yet exercising a strange all-determining power of influence upon both man and society. How this was possible was never explained, since, as has been said, culture was thought of as a closed system and the culture-historical method was incapable of explaining the origin and function of culture itself.

By adopting the theory of causality suggested here, namely, the position that an effect is an emergent of certain necessary and sufficient conditions, but cannot be deduced from a knowledge of the cause, one could safeguard the uniqueness and relative autonomy of cultural phenomena without subscribing to an absolute cultural and historical determinism. Human freedom and cultural determinism may then be reconciled, since it will no longer be necessary, in the interests of a science of culture or of society, to exclude man and society from an active, efficient role as the originators and directors of the cultural process.

II. CULTURE AS ENTITY AND AS PROCESS

The essential point of my criticism of the concept of culture as an autonomous level of reality centers in the argument that culture is not a new kind of entity or substance, but only a new kind of process and human achievement. Ontologically speaking, only substances, that is, things capable of independent existence, are to be conceived as constituting an order, or level, of reality. One may speak of an inorganic level of reality precisely because one may postulate an order of physical things and forces. Similarly, one may refer to an organic level of reality because empirically one may perceive biological entities or organisms capable of existing as distinct entities. Among biological or organic substances one may distinguish between those capable of conceptual thought and language, such as man, and those incapable of it. Ontologically and empirically, there is no distinct psychological or cultural level of reality, although there are distinct psychological and cultural phenomena. In other words, one must

not assume a complete identity between empirical phenomena and ontological reality. A psychological process is something we attribute to some
kinds of animals, but psychological processes, the so-called minds of
animals, are not empirically observable entities comparable to those of
physics or biology. Similarly, cultural processes may be attributed to man
in particular, but cultural phenomena, like mental processes in general,
do not have a substantial, independent reality of their own. Culture is an
attribute of man, a mode of acting and thinking which we attribute to
human enterprise; it is not, therefore, to be thought of as an entity *sui
generis*.

Historically, the notion that social or cultural facts constitute a new,
autonomous level of reality was introduced by Comte and Durkheim.
They argued that because society was to be regarded as a new kind of
entity, not reducible to the aggregates of individuals who compose it,
therefore the products of society and social life, namely, what Durkheim
called collective representations and social facts, were also new kinds of
entities, comparable to those of physics and biology and subject to their
own laws of development. By confusing society with the cultural achievements of society, they were led to postulate natural evolution of social
facts according to fixed laws corresponding to the evolution of society.
This tendency to identify the social and the cultural was continued by the
anthropologists, long after they isolated the culture concept for independent treatment. The notion that there were distinct levels of social
reality and autonomous sciences corresponding to them seemed very
plausible and attractive, as this thought opened the way for the founding
of new sciences. At last man and his culture were to be brought under the
scope of natural science, and the last citadel of animism and freedom
would be brought under orderly control and subjected to the laws of
nature. Any one who opposed this mechanistic, naturalistic approach was
henceforth regarded as a reactionary and an opponent of scientific
evolution.

Thus, the last ghost of "animism" was laid low. Animistic forces were
eliminated not only from the sphere of physical and biological nature but
from the human sphere as well. Not only nature but man also was deprived
of his "anima." The notion that man was a being capable of freedom of
initiative or that the development of his culture had anything to do with
what men in particular thought and willed and strove for was henceforth
dismissed as the outmoded delusions of reactionaries such as philosophers
and theologians; social scientists knew better.

While the impetus to investigate the origin and function of human
institutions and other cultural achievements has added greatly to our
store of knowledge and has provided modern scholars with a degree of
insight into human affairs which previous generations lacked, the gran-
diose claims on behalf of the science of society and the science of man

have not been realized. After some one hundred years, the alibi is still brought forward that the "youth" of these new sciences is responsible for their failure to obtain the necessary information and to establish the promised "laws," but that given sufficient funds and encouragement the promised results would certainly be forthcoming.

The fallacious assumptions underlying these universal claims were scarcely examined. The absurdity of eliminating man himself from the social and cultural process did not seem to matter; all that mattered was the establishment of the reign of law in human affairs—at least in theory. Meanwhile, human affairs have gone from bad to worse, and the past few generations have witnessed a series of social crises such as the apostles of social law and social progress hardly dreamed of in their theoretical utopian schemes.

It is high time, therefore, that we stop to reconsider the foundations of modern sociology and anthropology and to expose some of the fallacious assumptions and arguments upon which they are based. The foremost theoretical assumptions are that every new kind or type of phenomena requires the postulation of an autonomous level of reality subject to its own laws of development and that social and cultural facts require the postulation of distinct levels of social and cultural reality, each of which is *sui generis.* In opposition, I have urged, first, that neither social nor cultural facts constitute a distinct ontological level of reality intelligible through itself alone. As processes involving the active efforts of man in society, cultural and social facts are not ultimately intelligible through themselves alone, although they may be treated as abstractions for immediate purposes of investigation. The notion of the autonomy of social and cultural reality is based on nothing more than a misleading analogy initiated by Comte and Durkheim, whereby empirical kinds of phenomena and levels of reality became identified and confused. This had led to the fallacy of misplaced concreteness, whereby a category of phenomena properly termed an attribute of man in society became converted into an autonomous entity independent of man and society.

Ultimately culture is not intelligible by itself, for the simple reason that culture is a correlative phenomenon, always involving some reference to nature, including man and his geographical environment. One may distinguish at least four variables in the cultural process, namely, human nature, society, geography, and historical experience. Any cultural explanation is an attempt to indicate the limiting conditions of a given cultural phenomenon or pattern by reference to the interrelations of these factors. Because psychology, biology, sociology, or geography may be insufficient, when taken separately, to account adequately for the specificity of a given cultural phenomenon, it does not follow that therefore culture or culture history is sufficient by itself to account for it. In practice, the so-called autonomy of culture leads to the elimination of all explanation of cultural phenomena other than historical. Thus the present meaning of a given cultural process is always "reduced" to the past in an indefinite regress. This leads to an impoverishment of cultural explanation, since the origin of culture itself still remains a mystery not subject to explanation by the historical method. By contrast, I should say that cultural phenomena constitute the most open of systems, requiring complex explanations in terms of many variables. All one can do under given circumstances is to indicate the most probable conditions which underlie a given process or institution, including historical conditions, when these are known or can be reconstructed. Under no circumstances will one be able to "deduce" cultural phenomena from these limiting conditions, but only to indicate how they emerged. The assumption of a universal determinism leads to the confusion of explanation with deduction and prepares the way for a monistic "reduction" of effects to their causes. On the other hand, the postulate that culture is an emergent product of human freedom and creativity in relation to human potentialities and those of the geographical environment precludes the danger of reductionism and deductionism and leaves us free to take into consideration all the complex and variable factors which condition the cultural process.

3

The moral issues

Rationale of this section

We turn our attention now to the impact of the idea of evolution on the socio-cultural life of mankind as men go about the affairs of everyday existence. Here there seem to be three principal dimensions which must be indicated: the notion of morality which seems to be imposed on our thought by the discovery of man's evolutionary origins and natural state (the reading on "Evolution and Ethics" by John N. Deely); the absence of morality in all sectors of the known universe save the human sector, and the ontological or structural reasons inscribed in the nature of things for man's singularity and isolation as the ethical being (the reading on "The Consequences For Action" by Mortimer J. Adler); and the peculiar problems posed for the *homo humanus,* from the standpoint of evolutionary science, by *homo biologicus* (the reading on "Man in Evolution: A Scientific Statement and Some Ethical Implications" by F. J. Ayala).

Evolution and ethics

John N. Deely

I. THE POSSIBLE RELATIONSHIPS

It is often maintained that the idea of evolution either "has very little indeed to say to Ethics,"[1] or serves only to give the lie to any notion of moral absolutes by manifesting the brevity and above all the relativity of whatever we may hold to be of value.[2] More recently, Hans Jonas has pointed out that inasmuch as the idea of evolution forces on us a revision of the idea of nature, "through the continuity of mind with organism and of organism with nature, ethics becomes part of the philosophy of nature."[3] If the 'revision' of our concept of nature required by evolutionary science is indeed such as to make ethics part of the philosophy of nature, there would result "a principle of ethics which is ultimately grounded neither in the autonomy of the self nor in the needs of the community, but in an objective assignment by the nature of things (what theology used to call the *ordo creationis*)—such as could still be kept faith with by the last of a dying mankind in his final solitude."[4]

Without saying so expressly, and apparently without even realizing it,[5] Jonas' suggestion of an essential connection between *ought* and *being* entailed in the reality of mankind evolving is nothing more or less than a suggestion that the moral issues raised by evolution are identical with the issues at the core of the classical tradition of natural law ethics, wherein

---

This essay was specially prepared for the present volume. A variant of it was presented at the 1969 ACPA Convention.

the moral order is understood not as an aggregation of norms, but as an order of tendential existence bound up on every side with the deficiencies of human insight into the conditions and consequences of free and therefore responsible action.

We have then three rather different assessments of the relation between the discovery that mankind has and is evolving, and the questions as to what is right (or wrong) and just (or unjust) in the conduct of human affairs. There is the widely accepted stance of G. E. Moore, that evolutionary science has little or nothing to contribute to moral philosophy, in virtue of the principle that the way things *are* tells us in and of itself nothing at all of how things *ought* to be. Secondly, there is the position of dialectical materialism, of existentialism, and of "situation ethics," which contends that the principal import of the idea of evolution for moral philosophy lies in the realization that values are relative to mankind's *historical* situation (and to the individual's biographical situation) rather than to man's *natural* situation, with the consequence that "justice is based entirely on society."

It arises in the course of men striving to work out a common life, and, if it were possible for men to live apart from society, there would be no justice and no morality at all. Justice, like the morals of society, grows out of the mores of men. It is evolved by them through their efforts to meet the demands of living together.[6]

(This is so because assuming one sought to give an account of justice exclusively in terms of the historicity of man, one inevitably "places emphasis upon the social situation of man and its needs, and not upon man as such," i.e., as individually human, "in any social situation. The sociality, so to speak, that gives rise to justice is a property of special conditions, not a property belonging in any way to man as such. An individual is in no way naturally just."[7] Such an account, in short, inevitably reduces justice to a socio-cultural determinism or convention, "if the conventional is understood to refer to what is entirely the work of society without any natural basis."[8]) Finally, there is the position that "refers practical reason to knowledge of human nature, which is a matter of theoretical reason,"[9] by contending that it is possible for reason reflecting on human experience to discern criteria "which would make it possible to decide whether a certain ethical system of values is in some definite and important sense preferable to another,"[10] that "there are certain ways of behaving which are appropriate to man simply by virtue of the fact that he is a human being,"[11] in the sense that "it is possible to find, by considering the nature of man, an order of goods that are due him";[12] and this is the essential position of natural law ethics. "It presupposes that it makes sense to speak of 'human nature,' that man has a nature as well as a history,"[13] that "there is more to justice than can be caught by the notion of [positive] law or of the social good . . . a right that rests ultimately . . . . on the nature of man . . . underlying both positive law and the social good, and supplying a criterion for them." [14]

It should be clear that the decisive issue between these positions is that of the nature of obligation, or whether there is an intrinsic relation between factual description and moral prescription. [15] It is a question of whether, in addition to being distinct from ("opposed to" in this sense) the *conventional*, the *natural* is further opposed to the *moral* in such a way as to exist indifferent to the good and apart from the order of just action taken morally in itself; for the idea of evolution cannot be separated from the idea of nature without deracination, whence it can have no significant bearing on the idea of ethics if that idea is itself unrelated in every significant sense with whatever idea of nature the data of evolution may be seen to require. It is in just this respect that Jonas' insight is so penetrating.

Ontology as the ground of ethics was the original tenet of philosophy. Their divorce, which is the divorce of the "objective" and "subjective" realms, is the modern destiny. Their reunion can be effected, if at all, only from the "objective" end, that is to say, through a revision of the idea of nature. And it is becoming rather than abiding nature which would hold out any such promise. [16]

On the basis of any extended survey of the evolutionary evidences presented by the full range of contemporary studies—not only those of cultural anthropology and ethnology but also of physical anthropology, not only those of sociology and social and depth psychology but also those of rational psychology, not only those of genetics but also those of embryology and developmental psychology, and as well those of biogeography, ecology, comparative anatomy and physiology, cytology, biochemistry, paleontology, archeology, etc.—in short, on the basis of a cross-sectional sampling of the available materials relevant to understanding man, it is my view that we may fairly dismiss from further consideration all those ethical positions (notably the variants of the existentialism of Sartre or Ortega y Gasset) which rest on the flat assertion that man has no nature, but only a history. In fact, he has both. Therefore, in assessing the moral issues raised by the idea of evolution, or in deciding that this idea of itself raises no moral issues, everything depends on the idea of nature which is before our mind when we pose the question of morality.

The validity of . . . argument that one cannot logically proceed from an 'is' to an 'ought' depends entirely on what is the content of the notion conveyed by 'is'. . . . In fact, any invocation of 'is' other than as a logical copula, involves an epistemology, and it is impossible to reduce the relation of 'is' to 'ought' to a matter of pure logic. [17]

II. EVOLUTIONARY SCIENCE AND THE IDEA OF NATURE

What then, in the present state of evolutionary science, is the philosophical content of the notion conveyed by "is", i.e., what is the idea of nature which the data of evolution tend of themselves to require? Elsewhere, I have tried to manifest that the essential historical thrust of the rise of evolutionary science has been, from an epistemological point of view, toward a restoration and renovation of the Aristotelian explanatory mode, understanding by this the notion of science as a reasoned and not simply mathematized fact, as a fact, that is, seen in the light of its proper causes and not just in the light of a mathematical formula or set of equations.[18] If this thesis is correct, then it is not surprising that evolutionary science seems also to postulate as the ground of its very explanatory power, the Aristotelian notion of "nature" as involving *plurality* (a many each of which is itself one), *teleonomy* (a plurality of "epigenetic systems" or developmental unities), and *relation* between beginning and end (fertilized ovum or "genotype" and mature adult or "phenotype", for example). "Nature, in the physics of Aristotle, signifies entity, essence, whatness, quiddity with a constitutional relation to action, operation, movement, growth, development. A nature is a way of being which does not possess its state of accomplishment instantly but is designed to reach it through a progression *(Phys.* 2.1. 192b, *Met.* 5.4 1014b)."[19]

Who really doubts, asks Simon, that an acorn and an oak tree are related as nature folded and nature unfolded, as a nature in its initial condition and in its accomplished condition?[20] And this opposition of beginning and end is relevant in all consideration of nature. When we say of something that it is "natural," we may be referring primarily to an incipient state of affairs where there is little more than a tendency toward a state of accomplishment; or we may have chiefly in mind the mature state of accomplishment itself; or again we may be referring principally to the dynamic relation between these two states, the ordering of the one to the other, i.e., the transitional tendency itself.

A. Nature and the good

This notion of nature, moreover, which has been or is being largely restored by evolutionary science, just because it essentially implicates plurality, teleonomy and relation, carries with it a particular conception of good which can only be gainsaid if the notion of nature which comports it is itself gainsaid; and it is on this notion of the good, which is co-imposed on our mind with an affirmation of the explanatory validity of the

Aristotelian idea of nature, that the affirmation of an essential connection between ought and being (and therewith the importance of evolution for ethics) depends.[21]

Moral good . . . can only be a species of the good in general. The question then is: When do we call things good and when bad? . . . Good can signify the fitness of a thing for a special purpose, for example, when we speak of a good horse, meaning one fit for pulling loads or one willing and not recalcitrant. But *first and foremost* a horse is good if its organism functions well in all ·respects (including of course the normal intelligence of a horse). This is obviously the case if its nature fulfills the ends inherent in its essential functions, for instance, those of the digestive apparatus, of sight, of hearing, of the nervous system; otherwise it is defective in one way or another and is a bad horse. Thus, *we call things good or bad according to their fitness for carrying out the functions which constitute their nature.* Good in general is, therefore, the perfection proper to a thing . . . understood in an ontological sense. . . . Thus, the good is a mode of being and so a quality of a particular kind.[22]

To understand the manner in which this ontological sense of "good" provides, in the case of human nature, a criterion of morality and a ground of obligation, it is necessary to take explicit note of three of its distinctive features.

First of all, this notion of good is essentially bound up with the dynamism distinctive of the Aristotelian idea of nature. "Good" signifies immediately and of its very nature, or intrinsically and directly, suitability for a given function, aptness to perform. Thus it denotes relation to an end; and while this end by reference to which a thing is said to be more or less good may sometimes be an extrinsic one imposed by human con- sciousness (as when we seek out wood suitable for burning, or for building, etc.), such extrinsic finality is not the primary and ontological sense of "good". In and of itself, "first and foremost," as Messner says, "good" refers to the functioning of a natural unit in relation to its *own* maturation, so that while goodness always comports relation to an end, natural good- ness is distinct from artificial goodness in that the former denotes relation to an immanent end while the latter denotes relation to an extrinsic end set up by consciousness. Natural goodness denotes the spontaneous tendency of a nature to progress towards a state of accomplishment, and this is the primary sense of "good"; artificial goodness denotes the sub- ordination or subversion of nature's spontaneous unfolding to some human design, and this is but a secondary sense of "good."

The second point to be noted with regard to this conception of the good is that it transcends the order of voluntary action. It transcends the distinction between physical and moral.

In other words, the "natural just" of Aristotle—and the same remark holds, by all means, for the natural right and natural law of Thomas Aquinas—ignores the particular meaning that we often attribute to the contrast between the physical and the moral worlds. Between the two communication is insured not only by the notion of nature but also by that of justice, which fundamentally [as the fundamental mode of goodness] signifies adjustment.[23]

This was what Messner had in mind when he observed, following Aristotle and Aquinas, that moral good can only be a species or particular aspect of the good in general. Goodness as an ontological state of nature (tendential and functional, and sustained by a network of relations, be it remembered) is more primary than goodness in the moral sense, inasmuch as moral goodness is possible only on the basis of natural goodness as a specification thereof.

This brings us to the third distinctive feature of the ontological sense of "good," namely, that it varies in every instance. It defines itself differently in different cases, and in every case according to the mature state toward which the natural entity in question of itself tends. Exactly because "the goodness of any given thing consists in its being well disposed according to the mode of its nature,"[24] the ontological sense of the good subdivides in the world of natures into physical and moral goodness, according as a cosmic event is distinguished from a human action by the absence of reason. And according as a cosmic event is distinguished from a human action, the rule of nature's laws is divided into a physical realm and a moral realm.

B. Physical and moral goodness

Why do we sometimes dismiss the case of a man who has killed a fellow human being, asks Simon?[25] Why do we feel that an appalling injustice is perpetrated when a man who seems to have acted under the compulsive power of pathological feeling is declared guilty and punished accordingly? Again, why do we distinguish moral issues from legal issues? Why do we hold men responsible for the rejection of unjust legislation and the personal repudiation of abuses of authority, as was done for example in the proceedings at Nuremberg?

In distinguishing a moral issue from a physical issue on the one hand and from a legal issue on the other, we make reference to the same fact: wherever it is a question of human action, it is to reason that primacy belongs; whereas wherever it is a question of non-human action (many actions of men are not human actions, and no actions of other than rational natural entities are human actions), the primacy goes merely to the functional relationship between the individual and its environment. Thanks to the presence of reason in human life, morality transcends the prime requirement of physical nature to establish the smoothest possible functional relationship between the individual and the environment, and we hold that man to be living below the threshold of his humanness who merely proclaims the reality of the customs and sentiments prevailing in his community and then conforms to them without qualification. "Such blind submission is the docility of a slave, the pusillanimity of a domesticated animal. The *mores* themselves must be judged. They are not the standards of judgment. Rather they are measured by conformity to reason and the practical principles it dictates,"[26] under the guidance of the sole norm absolutely binding human action in any formal or a-priori fashion transcendent to history: to seek the human good—"good is to be done and evil avoided."[27] The morality principle, then, must be the rule of reason in human life.

"Thus by reflecting upon the rational character of what is recognized as 'human action' we come to understand that ruling human action primarily pertains to the reason,"[28] and we find ourselves presented, in the essential tendency proper to human nature to reach its state of accomplishment through a developmental process mediated by reason, with 'a principle of ethics which is ultimately grounded neither in the autonomy of the self nor in the needs of the community, but in an objective assignment by the nature of things.'[29]

It is clear then that the principal subdivision of the ontological good is into moral good and physical good, the former being that tendential or functional excellence proper to human actions, the latter being that functional perfection proper to whatever actions in this universe realize one or another natural tendency in its own order and apart from the concourse of reason. This is the profound sense of Messner's declaration that "natural law is nature for man";[30] and of Julian Huxley's declaration that "for a science of man, the problem is not whether or not to have anything to do with values, but how to devise satisfactory methods of studying them and discovering how they work," since "as soon as it is applied to man," science "finds values among its data."[31]

When G. E. Moore tells us that "there is no evidence for supposing Nature to be on the side of the Good,"[32] we may wonder exactly what notion of "Nature" stands before his mind; but we may be certain that it is not the Aristotelian notion of a fundamental natural unit with a constitutional relation to action, operation, movement, growth, development, and the realization generally of its richest potentialities.[33]

C. The "naturalistic fallacy"

On the other hand, when Moore contends that "what the highest good is Nature cannot determine,"[34] he effectively exposes the sophism of those who would understand human nature in purely physical terms, and he appropriately dismisses their standpoint as a "naturalistic fallacy," which might equally appropriately be termed a "physicalistic (or even 'deter-ministic') fallacy." When we say that "natural law is nature for man," we must have in mind that the play of physical causality, though necessary for human fulfillment, of itself cannot bring about that fulfillment. "Which one is more natural for man," queries Simon: "The nasty and brutish individualism which would follow the collapse of social structures or the relative social integration that we enjoy in a city where no more than about one person a day is shot down?"[35] (He is thinking of Chicago in the late fifties.) By reason of the law of development incarnate in *any* nature *as* nature we must acknowledge:

No doubt, a state of accomplishment is the most natural condition of a nature, for it is that toward which nature has been striving from the beginning and by reason of its identity with itself. Yet, in human affairs principally, the condition that nature is striving toward is not brought about by nature alone but by such causes as under-standing, crafts, arts, sciences, techniques, and above all, good will and wisdom.[36]

Because nature cannot realize its spontaneous inclinations in man except through the mediation of reflective thought (human nature cannot reach its distinctive state of accomplishment apart from the sustained exercise of reason), there is no possibility of equating without further qualification the latest products of biological or socio-cultural evolution with what is best: there is no necessity that what is more evolved in the temporal sense—e.g., Parson's College as contrasted with Plato's Academy—should also be higher and better in an ethical sense, or even (before the bar of reason) higher and better in *any* significant respect. In this sense, Moore is perfectly correct in noting that there is no evidence for sup-posing that nature is on the side of the good.

In another sense, however, there is all the evidence in the world to manifest that nature is on the side of the good, and particularly in the case of man; for "all the dynamism of nature would be missed if our language did not remind us of the relation between the initial and the terminal, the rudimentary and the accomplished, the natural in the sense of that which is just given by nature antecedently to knowledge, craft, and wisdom, and the natural as that which implies the work of intelligence, experience, good will, wisdom, society."[37]

D. Being and the ought

Yet just because the proper excellence of the human person is a work of reason disciplining and ordering the diverse tendencies of his physical being, although it is certain that nature is on the side of the good, what that good is toward which it inclines man, physical nature cannot deter-mine. There we have the connection between being and the ought, and the profound reason why only the preamble and principle of the natural law (good is to be done) and *not the natural law itself* is formally and absolutely speaking transcendent to history.[38] Each man finds himself, in the word of Heidegger, *thrown* into a situation of concrete possibilities, some inherited, some imposed by ·circumstance, some chosen by the individual himself. In the center of this situation, moreover, each one finds himself not allowed to rest, but constantly prodded by inner as well as outer exigencies to action, not just to the action sufficient to achieve the symbiosis of adjustment and survival, but to action tending to the full realization of his humanity so far as conditions and inner resolve will sustain him. "What this implies is that the 'ought' judgment never reduces to a simple 'is' assertion," simply in view of the fact that "basically there is an initial obligation arising from our natural bent to good (natural not in any sense contrasted with rational, for it is not blindly impulsive, but natural in the sense of being an ineluctable tendency of our human.make-up)."[39]

Man possesses one of those ways of being which does not secure its state of accomplishment instantly but through a progression, i.e., he has a nature in the Aristotelian sense. This nature is unique in the respect that it depends for its full flowering in every given situation on self-determination. "Because it is dependent on his reason and his will (self-determination) the conduct required of man by the full actuality of his nature takes on for him the nature of morality,"[40] that is to say, because *through self-determination* a man can stultify or even effectively warp the constitutional relation between what is most natural to man as being most native, most primitive, and what is natural as most just, most adequate to his nature; because one can through self-determination turn the dynamism of one's nature aside from the human good to serve the regime of rivalry and opposition which is "natural" to physical individuality—because of all this, the natural ordering of man to full humanity is governed by natural *moral* law as well as and superordinate to natural *physical* law; because he is necessitated in the moral order only initially but never terminally and circumstantially, because human action depends for its quality on the exercise of judgment, the laws structuring this order differ from physical laws structuring the physical order or from positive laws structuring the social order in that they express not what *will* happen or necessarily what government *wants* to happen under given conditions, all things being equal, but only what *should* happen, if the will is right and wisdom be not wanting. "Otherwise the moral rule would not be a basis for judging men as good and bad, right and wrong in their actions, according as they conform to or transgress the rule,"[41] i.e., the rule of practical reason, "of being *bound* by the good that I *see*," or the constraint of vision upon the will. That is why mankind has not only a nature, in the biological sense, but as well a history, in the cultural sense; "for that same man who is a person, and subsists in his entirety with the subsistence of his soul, is also an individual in a species and dust before the wind."[42]

E. The mutual implication of "natural" and "positive" law

Moreover, it is essential to keep in mind that in speaking of natural law we are in the dimension of ethics or morality rather than in the field of legality in the sense of positive law. It would be a mistake and a complete misunderstanding to confuse the functions of natural law philosophy with the functions of positive law, although the problem of natural law does in a sense begin with the interpretation of positive law.[43] "The theory of natural law is a theory of what makes laws laws, not an easy substitute for making laws, for legislating."[44] "In speaking of natural law as 'law' we should think of it not as lying alongside of, and somehow superior to, all other laws, but as that which is at the heart of, and constitutes the possibility, indeed the obligatory character, of every other law,"[45] provided only it be just; whereas, in the case of unjust laws, "the importance of the theory of natural law is that it affords the possibility of rebellion, it provides a court of appeal, and without it there is no court of appeal beyond the edicts of men."[46]

In this perspective, illumined indeed by the prehistory as well as the history of the human species, it can be seen that, "in the toilsome development of mankind and of reason, in proportion as the normal aspirations of the human personality succeed, under whatever given conditions, in more or less perfectly achieving reality, to the same extent the natural law—taken not as an abstract code but in its historical growth, which is itself natural—tends progressively to make explicit the potential requirements contained within its principles, and the positive law tends in its own sphere to open itself more to the influence of nature;"[47] so that even as moral principles rooted in nature come to fruition only through history, just so "all our concepts of the practical order have a signification that is both rational and historical."[48]

We are brought back again to the suggestions of Hans Jonas on the question of nature and ethics, with which suggestions we began:

True it is that obligation cannot be without the idea of obligation, and true that within the known world the capacity for that or any idea appears in man alone; but it does not follow that the idea must therefore be an invention, and cannot be a discovery. Nor does it follow that the rest of existence is indifferent to that discovery: it may have a stake in it, and in virtue of that stake may even be the ground of the obligation which man acknowledges for himself. He would then be the executor of a trust which only he can see, but did not create.[49]

III. THE ACTUAL RELATION BETWEEN EVOLUTION AND ETHICS

This discussion enables us, I think, to clarify in a decisive way the relation of the idea of evolution to legal and moral philosophy or ethics. In the first place, the morality principle is the rule of reason in human life. This "rule of reason," however, must be seen to be always bound up with and essentially dependent upon the constitutional dynamism of the epigenetic system, through which primitive nature (be it of the individual or the society) is intrinsically ordered to fuller realization. "The great problem of the relations between the conscious and the unconscious," both biological and culturally determined, "will be one of its principal concerns." [50] The "overwhelming importance" of the intellect or reason in this regard "is just that it is an instrument for going beyond the immediate present." [51] On the one hand, with respect to the unconscious workings of the mind, intellect "is misused, or under-used, if it is only employed to provide a verbal justification for some conclusion which has been previously arrived at by non intellectual means;" [52] on the other hand, with respect to the conscious life of the mind, "the intellect also remains under-used if it does not grapple with the major problems which its milieu offers." [53] On both these counts, then, "what is demanded of each generation is a theory of ethics which is neither a mere rationalization of prejudices, nor a philosophical discourse so abstract as to be irrelevant to the practical problems with which mankind is faced at that time." [54]

In the second place, inasmuch as the idea of evolution (the idea that all beings in the physical universe are subject to radical transformations in and through interaction) imposes itself on us as a properly philosophical fact, it must be acknowledged forthrightly that "the idea of evolution is inseparable from the nature and the natural law of man." [55] From this follows an essential dependency of moral and legal philosophy with regard to metaphysics for an ontological understanding of the good in general, and with regard to experience for a practical and normative understanding of the good for man in particular.

In the third place, inasmuch as the mores, the customs and prevailing attitudes of a given society—be it 4th century B.C. Greece or 20th century A.D. America—are not the standards of judgment but must themselves be judged, "in order to obtain a definition of the 'real good' [for man] (i.e., the criterion, 'wisdom') we have to look at the experience of the human world as a whole during its evolution rather than at any particular society;" [56] for "that is 'natural' to man which constitutes him not merely in isolation, but in relation to the whole world-for-man which he creates around him (the highly artificial world of civilization), and in relation to other persons who stand not simply as objects but as other subjects around him."[57] Within this perspective, a threefold tendency or finality seems clearly marked out in the evolution of human nature: one, covering the loftiest ambitions of science, is the mastery of nature by man, the progressive liberation of the human person from the control exercised over him by the vagaries of chance and the inexorability of physical laws; the second, remaining within man himself as the foundation of all that is human, is the development of knowledge in all its degrees and principal dimensions—the rational (where cognitivity is supreme), the artistic or "aesthetic" (where creativity and beauty are supreme), and the ethical or moral (where the control of human action and the direction of human life are the key concerns); and the third, resulting from the impact of culture on biological existence itself and man's material frame, the gradual unveiling of all the potentialities of human nature.[58]

This triple relationship which obtains between evolution and ethics is clear vindication of Jonas' proposition that, from the standpoint of its specific character, and in view of the revision of the idea of nature which the rise of evolutionary science imposes upon us, ethics becomes part of the philosophy of nature.

But it is necessary to realize how unexpected this turn of events really is.[59] It has long been supposed that evolutionistic notions cause a situation strongly unfavorable to the theory of natural law. And long they have (indeed, the idea of evolution is one which can exercise a dissolvent power on the intellect, and insofar merited Goethe's denunciation). Yet with the accumulation of further data, the achievement of new insights and the deepening of initial ones, it turns out that the moral issues posed by the realization of the evolution of man within the natural world are substantially identical with those underlying the classical tradition of natural law ethics. This surprising eventuality has in the evolutionist writings to date reached clearest formulation perhaps under the pen of C. H. Waddington:

The manner in which I have suggested that one can establish a connection between the evolutionary process and ethical beliefs does not involve a circular argument. . . . I argue that if we investigate by normal scientific methods the way in which the existence of ethical beliefs is involved in the causal nexus of the world's happenings, we shall be forced to conclude that the function of ethicizing is to mediate the progress of human evolution, a progress which now takes place mainly in the social and psychological sphere. We shall also find that this progress, in the world as a whole, exhibits a direction which is as well or ill defined as the concept of physiological health. Putting these two together we can define a criterion [for judging between ethical values, for "discussing in a rational manner whether, for example, an ethic which attaches high value to every individual life is preferable to one which condones or approves head-hunting"] which does not depend for its validity on any recognition by a pre-existing ethical belief [in the sense of *mores*].[60]

IV. SUMMARY AND CONCLUSION

We are now in a position to summarize the foregoing discussion of the moral issues which present themselves against the backdrop of the idea of evolution.

Is the underlying moral issue, the foundation of ethical knowledge, unearthed by research into the question "What is good," in the sense of how good is to be defined; or by research into the question, "What is good for man," in the sense of how is the human good to be defined? Yet can the latter be answered if the former is not? "The difficulty at this point derives in large part from the difficulty of ascertaining how an answer can be found to the question being asked. . . . At this point in the controversy, we face the fundamental question of what is to count as moral reasoning and good evidence in deciding questions of value."[61]

If however good is defined differently in different cases, and in every case according to the end in view, the human good will be defined according to the ends of human existence; and on whether there is such an end or ends the reality of ethical knowledge as distinct from theoretical understanding and wishful thinking alike depends.

The question of ethics is the same therefore as the question of whether there are other than accidental goals immanent in the evolutionary unfolding of human nature. Is there anything towards the realization of which human life of itself tends?

This does not however mean that it is necessary to assign an end for the total evolutionary unfolding, or to regard man as the "leading shoot of a cosmic epigenesis toward Point Omega,"[62] before any real answers can be essayed to the question of what is good for man. If that were the case, there would be no ethical knowledge, just as there would be no scientific understanding if this world were one single nature rather than many individuals varied in nature and forming distinct and relatively autonomous unities or wholes.[63] Indeed, just because science is able to provide us with some essential knowledge of various kinds of things—including our own constitution—which lie within the orbit of experience (even though the totality of nature does not lie within that orbit), we may hope to discern some essential inclinations or tendencies which are proper to human nature as such, even though we have not yet succeeded and may never succeed in discerning by reason that God could not have created a very different universe, nor whence this universe has come, or why it has its peculiar history, or just what its future will be.

Let it be said . . . that the eagerness to believe in natural law which is evident in a part of our society may occasion oversimplifications, unwarranted generalizations, and all sorts of illusions quickly conducive to skepticism. . . . Any proposition causes confusion and error as soon as it becomes a cliché. Clichés are obnoxious for a number of reasons but principally because they make things look clear and easy, because they render people unable to perceive the depth of difficulties. In brief, they kill the sense for mystery. Similarly, propaganda at the service of natural law is apt to make people believe that the things right or wrong by nature are easily determined and explained. One of the social functions of philosophers, when they speak of natural law, is to remind men that their own nature, the universe of morality, is no less mysterious than this physical universe.[64]

In one way or another, this point is forcefully pressed by the evolutionists, and most of them today would go at least this far with Simpson: "Ethical standards . . . are relative to man as he now exists on earth. They are based on man's place in nature, his evolution, and the evolution of life, but they do not arise automatically from these facts. . . . Part of their basis is man's power of choice . . . ; after all, if mankind does pursue the ethic of knowledge it should be able progressively to improve and refine any ethical system based on knowledge."[65] Within this line of sight, "the present chaotic state of humanity is not, as some wishfully maintain, caused by lack of faith but by too much unreasoning faith and too many conflicting faiths within these boundaries"—the boundaries of what can be achieved by perception and by reason—"where such faith should have no place. The chaos is one that only responsible human knowledge can reduce to order. This makes rational the hope that choice may sometime lead to what is good and right for man. Responsibility for defining and for seeking that . . . belongs to all of us."[66]

NOTES

[1] George Edward Moore, *Principia Ethica* (London: Cambridge, 1960), p. 58. According to Everett W. Hall in his study of *Modern Science and Human Values* (New York: Delta, 1966), pp. 445–446, "it is of course obvious that the idea of biological evolution, in its Darwinian or in any other specific form, has absolutely no implications for ethics. The obvious, however, is not always seen."

[2] Thus Engels, speaking for both himself and that Karl Marx who wished to dedicate his major writing to Charles Darwin, wrote: "We therefore reject every attempt to impose upon us any moral dogma whatsoever as an eternal, ultimate, and forever immutable moral law on the pretext that the moral world too has its permanent principles which transcend history and the differences between nations."—*Anti-Dühring* (New York: International Publishers, 1935), p. 109. For analogous moral perspectives, cf. Wilhelm Dilthey, *Gesammelte Schriften*, B. Groethysen, ed. (Stuttgart: Teubner, 1958), Vol. III, p. 173; Ruth Benedict, *Patterns of Culture* (New York: Mentor, 1959), pp. 220, 239–240. Further references in Clyde Kluckhohn's study of "Ethical Relativism, Sic et Non," *Journal of Philosophy*, 52 (November, 1955), pp. 666–677.

The existentialist ethics, as is well known, generally seeks its ground in the Sartrean premise that, "to begin with . . . there is no human nature" (Jean-Paul Sartre, "Existentialism," in *The Age of Reason*, ed. by Morton White [New York: Mentor, 1955], p. 124); so that when Simone de Beauvoir contends that "the individual is defined only by his relationship to the world and to other individuals" (*The Ethics of Ambiguity* [New York: Citadel, 1967], p. 156), the word "world" is used not at all in the sense of natural science. Elsewhere she will explain: "It is impossible to propose a morality to man if we define him as a nature, as something given." (Cited without reference by Ignace Lepp in *The Authentic Morality* [New York: Macmillan, 1965], p. 49.)

Similarly for "situational" or "contextual" ethics, although we are assured that it is "*not* existential . . . in the philosophical sense that a Kierkegaard or a Sartre would suppose," and that Pius XII confuses the issue by coupling "situational" with "existential" (in *Acta Apostolicae Sedis*, 44 [1952], pp. 413–419), while Karl Rahner (in *Theological Investigations* [New York: Helicon, 1963], Vol. II, pp. 217 ff., "On the Question of the Formal Existential Ethics") prolongs this confusion, nonetheless—and this is all that is needed for location in the present context—it (situationism) "simply refuses to accept *any* principle, other than neighbour concern, as always binding": Joseph Fletcher, *Moral Responsibility; Situation Ethics at Work* (Philadelphia: Westminster, 1967), pp. 74–75.

[3] Hans Jonas, "Epilogue" to *The Phenomenon of Life*, on "Nature and Ethics" (New York: Harper, 1966), p. 282. See fn. 59 below.

[4] *Ibid.*, p. 283.

[5] This must be said, for logically (if not psychologically) it is an inconsistency and a contradiction to suggest on the one hand that ethics is a part of the philosophy of nature, and on the other hand to contend that "the mere factual discovery of evolution . . . signifies the final victory of nominalism over realism" (Jonas, p. 45), since, from the standpoint of logic, "existentialism is the most radical conclusion drawn so far from the unreservedly accepted victory of nominalism over realism" (*ibid.*, p. 48), and existentialism is not exactly a deepening of natural philosophy. Indeed, at a point in his own book prior to his "Epilogue" on "Nature and Ethics" Jonas had himself questioned whether, "if values are not beheld in vision as being (. . .) but are posited by the will as projects" (p. 232)—which seems to result from any "absolute victory of nominalism over realism" (*ibid.*)—it is still possible to find a road "by which the dualistic rift [of Cartesianism] can be avoided and yet enough of the dualistic insight saved to uphold the humanity of man" (p. 234).

[6] From Otto A. Bird's exposition of "The Social Good Theory of Justice," Ch. 4 of *The Idea of Justice* (New York: Praeger, 1967), p. 86.

[7] *Ibid.*

[8] *Ibid.* See also p. 27. Thus it is necessary to point out here the capital distinction between the *conventional* and the *arbitrary*. One can deny that justice is arbitrary inasmuch as it is necessary for social life, and affirm that it is nonetheless conventional inasmuch as, for conventionality, "what is essential is the subordination of justice to society *in such a way that* all its principles are made ultimately dependent on that society and its needs" (*The Idea of Justice*, p. 90, my emphasis); so that what is "natural" in the sense of non-arbitrary about justice is its extrinsic source or "stimulant," its material aspect, whereas what is intrinsic and formally constitutive of the notion remains through and through conventional and a matter of socio-cultural determinism pure and simple. Cf. the assessments of Leslie A. White in *The Science of Culture* (New York: Grove, 1949), esp. Part II, pp. 120–281 and 330–361.

[9] Johannes Messner, *Social Ethics: Natural Law in the Western World*, trans. by J. J. Doherty (rev. ed.; St. Louis: Herder, 1965), p. 41. One should read however Yves Simon's *caveat* to the effect that "the ideal of a social science which would, in each particular case, procure a rational solution and render governmental prudence unnecessary is thoroughly deceptive" (*The Tradition of Natural Law*, p. 85).

[10] C. H. Waddington, *The Ethical Animal* (New York: Atheneum, 1961), p. 55.

[11] "Foreword" by John H. Hallowell to Yves R. Simon's study of *The Tradition of Natural Law* (New York: Fordham, 1965), p. viii. Cf. John Cogley's introductory essay in *Natural Law and Modern Society*, ed. by John Cogley (New York: Meridian, 1966), pp. 11–28. But note the *caveat* in fn. 15 below.

[12] From the exposition of "The Natural Right Theory of Justice," Ch. 6 of *The Idea of Justice*, p. 149.

[13] Hallowell's "Foreword" to Simon's *The Tradition of Natural Law*, p. viii. It is instructive to read in this connection the following passage from the major study of Theodosius Dobzhansky, *Mankind Evolving* (New Haven: Yale, 1962), p. 18: "Ortega y Gasset has epitomized the point of view of many social scientists and humanists as 'Man has no nature, what he has is history' (quoted in C. Kluckhohn, *Mirror For Man* [New York: McGraw-Hill, 1949]). The polar

opposite is the view of Darlington (*The Facts of Life* [London: Allen & Unwin, 1953]): 'The materials of heredity contained in the chromosomes are the solid stuff which determines the course of history,' and 'the structure of a society rests on the stuff in the chromosomes and on the changes it undergoes.'

"The thesis to be set forth in the present book is that man has both a nature and a "history." Human evolution has two components, the biological or organic, and the cultural or super-organic. These components are neither mutually exclusive nor independent, but interrelated and interdependent. Human evolution cannot be understood as a purely biological process, nor can it be adequately described as a history of culture."

[14] "The Natural Right Theory of Justice," in *The Idea of Justice*, p. 127.

[15] Cf. "The Fundamental Issues of the Controversy," Ch. 2 in *The Idea of Justice*, esp. pp. 27–29: "No theory of justice would deny that questions of justice and injustice may concern matters of conventional agreement among men in society—how they may best distribute their burdens and rewards, what is fair treatment in their common life, and the like. Disagreement does arise, however, when we ask whether justice is ever anything more than conventional, whether it has a natural basis apart from convention or agreement, and whether it makes sense to think of justice apart from leading a common life in organized society.

"The question of natural right thus provides us with a fundamental issue . . . In asking whether justice is based on natural right, it raises the question whether justice as a social norm has a natural basis. The issue is phrased in terms of right, which is a subject on which all theories of justice have something to say, and it asks whether the rights men have are entirely and exclusively conferred by society or whether they also have a natural basis. So understood, this issue serves to divide all theories of justice into two groups: those who deny and those who affirm that justice is based on natural right. The affirmative position identifies the Natural Right theory of justice. It includes the theories that base justice on natural law, although, as we shall see, some theories adopt the Natural Right theory of justice without asserting the existence of natural law." Thus "men may differ about natural law and still agree that justice is in some sense natural and is not entirely conventional."

In this reference, the present essay makes two critical suggestions. The first pertains to the problem of "The obligatoriness of justice" as it poses itself within the Natural Right theory (see *The Idea of Justice*, pp. 151–157). No doubt the affirmation of natural right depends on a general theory of nature which affirms the reality of differences in kind among things as well as differences of degree. It seems to me, however, that recognition of the reality of differences in kind, and in particular, of man as the most radically different and highest kind is not by itself a sufficient account of the source of obligation: see Bird's discussion of the "naturalist fallacy" in the context of the Natural Right theory, *op. cit.*, pp. 153–157; and M. J. Adler's analysis of "To Whom It Makes A Difference," Ch. 16 of *The Difference of Man and the Difference It Makes* (New York: Holt, Rinehart and Winston, 1967), pp. 255–258, notes pp. 355–359.

What I am suggesting is that it is not possible simultaneously to affirm the Natural Right theory of justice and to deny the existence of natural law without falling into the "naturalist" or "naturalistic" fallacy, "best described as the fallacy of deriving an *ought* statement from premises consisting of *is*-statements" (*The Idea of Justice*, p. 32). Bird rightly points out both that and how "the Natural Right position as such . . . is not necessarily guilty of the fallacy" (see pp. 153–157), and he also seems to agree that those Natural Right authors making the source of obligation "depend directly on the kind of being that man is" (p. 154) in such a way as "derives the ethical proposition that men have rights that ought to be respected from the descriptive and non-ethical observation that men have such-and-such a nature" (*ibid.*) are "guilty" of the fallacy in question. My first suggestion then is that the quarrel among the proponents of the Natural Right theory of justice over the obligatoriness of justice and over the nature and source of moral obligation generally is at bottom a quarrel over natural law, and that those proponents of natural right who at the same time deny natural law are in an inconsistent, i.e., untenable, philosophical position (not that all consistent positions are *ipso facto* tenable).

The second critical suggestion the present essay brings forward bears directly on "The Three-Sided Controversy About Justice" (Ch. 7 of *The Idea of Justice*, pp. 163–180) precisely at "the point at which the controversy may be said to come to a standstill. Up to now," comments Bird, "we have been faced with issues on which arguments can be given pro and con. But we have now come to what seems to be more like a question of fact, but a fact of which fundamentally discordant accounts are given . . .All three theories admit, for example, that man has a right to life. The question then arises—Why does he have a right to life? What is its basis? The one theory appeals to natural right, while the other two refuse to admit such an appeal." They maintain "there is only the right that men have come to recognize as a result of legal or social achievement." "The Positive Law theory tends to take the position that . . . moral judgment is . . . finally entirely a question of personal choice and of subjective feeling. The Social Good theory holds that moral judgments have an objective basis, but that this basis is provided, not by the nature of man but by what is needed for men to live together to achieve their social good." (*The Idea of Justice*, pp. 177–178, *passim.*).

One theory appeals to natural right, the other two join in denying the validity of the appeal: "Between these two answers," asks Bird (*The Idea of Justice*, p. 178), "what way is there to decide?" He goes on to observe that "at this point in the controversy, we face the fundamental question of what is to count as moral reasoning and good evidence in deciding questions of value." I suggest in line with this observation that the three-sided controversy about justice comes to a standstill at just this point because it is just at this point that it ceases to be a controversy about justice: in other words the level at which the Natural Right theory of justice joins issue with the Positive Law and Social Good theories is not the level at which the problem of the nature of "moral reasoning and good evidence in deciding questions of value" can be raised as an issue "on which arguments can be given pro and con;" so that just as the philosophical tenability of the Natural Right theory of justice depends on the existence of natural law, so does the three-sided controversy about justice turn in the end on this same underlying issue.

I will return to this point in Section IV of the present essay, its Summary and Conclusion.

[16] *The Phenomenon of Life*, p. 283. See fns. 29 and 55 below. In my opinion, this insight of Jonas is both sharpened and deepened by drawing the contrast between "temporal natures" and "static essences" rather than between "becoming" and "abiding nature," though neither of these sets of contrasting terms are mutually exclusive. (As Dondeyne points out in his assessment of *Contemporary European Thought and Christian Faith*, Pittsburgh: Duquesne, 1963, p. 36: "The sense of becoming, or more precisely the historical dimension of things (the two terms are not synonymous), is perhaps the most characteristic trait of our times." See John N. Deely, "The Philosophical Dimensions of the Origin of Species," *The Thomist*, XXXIII (January and April, 1969), esp. Section I of Part I and Section VIII of Part II; and Raymond J. Nogar, "Evolution: Scientific and Philosophical Dimensions," in *Philosophy of Biology*, V. E. Smith, ed., New York: St. John's University Press, 1962, pp. 23–66, esp. pp. 49–63).

[17] Waddington, *The Ethical Animal*, p. 54.

[18] In addition to Part I of this volume, "The Impact of Evolution on Scientific Method," see "The Philosophical Dimensions of the Origin of Species," esp. Sections III, IV, and VII. See also the essay by Benedict Ashley on "Change and Process" in Part II, Section IV, of this volume, The Metaphysical Issues.

[19] Yves Simon, *The Tradition of Natural Law*, pp. 42–43.

[20] Cf. *ibid.*, p. 47; also pp. 51–54.

[21] "Wherever there is nature there is direction toward a state of accomplishment, and in order to get rid of teleological considerations mechanism has first to replace nature by something else, e.g., extension . . . When we are confronted by a denial that is as stubborn as it is para-doxical, a denial that is unflinchingly maintained although no one can live up to it either in action or in thought, it is always enlightening to inquire into its reasons. The reasons why teleological notions are held suspicious by the scientific mind are numerous. One of the most profound is already familiar to us: there are no natures and no final causes in mathematics. When we watch a geometrical figure or an equation develop its properties, we are aware that it is not in order to achieve a better state of affairs that this equation or this figure is effecting this development. Indeed, "effecting" is here purely metaphorical. The properties of a mathemati-cal essence are not effected by this essence, they are identical with it and all the development takes place in our mind. Accordingly, whenever the interpretation of nature is mathematical, and insofar as it is mathematical, final causes are out of the picture. This is not an accident, and no misunderstanding is involved. The exclusion of final causes from every science where mathematical forms predominate follows upon the laws of mathematical abstraction and intelligibility.

"It is easy to see what the consequences are for a problem like that of natural law. When the Cartesian universe displaces the universe of Aristotle, when a universe made of natures is displaced by a single huge thing, extension, whose parts and their arrangements and re-arrangements lend themselves beautifully to mathematical treatment, we have to deal with a world picture in which teleological considerations are as irrelevant as considerations of color and taste would be in geometry. Of course, we are here supposing an ideal condition that mechanistic science has never actually attained . . . In its factual development, the modern science of nature—in all its parts but especially when it has to deal with living things—has continued to accept a few principles which have nothing to do with mathematics, principles connected with the notion of nature such as it was worked out by the Greeks and best expounded by Aristotle. With all our mechanistic good will, a chemical remains a thing ready to bring about definite effects under definite circumstances. Do you recognize a discreet expression of finality in this notion of readiness? This is how we keep arguing about teleology."

"It goes without saying that there cannot be such a thing as natural law in a thoroughly mechanistic universe. When mechanism is associated with idealism, as it is in Descartes and in most modern philosophers—again, whether outspoken or not—we have values instead of natural laws. Apparently, it is after having played a role of enormous importance in the work of the economists that the notion of value has reached the foreground, the most brightly lighted place in ethical philosophy. A realistic notion of value is not impossible; in a recent book Jacques Maritain did much to show what it would mean (*Neuf leçons sur les notions premières de la philosophic morale*, Paris: Desclée, 1951, in particular pp. 33, 38–66). But in the actual history of modern and contemporary philosophy, values have generally been conceived as placed in things, imposed upon them, forced into them by the human mind. Assuming that we still retain a sense for the distinction between the right and the wrong, what else can we do if things have no nature and no finality of their own? The idealism of the value theory is generally subjectivistic; this is the case, especially, when the ethical theory of values is influenced by the speculations of the economists. In schools of economics it is commonly held that the value of a thing is determined not at all by its relation to good human life but entirely by the willingness of men to pay a certain price for the possession or use of that thing. From a certain standpoint it could be held very reasonably that food rich in carbohydrates and proteins is more valuable than, say, alcohol. Yet it seems to be a lasting convention among economists that the greater value simply coincides with the greater eagerness on the part of the consumer, so that if the majority is ready to sacrifice their biologically normal ration of carbohydrates and proteins in order to procure their full ration of vodka then all we can say is that vodka, in this particular district, has the greater value. Still, in order to understand the history of modern thought, and a few philosophic subjects, we must be aware that there is such a thing as a non-subjectivistic idealism. Working out such an idealism was the task to which Kant dedicated his life, at least from the time he discovered the principles of criticism (see Kant's *Critique of Practical Reason and Other Works on the Theory of Ethics*, trans. by T. A. Abbott, 6th ed.; London: Longmans, Green and Co., 1906). Subjectivistic interpretations of Kant are common —and plausible enough—but actually erroneous. Kant was too much of a philosopher and too honest a man to produce another system of subjectivistic idealism; he dedicated the best of his efforts to reinterpreting the notion of scientific object. Whether he succeeded is another question. At any rate, when we hear today of moral values, esthetic values, social values, political values, spiritual values, etc., we should know where these come from. They come from the mind, they come from outside the things, and they are not embodied in entities, in nature. Thus, 'this has value' does not mean that by reason of what the thing is it is adjusted to something else, to some operation or to some relation: its value is something assigned to it by the mind while, in itself, it remains without value, without nature." (Yves R. Simon, *The Tradition of Natural Law*, pp. 47–48 and 50–51, *passim*). See further Ernest Barker's "Introduction" to *The Politics of Aristotle* (Oxford, 1960), esp. pp. 80–82.

[22] Messner, *Social Ethics*, pp. 15–16.

[23] Simon, *The Tradition of Natural Law*, p. 42. This partially indicates the profound reason why, "within the Natural Right position, the tendency is sometimes felt to identify justice with the whole of virtue." (Bird, *The Idea of Justice*, p. 34). See also Bird's presentation of the tri-aspectual definition of justice accepted as a *datum primitivum* within the tradition of natural law, pp. 124–126; and his presentation of the *debitum*, pp. 150–151; and of justice as a distinctly and specifically moral virtue for the upholders of natural right, pp. 158–159.

[24] Thomas Aquinas, *Summa theologica*, I–II, q. 71, art. 1.

[25] *The Tradition of Natural Law*, p. 78.

[26] Mortimer J. Adler, *What Man Has Made of Man* (New York: Ungar, 1957), pp. 232–233. See David Bidney's penetrating statement in this connection on "The Problem of Human Freedom," in his book *Theoretical Anthropology* (New York: Columbia, 1953), pp. 9–14. Cf. the statement by Charles Darwin "notwithstanding many sources of doubt" on the distinction between "higher" and "lower" moral rules, in *The Descent of Man* (New York: Modern Library edition G27), p. 491.

[27] Cf. Jacques Maritain, *Man and the State* (Chicago: University of Chicago Press, 1951), pp. 89–90: "Natural law is not a written law. Men know it with greater or less difficulty, and in different degrees, running the risk of error here as elsewhere. The only practical knowledge all men have naturally and infallibly in common as a self-evident principle, intellectually perceived by virtue of the concepts involved, is that we must do good and avoid evil. This is the preamble and the principle of natural law; it is not the law itself." See further Columba Ryan, "The Traditional Concept of Natural Law," in *Light on the Natural Law*, ed. by Illtud Evans (Baltimore: Helicon, 1965), esp. pp. 26–29; Thomas Aquinas, *Summa theologica*, I–II, q. 94, esp. arts. 2 and 4. In particular, see St. Thomas' thesis in the *Summa contra gentiles*, III, Ch. 121, as cited in fn. 66 below.

[28] Simon, *The Tradition of Natural Law*, p. 79.

[29] Messner explains this subtle and touchy point in this way: in natural law ethics, "the immutability of natural moral law" means that "there is a *constant* in human nature and existence which remains throughout all historical and cultural change," in the very precise and strictly delimited sense that as soon as man endowed with reason, the *animal rationale*, appeared, "there could no longer be any change in what is fundamentally good or evil . . ." (*Social Ethics*, p. 75: but see the further citation of this passage in fn. 55 below. On the question of a "constant" in human nature raised in the context of contemporary evolutionary theory—the so-called "synthetic theory" of biological evolution and origins—see John N. Deely, "The Emergence of Man: An Inquiry into the Operation of Natural Selection in the Making of Man," *The New Scholasticism*, XL (April, 1966), pp. 141–176. See also the discussion of "The two hierarchies," Section VIII of "The Philosophical Dimensions of the Origin of Species").

John Cogley, in the book *Natural Law and Modern Society*, pp. 19–20, puts the same issue in terms of a threefold proposition: "Fundamentally, the idea of natural law . . . is based on a belief that there exists a moral order which every normal person can discover by using his reason and of which he must take account if he is to attune himself to his necessary ends as a human being. Three propositions, then, are included in the definitions: 1) there is a nature common to all men—something uniquely human makes all of us *men* rather than either beasts or angels; 2) because that "something" is rationality we are capable of learning what the general ends of human nature are; and 3) by taking thought we can relate our moral choices to these ends." (For a thorough discussion of what is involved in the first and most basic of these three propositions, see M. J. Adler, *The Difference of Man and the Difference It Makes*).

[30] *Social Ethics*, p. 44.

[31] *Evolution In Action* (New York: Mentor, 1963), p. 117. See further fn. 37 below.

[32] *Principia Ethica*, p. 57.

[33] See references in fn. 37 below. Cf. also Jacques Maritain's remarks "Concerning the World Considered in Its Natural Structures," pp. 38–40 of *The Peasant of the Garonne* (New York: Holt, Rinehart and Winston, 1968). Also Bird's assessment of "The Natural Right Theory of Justice" in *The Idea of Justice*, pp. 118–160.

[34] *Principia Ethica*, p. 45. See J. J. C. Smart, *Philosophy and Scientific Realism* (New York: Humanities, 1963), and comments thereon by M. J. Adler in *The Difference of Man and the Difference It Makes*, fn. 2 pp. 357–358.

[35] *The Tradition of Natural Law*, p. 52.

[36] *Ibid.*, p. 53. Hence the force of G. E. Moore's contention (*Principia Ethica*, p. 44) that "we must not, therefore, be frightened by the assertion that a thing is natural into the admission that it is good [in any sense that has an import for ethics: see p. 42;] good does not, by definition mean anything that is natural; and it is therefore always an open question whether anything natural is good," again, in the particular sense which has ethical import. Cf. A.-D. Sertillanges, *Le problème du mal*, Vol. I, *L'histoire* (Paris: Aubier, 1948), Vol. II, *La solution* (Paris: Aubier, 1951); J. Maritain, *St. Thomas and the Problem of Evil* (Milwaukee: Marquette, 1942); Charles Journet, *The Meaning of Evil* (New York: Kennedy, 1963).

[37] Simon, *The Tradition of Natural Law*, p. 53. Huxley argues (*Evolution In Action*, pp. 73 and 116–117) that "advance and progress are possibilities of the evolutionary process, and have

been realized to a remarkable extent during the history of life. This is the one major fact which links biological evolution with human values." "It is not true that the nature of things is irrelevant to the interests of man, for the interests of man turn out to be part of the nature of things. Nor is it true that science cannot be concerned with values. Science is a method of inquiry which can be applied in all kinds of fields. In any particular field, it has to deal with the subject matter it finds there. [See "The logic of rational understanding," Section III of "The Philosophical Dimensions of the Origin of Species."] In biology it can do something toward explaining the origins of conscious evolution. But as soon as it is applied to man, it finds values among its data; you cannot either understand or control human affairs without taking them into account. . . .

"And so, in human life, the fact of progress is linked with the problem of destiny, in its dual sense of something to be obeyed and something to be fulfilled. Man alone is conscious of destiny; human organization is so constructed as to make men pose the problem of existence in this form. Ever since he first began, man has been groping to discern the features of his destiny more clearly. In the light of the evidence now available, he could come to the realization that his destiny is to participate and lead in the creative process of evolution, whereby new possibilities can be realized for life.

"To draw this general conclusion appears to me as a real advance in thought. *But it still remains to explore its implications and to study the means by which we might achieve in practice something of this destiny which is being revealed to us.*" (My emphasis.)

Cf. George Gaylord Simpson, *The Meaning of Evolution* (New Haven: Yale, 1949), esp. Ch. XV, "The Concept of Progress in Evolution," pp. 240–262, and Part III, "Evolution, Humanity, and Ethics," pp. 280–337.

[38] See fn. 27 above; and Maritain, *Neuf leçons* . . . p. 23.

[39] C. Ryan, "The Traditional Concept of Natural Law," p. 26. See Bird's presentation in this regard in his discussion of the "naturalist fallacy" in the context of the Natural Right theory of justice: *The Idea of Justice*, pp. 153–157. See also the remarks in fn. 15 above.

[40] Messner, *Social Ethics*, p. 18.

[41] Mortimer J. Adler, *A Dialectic of Morals: Towards the Foundations of Political Philosophy* (New York: Ungar, 1941), p. 19.

[42] Jacques Maritain, *The Degrees of Knowledge*, trans. from the 4th French ed. under the general supervision of Gerald Phelan (New York: Scribner's, 1959), p. 232.

[43] "This much," Bird notes (*The Idea of Justice*, p. 25), "would be generally admitted." See Jean Piaget, *The Moral Judgment of the Child* (London: Routledge, Kegan Paul, 1932). Maritain seems to think that the thought of St. Thomas would introduce an essential qualification here: see his analysis of "The Immanent Dialectic of the First Act of Freedom" in *The Range of Reason* (New York: Scribner's, 1952), pp. 66–85.

In any event, as Simon points out, practically speaking, "the problem" of whether the ground of obligation lies in the common good (but understood as the essential demands of human nature) or the majority will "actually begins with the interpretation of positive law. When we are told that 'this is the law of the land,' we may be satisfied with the practical signification of these words and conclude that we have to conform or risk trouble. But if we care to go beyond such a behavioristic notion of conduct, we have to determine whether, by the law of the land, we primarily mean a rule worked out rationally, which always should be entirely reasonable and which falls short of its nature insofar as it fails to achieve complete reasonableness, or an act which holds because it is born of sovereign will and which, in order to hold, needs no other grounds than the sheer fact that it has been elicited by a sovereign will. The question is whether by the law of the land, we primarily mean a work of public reason or an act of will elicited by the sovereign (whether king or people makes little difference)."— Simon, *The Tradition of Natural Law*, p. 61.

Cf. Abe Fortas' discussion *Concerning Dissent and Civil Disobedience* (New York: Signet, 1968).

[44] Ryan, "The Traditional Concept of Natural Law," p. 19.

[45] *Ibid.* See also pp. 33–34: "At the end of this . . . statement . . . of natural law, it may be asked whether I have not so diminished it as to make it of little value. I have argued that it is not 'law' in the same sense as positive law; that its primary injunctions are purely formal, amounting to little more than discrimination between good and evil; that its derivative precepts are either so general as to provide little guidance to conduct or else so disputable as to win no general consent; and that in its more detailed applications it may be subject to change and exception.

Is there anything left? I think the suggestion that there is not comes from asking and expecting too much of the natural law (as sometimes men ask too much of the existence of God, as if he were there . . . to solve all problems). The natural law does not provide a ready-made yardstick by which to measure other laws; it is not an alternative code which may be consulted to find whether it contains the laws made by men. If it were, there would be no need for the laws of men. . . . But it does provide the moral background or rather the immanent structure whereby good laws may be seen to be good laws, in their making and in their acceptance; it represents the pattern of law as law, discoverable, as patterns are, by those who inquire diligently; it might even be described as the special logic of law, as ever present and necessary to law as logic is to argument, but as little obvious to those who are not trained in it (even when they use it) as logic is to most men when they are engaged in argument. And as we may justify or invalidate an argument by appeal to logic, so we may justify or invalidate the laws of men by appeal to the natural law."

[46] *Ibid.*, pp. 17–18. See Thomas Aquinas, *Summa theologica*, II–II, q. 60, art. 5 (on the source and extent of the binding authority of human legislation; I–II, q. 96, art 4 (on the criteria legislation must conform to in order to be just); I–II, q. 95, arts. 2 and 4 (on the twofold relation of positive to the natural law). Cf. Hegel's *Philosophy of Right*, trans. by T. M. Knox (Oxford, 1945), esp. No. 3. See also Bird's discussion of "The criterion of law" within "The Natural Right Theory of Justice" in *The Idea of Justice*, pp. 137–144.

[47] Jacques Maritain, essay on "Human Equality," in *Redeeming the Time* (London: Centennary Press, 1946), p. 21.

[48] Jacques Maritain, *Neuf leçons . . .*, p. 23.

[49] *The Phenomenon of Life*, p. 282.

[50] Jacques Maritain, *Moral Philosophy* (New York: Scribner's, 1964), p. 451. Also see his essay on "The 'Natural' Knowledge of Moral Values" in *Challenges and Renewals* (South Bend, Ind.: University of Notre Dame Press, 1966), pp. 229–238; and *Creative Intuition in Art and Poetry* (New York: Pantheon, 1953), esp. Ch. III, "The Preconscious Life of the Intellect," pp. 71–105.

[51] Waddington, *The Ethical Animal*, p. 19.

[52] *Ibid.*

[53] *Ibid.*

[54] *Ibid.*

[55] Messner, *Social Ethics*, p. 76. Thus Messner comments (and this is the necessary completion to his remarks as cited in fn. 29 above) that "there exists, notwithstanding the immutability spoken of . . . a further range of *mutability*. We have referred to the 'constant' in human nature and existence: it concerns the fundamentals of the order of truly human existence and the apprehension of the fundamental moral principles of this order. But the fact that the nature of man, of society, and of civilization is subject to far-reaching changes, and involves a very extensive *variable*, can be seen from a glance at the course of development from Neanderthal man to the present day in the region where he once moved. He was unaware of a considerable proportion of the more advanced moral principles which are today the common property of civilized nations. As civilization develops, new demands of natural law arise, so that natural law is itself changeable in its operation. [One of the best and most concise explanations of this first dimension of mutability in the natural moral law is contained in George E. Gerharz' article, "Natural Law and Self-Realization," in *Listening*, 2 (Autumn, 1967), pp. 184–193.] From this *first* kind of mutability of natural law arising from the development of civilization there is to be distinguished a *second* kind, owing to the fact that as circumstances change the application of the same principle leads to different conclusions. [A powerful illustration and explanation of this second dimension of mutability in the natural moral law is provided by Jacques Maritain in his analysis of "The Problem of Means in a Regressive or Barbarous Society" in *Man and the State*, pp. 71–75.] . . . Thus, we come back to the idea of evolution, and at the same time to a *third* kind of mutability in the natural law: the moral consciousness of individual peoples and of mankind as a whole underlies evolution. Even though individual moral 'geniuses' may provide stimuli in this direction, such stimuli have also proceeded from the experience linked with the desire for happiness as driving force in society, from the inquiring mind and from philosophical ethics, from the moral sciences and social movements. In this process, truth and falsehood, justice and injustice have generally been mixed. But, as in all domains of the human spirit, in the realm of moral law mankind progresses through the struggle of truth and of error. And it is not the least proof of the potency of natural law that in

this struggle it always asserts itself, not continuously and without reverses, to be sure, yet in the traceable history of mankind it operates in the formation of social systems more in accordance with the demands of true humanity." See further Maritain, *Man and State*, esp. Chs. 4–7; *The Rights of Man and Natural Law* (New York: Scribner's, 1943); and the UNESCO edited Symposium with an Introduction by Maritain, *Human Rights* (New York: Columbia, 1949).

Cf. Ignace Lepp, *The Authentic Morality* (New York: Macmillan, 1965), esp. Part I, "Fundamental Principles," pp. 3–97; and J. H. van den Berg, *The Changing Nature of Man; Introduction to a Historical Psychology* (New York: Delta, 1964).

[56] Waddington, *The Ethical Animal*, p. 57.

[57] C. Ryan, "The Traditional Concept of Natural Law," p. 23.

[58] Cf. the intriguing study of *Human Potentialities* (New York: Basic Books, 1958) essayed by Gardner Murphy around the thesis that "a cycle of interactions between a changing culture and a changing biological potential will . . . give us a spiral of transformations which in no sense grow directly from the patent trends of today," so that "new ways of acting upon man result in new kinds of humanness," and "cycles of new humanness are laid bare by probing more deeply into the latent structure which was never suspected before." "All the past and existing societies have arisen within a rather narrow range of possibilities in comparison with what can easily be imagined."

[59] And how closely it is linked by the strongest logical and conceptual bonds to the Aristotelian tradition of natural philosophy. These are laid bare most thoroughly perhaps in the classical writings of John St. Thomas (1588–1644). (Though a contemporary of Galileo and Descartes, his completely Aristotelian cast of mind on questions of natural science gives the whole of his work the noetic features typical of classical antiquity without any of those typical of modern times [see Section II of "The Impact of Evolution on Scientific Method," in Part I of this volume]: in the radical change of age which gave birth to the great mathematicians of nature, the last of the great classical philosophers of nature took his resolute stand—"the name of Descartes does not appear in John of St. Thomas' Courses of philosophy any more than the name of John of St. Thomas appears in the works of Descartes," significantly notes Simon). E.g.: "If moral knowledge is considered as animating action, it is seen to be identical with prudence, and within this perspective it does not pertain to the patterns of speculative thought but to those of practical thought. . . . On the other hand, if *moral knowledge* is considered *inasmuch as it establishes an understanding which provides a foundation for the direction of action*, that is to say, as it treats of the nature of virtue, within this perspective it does pertain to the patterns of speculative thought and *is a part of natural philosophy*, for it treats of the distinctive nature of man and consequently is obliged to deal with the moral actions which are consequent on that nature." (John of St. Thomas, *Logica* II. p. Q. XXVII, I, Reiser ed. pp. 826–827). See further Q. I. Art. IV, pp. 276–277.

The fact moreover that the ancients so included in the philosophy of nature all wisdom concerning the things of nature, right up to and including man and ethics, did not prevent them from realizing that from other points of view *scientia moralis* and *scientia naturalis* are irreducibly distinct: see Appendix VII to *The Degrees of Knowledge*, on " 'Speculative' and 'Practical'," pp. 456–464.

[60] *The Ethical Animal*, p. 59. Rephrasing one of his critic's comments, Waddington puts forward his essential position thus: " 'Waddington thinks the answer to the question: "What would it be wise for me to do and for what reason?" can be deduced from the answer to the question "What has the world at large been doing during its history and from what causes?" This is why he thinks that a causal account of how individual ethical judgments have come to be what they are can supply a criterion or rational ground for the judgment between different ethical beliefs which it is wise for us to make. These are two questions and the kinds of answer they seek are of different logical types. Waddington's argument for using the direction of evolution as the criterion for judgment between ethical beliefs rests on acceptance of this." (*Ibid.*, p. 58). Cf. with John Dewey's views in *Human Nature and Conduct* (New York: Random House, 1922).

[61] Bird, "The controversy over natural right" within "The Three-Sided Controversy About Justice," Ch. 7 of *The Idea of Justice*, p. 178. Perhaps then it may be hoped that the present essay will make some contribution toward bringing the debate over justice closer to a "realization of the ideal of rational debate concerning a great idea." (See M. J. Adler's "Foreword" to *The Idea of Justice*, esp. p. xii.)

[62] The view, it seems, of Teilhard de Chardin. See, *inter alia, The Phenomenon of Man* (New York: Harper, 1959), esp. pp. 225–267; *The Future of Man* (New York: Harper, 1964), esp. Chs. XIII, XX, XXII; and Part II of Piet Smulders excellent study, *The Design of Teilhard de Chardin* (Westminster, Md.: Newman, 1967), pp. 89–195.

Yet see Simon's brief remarks on "the connection of the problem of natural law with the problem of God," in *The Tradition of Natural Law*, pp. 62–63.

[63] The point is brilliantly made, if insufficiently developed, by Benedict Ashley in an essay titled "Does Natural Science Attain Nature or Only the Phenomena" in *The Philosophy of Physics*, ed. by V. E. Smith (New York: St. John's University Press, 1961), pp. 63–82.

[64] Yves Simon, *The Tradition of Natural Law*, pp. 39–40. "Our time has witnessed a new birth of belief in natural law concomitantly with the success of existentialism, which represents the most thorough criticism of natural law ever voiced by philosophers. Against such powers of destruction we feel the need for an ideology of natural law. The current interest in this subject certainly expresses an aspiration of our society at a time when the foundations of common life and of just relations are subjected to radical threats. No matter how sound these aspirations may be, they are quite likely to distort philosophic treatments. For a number of years we have been witnessing a tendency, in teachers and preachers, to assume that natural law decides, with the universality proper to the necessity of essences, incomparably more issues than it is actually able to decide. There is a tendency to treat in terms of natural law questions which call for treatment in terms of prudence. It should be clear that any concession to this tendency is bound promptly to cause disappointment and skepticism. People are quick to realize what is weak, or dishonest, in pretending to decide by the axioms of natural law, questions that really cannot be solved except by the obscure methods of prudence, and they gladly extend to all theory of natural law the contempt that they rightly feel toward such sophistry." (*Ibid.*, pp. 23–24). It is of some interest to know that contraception was one of the issues Dr. Simon had in mind while writing thus.

[65] George Gaylord Simpson, *The Meaning of Evolution*, p. 324.

[66] *Ibid.*, pp. 347–348. See Thomas Aquinas' thesis in his *Summa contra gentes*, Bk. III, Ch. 121, "That divine law orders man according to reason with respect to corporeal and sensible things," with the consequence that (*ibid.*, Ch. 122, n. 2) "we do not offend God except by acting counter to the human good" ("Non enim Deus a nobis offenditur nisi ex eo, quod contra nostrum bonum agimus, ut dictum est"—scil., in caput 121).

The consequences for action

Mortimer J. Adler

CONTEXTUALIZING COMMENTS

Dr. Adler is so well known a figure in contemporary intellectual life as to hardly need an introduction. As associate editor with Dr. Robert Hutchins of Britannica's 54-volume set, *Great Books of the Western World,* it was Adler who conceived and executed the assemblage of the two-volume *Syntopicon,* the first basic reference work in the great ideas of the Western tradition. Later, as Director of the Chicago-based Institute for Philosophical Research, he produced the two volume comprehensive survey of *The Idea of Freedom,* which remains the indispensable reference work on that topic.

In the present reading, selected from Dr. Adler's most recent book, we have an assessment of the effects upon conduct, and on the principles governing conduct, for one who consistently maintains one or another answer to the question faced in Section I of this book, the question about how man differs. We return thus to the root question of human uniqueness, of how man differs, but now with a new accent or emphasis: what difference does it make for the practical conduct of affairs, how man differs?

Since according as we answer in one way or another the factual question about how man differs, we sustain or abrogate the distinction between person and thing upon which rests the tenability of the three fundamental Western religions (Judaism, Christianity, Mohammedan-ism), the fundamental position of Western jurisprudence, the basic doctrine of the dignity of man and of what justice is, Dr. Adler has little trouble showing that the factual question as to how man differs makes a great deal of difference in every sphere for anyone who does not maintain that *ad hoc* justifications for conduct (rationalizations of what we want to do) are just as good as reasons of principle, that principled and unprincipled conduct are equally good: or, to put the matter positively, for those who maintain "that sound policies for the conduct of our relations with our fellow men and for our quite different treatment of other animals must be based on the nature of man, on the nature of other animals, and on the character of the difference between them."[1]

Since then everything in the discussion to follow depends on a clear grasp of the distinction between principled and unprincipled policies or conduct, it is incumbent on us in these preliminary remarks to make clear what Dr. Adler intends by this distinction.

Unprincipled conduct is conduct based on reasons of expediency; its code is that the end justifies the means. Principled conduct, by con-trast, rests on reasons of principle; it "requires us to subsume the facts of the case under a normative principle that applies . . . without regard to who the parties are and what their momentary purposes may be."[2] In the former case, "no consistency is to be expected in our policies. We can excuse ourselves for doing what we condemn in others on the ground that, even though the facts are the same in both cases, our conduct serves the purpose of the moment, whereas the conduct of others worked in the opposite way so far as we were concerned."[3] In the latter case, "opposite courses of conduct can be justified only by appealing to opposite norma-tive principles; and then the question of which of the conflicting principles is the right one must be faced."[4]

There we have the clear advantage and decisive superiority of conduct that is principled over conduct that is unprincipled: on any appeal to the facts of the case in the light of normative principles, it is impossible to *justify* opposite lines of conduct; whereas *ad hoc* justification shifts from time to time and from case to case, according to whim. The question of *right* cannot arise in *ad hoc* policy and conduct, and the rule of human action bears no inner relationship to reason (see the preceding essay, "Evolution and Ethics," on the distinction between a cosmic event and a human action).

[1] Mortimer J. Adler, *The Difference of Man and the Difference It Makes* (New York: Holt, 1967), p. 257.
[2] *Ibid.*, p. 358.
[3] *Ibid.*
[4] *Ibid.*

When our conduct is *principled*, we must be prepared to defend the soundness of the principles on which we act and if we act on a certain principle in one case, we cannot justify acting in an opposite way in another case in which the facts are the same and the same principle applies. But when our conduct is *unprincipled*, we may concoct a "justification" or explanation of our conduct, if one is called for, and we seldom find insuperable difficulties in the way of rationalizing opposite policies or courses of action, even when the facts are the same in the cases in which, to serve our purposes, we act in opposite ways.[5]

This is the same distinction in the moral sphere which we have already encountered in the theoretical sphere as the difference between *scientific* and *wishful* thinking (see fn. 46 of the Introductory Essay, "The Impact of Evolution on Scientific Method"). Thus, "I have used the phrase 'reasons of expediency' as it is often used in everyday speech to cover those *ad hoc* justifications of action that recommend the action solely on the ground that it serves the purpose at hand."[6] Conversely, "I have used the phrase 'reasons of principle' for those justifications of conduct that appeal to antecedent facts and principles, not merely to consequences (i.e., purposes to be served). Reasons of principle"—and this is why it makes an inescapable difference for reasoned action, how man differs—"never consist of principles alone, but always principles combined with assertions of fact; and the principles involved are always normative rules or prescriptions (i.e., statements about what ought or ought not be done)."[7]

[5] *Ibid.*
[6] *Ibid.*
[7] *Ibid.*, p. 355. "I have called such statements principles in order to distinguish them from the statements of fact with which we associate the given reasons for action, not in order to elevate them to the level of indubitable or incorrigible truth. This use of the word 'principle' conforms to everyday usage. When we say, in view of a person's conduct or character, that he is a 'man of principle', or when we say that 'it is not the money, but the principle that matters,' the kind of principles that we have in mind are moral principles, that is, normative or prescriptive statements about what ought or ought not to be done or sought. In line with this use of words, . . . when I refer to unprincipled conduct, I mean conduct that is justified only by *ad hoc* reasons of expediency (by reference only to the purpose the conduct serves or the desirable consequences one expects from it)." (*Ibid.*, pp. 355–356.)

The mistake of thinking "that the decision whether an entity under consideration (be it a man or a machine) is a person or thing depends entirely on how we act with respect to it, how we wish to treat it, or how we talk about it . . . totally ignores the possibility that the question can be decided entirely by an appeal to observed or behavioral facts" (*ibid.*, p. 356); such argument "proceeds without any regard for the logic involved in the process of determining whether two things differ in degree or in kind, and without any cognizance of the wealth of empirical evidence that is now available to decide how men differ from other things." (*Ibid.* For an exemplification of the type of argument criticized, see Amelie O. Rorty, "Slaves and Machines," *Analysis*, 22, No. 5, N.S. No. 89 [April, 1962], pp. 118–120. Further references in Adler, *loc. cit.*)

Within this perspective, then, "what is called the naturalistic fallacy in moral reason or argumentation consists in attempting to draw normative conclusions from assertions of fact. I commit this fallacy if the only grounds or reasons that I offer for my recommendation that this or that ought or ought not to be done consist of the views I hold concerning the nature of things, i.e., assertions I make concerning the way things are. The nature of things—the way

We saw in the preceding essay that the idea of evolution is conceptually bound up with the idea of ethics, of a moral order, through the tradition of natural law; and that, in this tradition, the reality of the moral universe is predicated on the presence of reason in human action as distinct from a simple cosmic event, since the subdivision of goodness into physical and moral depends in the world of natures on the reality of the difference between a rational nature and a simply physical nature.

If that difference, however, proves in the end to be *apparent* and *not real*, it is not possible to hold in principle with the traditional Western views of human equality and freedom, i.e., responsibility; while if that difference, in the end, while proving real, proves to be *superficial* and *not radical*, it is still not consistent with the idea of personal responsibility, however limited. These antinomies, more than any single factor in contemporary socio-cultural life, underlie the present malaise of Western democracy, our general sense of absurdity and anomie, and our irresolution in dealing with the crises of our age. For, on the one hand (as Adler manifests in the following pages), the revolutionary social and political ideal of human equality which has from the beginning guided the rise of the West and has insisted time and again on violence when its guidance was spurned, depends for its ultimate validity on a radical difference of man; while on the other hand, "in the last hundred years, the altered view of man [as not radically different] has come so generally to pervade the learned world that if now or in the future the immaterialist hypothesis"—the hypothesis that man differs radically in kind, and is not only a personal being, "one who speaks," but also a being capable of *malice* (not just maladjustment) and so of moral responsibility, worthy of praise and blame—"were to be falsified, few would be surprised, and fewer would suffer any serious embarrassment."[8]

Thus, when it comes to moral values and moral standards, the con-

---

things are—does not by itself validly support any normative conclusions, i.e., any statements about what ought or ought not to be done. I do not commit the fallacy of supposing that the nature of things leads to moral conclusions when I employ moral principles in my reasoning and combine these moral principles with statements of fact about the nature of things (i.e., when I combine an ought-premise with an is-premise to arrive at an ought-conclusion). (For further discussion of this point see Adler, *The Conditions of Philosophy* [New York: Atheneum, 1965], ch. 11, esp. pp. 188–195.)

"It is a misunderstanding of the naturalist fallacy to hold that facts have no normative consequences at all. While it is true that facts by themselves (unaccompanied by appeal to normative principles) do not have normative consequences, they do have such consequences when they are subsumed under normative rules; e.g., the fact that A ignored a red light leads to a normative judgment about A's driving only when it is subsumed under the rule that red lights ought to be heeded by drivers." (*The Difference of Man*, p. 357.)

On the question of the ultimate ground for maintaining an incompletely arbitrary character for positive law, and for the distinction between moral issues and legal issues generally (questions of right and questions of custom), see the preceding essay, "Evolution and Ethics."

[8] *Ibid.*, p. 294.

sideration of our present world authorizes us to cite the following remark:

It is a great misfortune that a civilization should suffer from a cleavage between the ideal which constitutes its reason for living and acting, and for which it continues to fight, and the inner cast of mind which exists in people, and which implies in reality doubt and mental insecurity about this same ideal. As a matter of fact, the common psyche of a society or a civilization, the memory of past experiences, family and community traditions, and the sort of emotional temperament, or vegetative structure of feeling, which have been thus engendered, may maintain in the practical conduct of men a deep-seated devotion to standards and values in which their intellect has ceased to believe. Under such circumstances they are even prepared to die, if necessary, for refusing to commit some unethical action or for defending justice or freedom, but they are at a loss to find any rational justification for the notions of justice, freedom, ethical behavior; these things no longer have for their minds any objective and unconditional value, perhaps any meaning. Such a situation is possible; it cannot last. A time will come when people will give up in practical existence those values about which they no longer have any intellectual conviction. Hence we realize how necessary the function of a sound moral philosophy is in human society. It has to give, or to give back, to society intellectual faith in the value of its ideals.[9]

In this regard, and by contrast to the science of the nineteenth century, the steady rise of evolutionary science in the present age strikes a note of optimism. The reality of human freedom, its possible dimensions and its relation to determinism is no longer a fringe issue for the scientific community, since the assumption that scientific knowledge is dependent on the unqualified and universal determinism of nature is no longer unquestioningly made (and not only in recognition of Heisenberg's principle of indeterminacy, itself a physico-mathematical notion: see the Introductory Essay, "The Impact of Evolution on Scientific Method," esp. sec. II—F). For, as a contemporary observer has astutely noted, "the evolutionary sciences speak about man with two quite different voices."[10]

On the one hand, the sciences of nature, joined most recently by the newer sciences of man and society, and by the technology of computers, have sketched out a picture of man—that chance selection of cumulative mutations, each one as blind as the proverbial bat (not so blind, after all, now that we know of radar)—as from stone age to present "an *effect* of nonrational determinants rather than the intentional initiator or *cause* of events."[11] And this is the first voice with which evolutionary science speaks about man: "man is a determined creature through and through, a result of invariant causal interrelations, a factor within a necessitating

[9] Jacques Maritain, *On the Use of Philosophy* (New York: Atheneum, 1965), first essay, "The Philosopher in Society," pp. 11–12.
[10] Langdon Gilkey, "Evolutionary Science and the Dilemma of Freedom and Determinism," *The Christian Century*, LXXXIV (15 March 1967), p. 339.
[11] *Ibid.*, pp. 339–340.

natural process; and consequently in the scientific account of man there is no longer any evidence of or any room for the traditional concept of freedom."[12]

On the other hand, these same sciences, or at least the scientists who practice them and give them voice, have begun now to tell us that the new knowledge of man through evolutionary biology, social and historical psychology, sociology and cultural anthropology can give us control over the destiny of the species and over our own lives. And this is the second voice, dissonant with the first, more practiced voice, with which evolutionary science speaks about man:

Heretofore evolution has been blind and its results have been contingent and unplanned . . . ; now, however, with human knowledge and its resultant possibilities for control over these determining factors, man himself and society, and so the course of evolution and history, are coming under human control and guidance. Thus we are faced at present with crucial possibilities that challenge our moral sense and our rationality, and that call for mature and responsible decisions from all of us. Shall we use this knowledge creatively to form a more rational, more secure, more peaceful and more democratic life for all of us, or shall we use it to destroy ourselves?[13]

"Very rarely," notes this same observer, "are these two views of man," the thoughts behind the voices, "brought together in any sort of conceptual unity; rather, they appear usually in different parts of a book or an article, the determined image dominating the section concerned with 'what we know about man,' and the free image taking over when the future uses of scientific knowledge are discussed."[14]

It is to the possibility of achieving such conceptual unity, then, and to the consequences of failure or inability to do so, that Dr. Adler would draw our attention in the reading to follow. For "the beliefs or disbeliefs of the learned eventually filter down and exert an influence upon the lives and conduct of their fellow men. But quite apart from the doctrines that prevail among the learned, the ultimate resolution of the question about how man differs from other things will make a difference—a serious

---

[12] *Ibid.*, p. 340.

[13] *Ibid.*, p. 340: "At the conclusion of most books and articles on man as understood by the new sciences, the issue of the responsible and creative *use* of our knowledge is pointedly raised over and over, in language which we can only call the traditional language of 'freedom.' It includes such key words as 'responsible,' 'rational,' 'moral,' 'choice,' 'decisions,' 'purposes,' and thus seems to paint a quite different picture of man than did the other voice. *As scientific and technological man,* man is here seen as an initiating moral and rational cause as well as a determined effect of natural and historical forces—which is what the category of 'freedom' has traditionally sought all along to say. To the cynical observer, it might almost seem that the category of freedom, if defended by theologians or philosophical idealists, is inadmissible in a scientific age, but that the same category, voiced by the scientific community, is thoroughly in tune with the aims and aspirations of that community."

[14] *Ibid.*

difference—to the future course of human affairs; for the image that we hold of man cannot fail to affect attitudes that influence our behavior in the world of action, and beliefs that determine our commitments in the world of thought." [15]

The consequences for action

(1)

As an initial step toward determining the consequences for action that flow from asserting or denying man's difference in kind, I propose to examine some contemporary views of the matter—the opinions of a number of scientists and philosophers who have faced up to this problem in one way or another.

Let me present first the warning given us by Dr. John Lilly with regard to the possibility that, in the not too remote future, we will be able to engage in a two-way conversation with the bottle-nosed dolphin. If and when this occurs, according to Dr. Lilly, we will have to attribute to dolphins the same kind of intellectual power that we attribute to men and deny to other non-linguistic or non-conversational animals. In other words, though men and dolphins may differ in the degree of their common intellectual power, they will stand on the same side of the line that divides animals that have such power from animals that totally lack it. Men and dolphins together will differ in kind from other animals.[1]

Would this possible state of facts, if realized, have any practical consequences? Dr. Lilly thinks it would. He writes:

The day that communication is established, the [dolphin] becomes a legal, ethical, moral, and social problem. At the present time, for example, dolphins correspond very loosely to conserved wild animals under the protection of the conservation laws of the United States and by international agreement, and to pets under the protection of the Society for the Prevention of Cruelty to Animals.

[But] if they achieve a bilateral conversation level corresponding, say, to a low-grade moron and well above a human imbecile or idiot, then they become an ethical, legal, and social problem. They have reached the level of humanness as it were. If they go above the level the problem becomes more and more acute, and if they reach the conversational abilities of any normal human being, we are in for trouble. Some groups of humans will then step forward in defense of these animals' lives and stop

---

[15] Adler, *The Difference of Man and the Difference It Makes*, p. 294.

their use in experimentation; they will insist that we treat them as humans and that we give them medical and legal protection.[2]

Let us consider next the view expressed by Professor Michael Scriven of the University of California in his Postscript to an article on "The Mechanical Concept of Mind." He is concerned with the question whether a robot that is successful at playing Turing's game[2a] can also pass the test that would require us to attribute consciousness to it. "With respect to all other performances and skills of which the human being is capable," Scriven writes, "it seems to me clear already that robots can be designed to do as well or better." But with respect to this special performance—the one that would be the test of the robot's consciousness— Scriven says that he was not certain at the time of writing the article; however, in the postscript which he added, he tells us that he is, "upon further deliberation, confident that robots can in principle be built that will pass this test too, because they are in fact conscious."[3]

We need not agree with Scriven's prediction about the behavior of some future robot in order to take account of his comment on the practical consequences of his prediction's coming true. On the outcome of his prediction depends, in his judgment, "not only the question of matching a performance, but . . . also the crucial ontological question of the status of a robot as a person and thence the propriety of saying that it knows or believes or remembers. . . . If it is a person," Scriven goes on to say, "of course it will have moral rights and hence political rights."[4]

I turn next to the reflections of Professor Wilfrid Sellars on what it means to be a person rather than a thing and on the criteria for drawing the line that divides persons from things. Sellars writes:

To think of a featherless biped as a person is to think of it as a being with which one is bound up in a network of rights and duties. From this point of view, the irreducibility of the personal is the irreducibility of the "ought" to the "is." But even more basic than this . . . is the fact that to think of a featherless biped as a person is to construe its behavior in terms of actual or potential membership in an embracing group each member of which thinks itself a member of the group.

Such a group, according to Sellars, is a community of persons. From the point of view of each of us as an individual, the most embracing community of persons to which we belong includes "all those with whom [we] can enter into meaningful discourse. . . . To recognize a featherless biped or dolphin or Martian [Sellars might have added, "or robot"] as a person is to think of oneself and it as belonging to a community"—the group of those who can engage in meaningful discourse with one another.[5]

I call the reader's attention to the criterion of being a person or a member of the community of persons. It is the same conversational test that Lilly and Scriven use for deciding whether dolphins and robots are persons or things. And that same criterion—conversational ability or ability to engage in meaningful discourse—also operates to differentiate man from brute. In other words, the same line that divides man from brute as different in kind also divides person from thing as different in kind. Furthermore, as Lilly, Scriven, and Sellars all point out, how we treat a particular entity depends on which side of that line we place it. These authors would, therefore, seem to be maintaining that a difference in kind has practical—legal, ethical, and social—consequences.

I would like, finally, to add the testimony of another philosopher, Professor J. J. C. Smart. Professor Smart, like Professor Sellars, is a moderate materialist. Each in his own way argues that conceptual thought can be entirely explained in terms of neuro-physiological processes. Hence, both would deny that man differs radically in kind from other animals or machines, and both would affirm the unbroken continuity of nature. But both also appear to maintain that man differs in kind rather than merely in degree from other animals, and that this difference, which is marked by the possession or lack of "conversational ability," also operates to draw a sharp line between persons and things, with the practical consequence of the differential treatment accorded persons and things.[6] Sellars makes all these points more explicitly and clearly than Smart, but it is, nevertheless, instructive to observe Smart moving in the same direction. He writes:

A scientist has to attend seriously to the arguments of another scientist, no matter what may be that other scientist's nationality, race or social position. He must therefore at least respect the other as a source of arguments and this is psychologically conducive to respecting him as a person in the full sense and hence to considering his interests equally with one's own.[7]

The moral obligation of one scientist to another, here recognized by Smart, can be generalized into the moral obligation of one person to another. The other to whom we owe respect, the other whom we ought to treat "as a person in the full sense," is here being defined as the giver or receiver of arguments. Interpreted broadly yet without violence to the essential point, the giver or receiver of arguments is one who can enter into meaningful—one might even say "rational"—discourse. Hence, the line that Smart draws between persons and things is the same line that differentiates man from brute; and, like Sellars and the others, he attaches definite moral consequences—respect and other obligations—to being on one side of this line rather than the other.

(2)

The foregoing reference to the opinions of Dr. Lilly and Professors Scriven, Sellars, and Smart indicates some practical consequences of opposed answers to the question about how man differs from other animals—in kind or in degree only. These writers all assume that the difference in kind that is established by man's having, and by all other animals' lacking, the power of propositional speech is only a superficial difference. They assume, in other words, that the power of conceptual thought can be adequately explained in neurophysiological terms, and that its presence in man and not in other animals can be explained by the size and complexity of the human brain, which is above the critical threshold of magnitude required for conceptual thought.

On this interpretation of the observed fact that linguistic animals differ in kind from non-linguistic animals, *is man a person rather than a thing?* The answer is affirmative if, as suggested by the above-mentioned writers, the line that divides persons from things can be drawn by such criteria as conversational ability, the ability to engage in meaningful discourse, and the ability to give and receive reasons or arguments. By these criteria, men are at present the only beings on earth that are persons. All other animals and machines are things—at least in the light of available evidence. The special worth or dignity that belongs exclusively to persons, the respect that must be accorded only to persons, the fundamental imperative that commands us to treat persons as ends, never solely as means—all these are thought to obtain on this theory of what is involved in being a person.

If in the future we should discover that dolphins, too, or certain robots, are persons in the same sense, then they too would have a dignity, deserve a respect, and impose certain obligations on us that other animals and other machines would not. However, if in the future we should discover that man differs from other animals *only in degree,* the line that divides the realm of persons from the realm of things would be rubbed out, and with its disappearance would go the basis in fact for a principled policy of treating men differently from the way in which we now treat other animals and machines.

Other practical consequences would then follow. Those who now oppose injurious discrimination on the moral ground that all human beings, being equal in their humanity, should be treated equally in all those respects that concern their common humanity, would have no solid basis in fact to support their normative principle. A social and political ideal that has operated with revolutionary force in human history could be validly dismissed as a hollow illusion that should become defunct. Certain anatomical and physiological characteristics would still separate the human race from other species of animals; but these would be devoid of moral significance if they were unaccompanied by a single psychological difference in kind. On the psychological plane, we would have only a scale of degrees in which superior human beings might be separated from inferior men by a wider gap than separated the latter from non-human animals. Why, then, should not groups of superior men be able to justify their enslavement, exploitation, or even genocide of inferior human groups, on factual and moral grounds akin to those that we now rely on to justify our treatment of the animals that we harness as beasts of burden, that we butcher for food and clothing, or that we destroy as disease-bearing pests or as dangerous predators?

It was one of the Nuremberg decrees that "there is a greater difference between the lowest forms still called human and our superior races than between the lowest man and monkeys of the highest order." What is wrong *in principle* with the Nazi policies toward Jews and Slavs if the facts are correctly described and if the only psychological differences between men and other animals are differences in degree? What is wrong *in principle* with the actions of the enslavers throughout human history who justified their ownership and use of men as chattel on the ground that the enslaved were inferiors (barbarians, gentiles, untouchables, "natural slaves, fit only for use")? What is wrong *in principle* with the policies of the American or South African segregationists if, as they claim, the Negro is markedly inferior to the white man, not much better than an animal and, perhaps, inferior to some?

The answer does not consist in dismissing as false the factual allegations concerning the superiority or inferiority of this or that group of men. It may be false that, within the human species, any racial or ethnic group is, as a group, inferior or superior. But it is not false that extremely wide differences in degree separate individuals who top the scale of human abilities from those who cluster at its bottom. We can, therefore, imagine a future state of affairs in which a new global division of mankind replaces all the old parochial divisions based upon race, nationality, or ethnic group—a division that separates the human elite at the top of the scale from the human scum at the bottom, a division based on accurate scientific measurement of human ability and achievement and one, therefore, that is factually incontrovertible. At this future time, let the population pressures have reached that critical level at which emergency measures must be taken if human life is to endure and be endurable. Finish the picture by imagining that before this crisis occurs, a global monopoly of authorized force has passed into the hands of the elite—the mathematicians, the scientists, and the technologists, not only those who make and control machines of incredible power and versatility, but also those whose technological skill has mechanized the organization of men in all large-scale economic and political processes. The elite are then the *de facto* as well as the *de jure* rulers of the world. At that juncture, what would be wrong *in principle* with their decision to exterminate a large portion of mankind—the lower half, let us say—thus making room for their betters to live and breathe more comfortably?

Stressing "in principle," the question calls for a moral judgment.

Validly to make a moral judgment in a particular case, real or imaginary, we must appeal to a defensible normative principle and one that is applicable to the facts as described. Can we do so in the case that we have been imagining? The facts include not only the scientifically measured ranking of individuals according to degrees of ability and achievement, but also the overarching fact that we have been taking for granted for the purpose of this discussion; namely, that it has been discovered that the psychological differences between men and other animals are all differences of degree. With exceptions that constitute a small minority, men have found nothing morally repugnant in killing animals for the health, comfort, sustenance, and preservation of human life. It seems reasonable to regard as morally sound those policies that have the almost unanimous consent of mankind, including its most civilized and cultivated representatives. By this criterion, we must acknowledge the moral validity of the policy that men have always followed with regard to the killing of animals for the benefit of the human race. If that policy is morally sound, it must reflect a valid normative principle. What is it?

It is indicated by the fact that, with the exception of relatively small numbers of scientists and philosophers, the members of the human race have always interpreted and still do interpret the observation that they alone of all animals have the power of speech as signifying not only a psychological difference in kind between themselves and the brutes, but also the psychological superiority of their own kind. Combining this fact with the policy that men have pursued in their treatment of animals, we can discern the normative principle underlying the action. It is that *an inferior kind ought to be ordered to a superior kind as a means to an end;* in which case there is nothing wrong about killing animals for the good of mankind. The same rule applies to other uses of animals as instruments of human welfare.

Now let us alter the picture by introducing into it the supposition with which we began this discussion—the supposition that it has been discovered that men and other animals differ psychologically only in degree. If, on that supposition, we still think it is a morally sound policy to use animals as means to our own good, including killing them, the underlying normative principle must be that *superiors in degree are justified, if it serves their welfare, in killing or otherwise making use of inferiors in degree.* But that principle, once it is recognized to be sound, cannot be restricted to the relation between men and animals; it applies with equal force to the relation between men of superior and men of inferior degree, especially to those who are at the top and at the bottom of the scale of ability and achievement, since the difference in degree that separates them may be as large as the difference in degree that separates the lowest men from the highest animals. Thus we appear to have reached the conclusion that, given only psychological differences of degree between men and other animals, and given a scientifically established ranking of individuals on a scale of degrees, the killing or exploitation of inferior by superior men cannot be morally condemned.

Is there a flaw in the argument? If there is one, it would appear to lie in the illicit substitution of the relation between superiors and inferiors *in degree* for the relation between a superior and an inferior *kind*. I am not prepared to say that the substitution is illicit, particularly if the superior and the inferior in kind are only superficially different in kind and hence in their underlying constitution differ only in degree. But if the normative principle that subordinates inferiors in kind to the good of their superiors in kind is defensible only when the superiors and the inferiors differ radically in kind, then we cannot validly convert that normative principle into a rule governing the action of superiors in degree with respect to inferiors in degree. Since, in the long history of man's reflective consideration of his action with respect to brute animals, the prevailing view of the difference between human beings and non-human animals has always been that it is not only a difference in kind, but also a radical difference in kind, I think it is reasonable to presume that the conscience of mankind has sanctioned the killing or exploitation of animals on this basis, and not on the view that the difference in kind is only superficial. The latter view, as explicitly formulated in this book, represents the position implicitly held by a relatively small number of scientists and philosophers in very recent times. It can hardly be regarded as generating the almost universal moral conviction that there is nothing reprehensible in the killing or exploitation of animals.

The conclusion that we have now reached has both negative and positive corollaries. On the negative side, the practical consequences may be very difficult to live with. If nothing less than the superiority of human to non-human beings that is based on a radical difference in kind between men and other animals can justify our killing and exploitation of them, we are without moral justification for our practices in this regard, should it turn out, as well it may, that the success of a Turing machine in the conversational game decisively shows that the difference in kind is only superficial. Two future possibilities—the one just mentioned or the possible discovery by psychologists that the difference between men and other animals is only one of degree—would leave us with what, after protracted consideration, might turn out to be an insoluble moral problem. We might have to concede that there is no clearly defensible answer to the question whether we ought or ought not to kill subhuman animals. We would then be forced to treat the problem as one of pure expediency, totally outside the pale of right and wrong. And in that case, would not the problem of how superior men should or should not treat inferior men also cease to be a moral problem, and become one of pure expediency? For those of us who still hold on to the traditional belief that moral principles of right and wrong govern the treatment of man by man, the contemplation of that eventuality is as upsetting as the possibility earlier envisaged—that with the discovery that men and other animals differ only in degree, it would be possible morally to justify a future elite in exterminating the scum of mankind in a global emergency brought on by population pressures that exceeded the limits of viability.

The positive corollary reveals that some of our traditional moral convictions rest on the supposition that men and other animals differ radically in kind. When we affirm the equality of all human beings in virtue of their common humanity, and subordinate to that equality all the differences—and inequalities—in degree between one individual and another, that affirmation involves more than simply asserting that all men belong to one and the same kind, which can be anatomically or physiologically identified. It involves the assertion that men differ from other animals in kind, not only psychologically, but also radically in their underlying constitution. Their superiority to other animals by virtue of such a radical difference in kind is that which gives their equality with one another as members of the human species its normative significance—for the rules governing the treatment of men by men as well as for the rules governing the treatment of other animals by men. The revolutionary social and political ideal of human equality is thus seen to depend for its ultimate validity on the outcome of the test that will decide which of the competing hypotheses about man is nearer the truth.[8]

(3)

We have seen that the line we now draw between men as *persons* and all else as *things* would be effaced by the discovery that nothing but differences in degree separate men from other animals and from intelligent robots. But can it be preserved if the difference, while one of kind rather than of degree, turns out to be only a superficial difference in kind and, therefore, one that is ultimately reducible to, or at least generated by, a difference in degree? Can the special dignity that is attributed to man as a person and to no other animal, and can the rights and responsibilities that are usually associated with that dignity, continue to be defended as inherently human if man is not radically different in kind from everything else?

Dr. Lilly and Professors Scriven, Smart, and Sellars have presented us with what I shall call a diminished view of what it means to be a person. For them, men are persons by virtue of their distinctive power of conceptual thought, manifested by propositional speech—and so also will dolphins and robots deserve to be ranked as persons if and when they, too, manifest their possession of conceptual thought by conversational or linguistic performances comparable to man's. In an etymologically warranted sense of the word "rational," talking animals are rational, and non-talking animals are brute; for the Greek word "logos" and its Latin equivalent "ratio" connote the intimate linkage of thought and word that is manifested in propositional speech. But if the difference between rational and brute animals solely and ultimately depends upon a difference in degree that places the talking animal above and the non-talking animals below a critical threshold in a continuum of brain magnitudes, such criteria as conversational ability, ability to engage in meaningful discourse, or ability to give and receive arguments may not suffice to establish men as the only persons in a world of things, with the dignity or moral worth that attaches to personality and with all the moral rights and responsibilities that appertain thereto. This began to become clear in our consideration of the hypothetical case that we explored dialectically in the preceding section. The argument there led us to the conclusion that the age-old prohibition against treating men as we have for ages treated animals, and the basic equality of men that rests not only on their all being the same in kind but also on their superiority in kind to animals, not just superiority in degree, cannot be defended—at least, not adequately—except on the ground that men differ *radically* from other animals and other things.

The reason why this is so can be made clearer by going back to the conception of personality as the bearer of moral worth, moral rights, and moral responsibility, which originated in classical antiquity with Plato and Aristotle and with the Roman Stoics, which developed under the influence of Christianity in the Middle Ages, and which, as reformulated in the eighteenth century, especially by Kant, prevailed in Western thought until very recently. As contrasted with the minimal or diminished view advanced by a number of contemporary writers, the traditional view conceived a person as a rational being with free choice. Rationality by itself—if that is nothing more than the power of conceptual thought as manifested in propositional speech—does not constitute a person. A dolphin or a robot would not have the moral worth or dignity that demands being treated as an end, never merely as a means; would not have inherent rights that deserve respect; and would not have the moral responsibility to respect such rights, if the dolphin or robot was nothing more than a talking animal by virtue of having the requisite brain power for speech. Nor would a man! On the traditional view, a person not only has the rationality that other animals and machines lack; he also has a freedom that is not possessed by them—the freedom to pursue a course of life to a self-appointed end and to pursue it through a free choice among means for reaching that end.

I think the traditional view is correct as against the minimal view that has recently been advanced. Man as a person belongs to what Kant calls "the kingdom of ends" precisely because the end he himself pursues and the means whereby he pursues it are not set for him but are freely appointed and freely chosen by himself. His moral rights and moral responsibility stem from the freedom that is associated with his rationality, not just from his rationality itself. If the power of conceptual thought that constitutes his rationality can, according to the identity hypothesis, be adequately accounted for in neurophysiological terms, then man's rationality does not carry with it the freedom of choice that is requisite for his having the moral rights and responsibility that comprise the dignity of a person. The power of conceptual thought elevates man above the world of sense, the world of the here and now; but the power that elevates him above the world of physical things and makes him a person is the power of free choice, which, as Kant puts it, involves "independence of the mechanism of nature."[9] Such independence can be man's only if the psychological power that is distinctive of man involves an immaterial or non-physical factor and can, therefore, operate with some independence of physical causes.

The freedom of free choice is properly called a "contra-causal" freedom when "contra-causal" is understood not as the total absence of causality, but as the presence of a non-physical causality. This does not mean total independence of physical causes; it means only that the act of free choice cannot be wholly explained by the action of physical causes. As will be pointed out in Chapter 18, one of the theoretical consequences of affirming the materialist hypothesis is the denial of free choice. If the brain is the sufficient condition of conceptual thought and if, therefore, there is no reason for positing an immaterial or non-physical factor as operative in man, then man may have other freedoms, just as brute animals do, but he does not have that freedom of choice which makes him the master of himself and of his destiny—the course he takes in life from beginning to end. Conversely, the affirmation of free choice presupposes the truth of the immaterialist hypothesis, which posits in man the operation of a non-physical factor, needed not only to explain his power of conceptual thought, but also to explain his contra-causal freedom of choice.

The proposition that man differs in kind, not just in degree, from other animals and from machines represents the conclusion that we have reached in the light of all the evidence that is at present available. This proposition may not be overturned by future findings, but if future experiments with Turing machines decisively show that man's difference in kind is only superficial, not radical, the practical consequences would be almost the same as they would be if future evidence showed that man differed only in degree. The distinction between men as persons and all else as things, and with it the attribution of a special dignity and of moral rights and responsibility to men alone, can be sustained only if man's difference is a radical difference in kind, one that cannot ultimately be explained by reference to an underlying difference of degree.

We saw, in the course of the preceding discussion, that the dignity of man as a person and his moral rights and responsibility rest on his freedom to determine the goal he pursues in life and on his free choice of the path by which to attain it. This throws light on the fact that we do not refer to other animals as engaged in the pursuit of happiness. Their goals are appointed for them by their instinctual drives, and the means they employ to reach these goals are provided either by fully developed instinctive patterns of behavior or by rudimentary instinctive mechanisms that require development and modification by learning. If man were just another animal, differing only in the degree to which his rudimentary instinctive mechanisms needed to be supplemented by learning, the pursuit of happiness would not be the peculiarly human enterprise that it is, nor would there by any ethical principles involved in the pursuit of happiness. There can be an ethics of happiness only if men can make mistakes in conceiving the goal that they ought to pursue in life, and can fail in their efforts by making mistakes in the choice of means. Lacking the power of conceptual thought, other animals cannot conceive, and hence cannot misconceive, their goals; only man with the power of conceptual thought can transcend the perceptual here and now and hold before himself a remote goal to be attained.

To this extent, a difference in kind, even if only superficial, is involved in man's concern with living a whole life well, not just with living from day to day. Other animals do not have this problem. This is just another way of saying that they do not have moral responsibility or moral rights. But if there were only one solution to the human problem of living well—the problem of how to make a good life for one's self—and if that solution were determined for each man by causes over which he had no control, then man would not be master of his life, would not be morally responsible for what he did in the pursuit of happiness, and could not claim certain things as his by right because he needed them to achieve his happiness—the happiness he has a right to pursue in his own way.

This last right, the source of all other rights, would not be the fundamental human right that it is, were man not master of his life, not only able to conceive a remote goal toward which to strive, but also able freely to choose between one or another conception of the goal to seek as well as freely to choose the means of seeking it. More than the power of conceptual thought is thus involved in the pursuit of happiness. Freedom of choice is also involved, and with it a radical difference in kind between men and other animals that have no moral problems, no moral rights, and no moral responsibility. Hence, should a Turing machine of the future succeed in the conversational test, as proponents of the materialist hypothesis predict that one will, the moral aspects of human life will be rendered illusory. Of course, unable quickly to shake off the habit of centuries, men may for some time hold onto the illusion that there are better and worse ways to live; but in the long run the truth will prevail, and men will give up the illusion that there is a fundamental difference between living *humanly* and living as other animals live.

This, in my judgment, is the most serious and far-reaching practical consequence of a decision in favor of the materialist hypothesis concerning the constitution of man, and with it a decision that man's difference in kind is only superficial. Only if the immaterialist hypothesis is confirmed by repeated trials and failures of Turing machines in the conversational test, only if man's difference from other animals and machines is a radical difference in kind, will the truth about man sustain a serious concern on his part with the moral problems involved in the pursuit of happiness— the problem of trying to find out what the distinctively human goods are and the problem of engaging by choice in one or another way of life aimed at a maximization of the goods attainable by man.[10]

(4)

One matter mentioned in the preceding discussion deserves further elaboration. It concerns the role of instinct in human life as compared with its role in the life of other animals. The view we take of the way in which man differs from other animals—in degree only, superficially in kind, or radically in kind—directly affects our understanding of the role of instinct in human life; and so, in the first instance, we are concerned with the theoretical consequences of diverse views of the difference of man. But there are practical consequences, too, though they are less immediate; for according as we understand the role of instinct in human life in one way or another, we may be led to adopt one or another practical policy with respect to the alteration or control of human behavior. A striking example of this is to be found in certain recent popularizations of the findings of ethology concerning the instincts of aggression and territoriality that are operative in fish, birds, and mammals. On the basis of those findings, interpreted in terms of the view that man differs only in degree or at most only superficially in kind, the thesis is advanced that the basic patterns of human behavior underlying the institutions of property, nationalism, and war are determined by these same animal instincts; and, being thus instinctively determined, the human institutions in question are unamenable to alteration or eradication as long as man remains the animal he is.[11]

To state the theoretical problem with clarity, a number of distinctions must be made. First, we must distinguish between that which is innate or unlearned, as indicated by its being species-predictable, and that which is acquired or learned, as indicated by its variable presence or absence in individual members of a given species. Second, we must distinguish between those completely formed instinctive mechanisms that operate effectively without the intervention of learning and those more rudimentary instinctive mechanisms that need to be supplemented by learned behavior in order to be effective in operation. And, third, we must distinguish between instinctive mechanisms, on the one hand, both those that are fully formed and those that are rudimentary, and instinctual drives, on the other hand.

The former are patterns of overt behavior; the latter are conative sources of behavior—sources of energy impelling toward certain biological results. Such are the instinctive drives of sexual or reproductive behavior, self-preservative behavior through feeding or flight, aggressive behavior, and associative behavior. These instinctual drives are innate in the sense of being species-predictable; when activated by specific releasing mechanisms, they impel the animal toward specific objects or conditions that constitute satisfactions of the drive and bring about its temporary quiescence. Though quiescent for a time, the instinctual drive remains as a potency to be aroused again, and when aroused it once again activates patterns of behavior seeking its fulfillment. The behavioral means of fulfillment (1) may consist of fully formed instinctive mechanisms, as in the case of the insects without brains or cerebro-spinal nervous systems, and also as in the case of the cerebro-spinal vertebrates with relatively small brains; or, (2) they may consist of rudimentary instinctive mechanisms supplemented in varying degrees by acquired or learned patterns of behavior, as in the case of the higher mammals with relatively large brains; or, (3) as in the case of man, the means of satisfying instinctual drives when they are operative may consist of overt patterns of behavior that are products of learning or intelligence.

While there seems to be no question that the instinctual drives found in the vertebrates and especially in the mammals are also present in man, the prevailing scientific opinion is that man has no fully formed instinctive mechanisms for the satisfaction of these drives, nor even rudimentary ones as in the case of other higher animals. The only species-predictable behavior in a mature human being consists of such simple reflex arcs as the pupillary, the salivary, the patellar, or the cilio-spinal reflex, together with such involuntary innervations as are produced by the action of the autonomic and sympathetic nervous systems. Men are impelled to overt behavior of certain sorts when in states of fear, anger, hunger, or sexual arousal. This overt behavior will be accompanied by visceral changes—in the glands and in the involuntary musculature—that are set in motion by the automatic and sympathetic nervous systems. But the behavior itself will consist of voluntary actions that have been learned, that are intelligently organized, and that may be directed to the immediate fulfillment of the drive, to a postponed fulfillment of it, or to its frustration. Such behavior will vary from individual to individual; and in any one individual, it will vary from time to time, though the instinctual drive may be the same and be of the same strength.

The foregoing description of the way in which instinctual drives operate in man as compared with the way in which they operate in other animals is more consonant with the view that man differs in kind than with the view that he differs only in degree; for the difference between the operation of instinctual drives and instinctive mechanisms in other animals and the functioning of instinctual drives in man appears to be one of kind rather than of degree. What other animals do entirely by instinct or by the combination of instinct and perceptual intelligence (i.e., the power of perceptual thought through which animal learning takes place), man does entirely by learning, through the exercise of his perceptual intelligence and especially his power of conceptual thought. The presence of the same instinctual drives in man and other animals does not lead to the same overt performances in man and in other animals when these same drives are operative; nor does the presence of the same instinctual drives in all members of the human species lead all men to behave in the same way when they are activated by the release of instinctual energies.

The power of conceptual thought in man enables him to devise alternative ways of dealing with his instinctual urges. But if all the driving power behind human behavior comes from the instinctual urges that man has in common with other animals, and if man's power of conceptual thought is merely the servant of his instinctual drives, then in its main outlines human behavior is instinctively determined, as animal behavior is to a greater extent and in more detail. For human behavior to be radically different in kind from animal behavior, with respect to the role that instinct plays, man must be radically different in kind from other animals. Not only must the power of conceptual thought enable man to devise diverse ways of dealing with his instinctual urges, but he must have psychic energy not drawn from instinctual sources in order to exercise mastery over them—to sublimate or divert them to non-animal satisfactions, to postpone their gratification for long periods of time, or to subdue and frustrate them entirely if he so chooses. No other animal manifests such mastery of its instinctual urges. In Freudian language, no other animal suffers the discomforts or pains that result from domesticating and civilizing its instincts. Both civilization and its discontents belong only to man: civilization with its technology, its laws, its arts and sciences, because man alone has the power of conceptual thought that produces these elements of human culture; the discontents of civilization, born of the frustration, prolonged postponement, or sublimation of instinctual urges, because man alone exercises some voluntary control over the instinctual drives that he shares with other animals.

Surprising as it may seem, Freud's account of the relation between man's intellect and his instincts presupposes that man differs radically in kind from other animals. I say this with full knowledge that Freud himself, if explicitly asked the question about how man differs, would give one of the opposite answers—either that man differs only in degree or that his difference in kind is only superficial. No other answer fits Freud's explicit commitment to the principle of phylogenetic continuity, and his equally strong commitment to a thoroughgoing determinism that precludes free choice. Nevertheless, when we read *Civilization and its Discontents*, we find many passages difficult to understand unless man is radically different from other animals that have the same instinctual drives; such as the following:

Sublimation of instinct is an especially conspicuous feature of cultural evolution; this it is that makes it possible for the higher mental operations, scientific, artistic, ideological activities, to play such an important part in civilized life. If one were to yield to a first impression, one would be tempted to say that sublimation is a fate which has been forced upon instincts by culture alone. But it is better to reflect over this a while. Thirdly and lastly, and this seems most important of all, it is impossible to ignore the extent to which civilization is built up on renunciation of instinctual gratifications, the degree to which the existence of civilization presupposes the non-gratification (suppression, repression or something else?) of powerful instinctual urgencies. This "cultural privation" dominates the whole field of social relations between human beings. . . . *It is not easy to understand how it can become possible to withhold satisfaction from an instinct.*[12]

I have italicized the last sentence quoted because I want to call attention to the question that must be answered. *How is it possible for us to withhold satisfaction from an instinct?* What power in us enables us to do so? Freud's answer to that question is, in my judgment, very revealing. It is given in the following passage:

We may insist as much as we like that the human intellect is weak in comparison with human instincts, and be right in doing so. But nevertheless there is something peculiar about this weakness. The voice of the intellect is a soft one, but it does not rest until it has gained a hearing. Ultimately, after endlessly repeated rebuffs, it succeeds. This is one of the few points in which one may be optimistic about the future of mankind, but in itself it signifies not a little. And one can make it the starting-point for yet other hopes. The primacy of the intellect certainly lies in the far, far, but still probably not infinite, distance.[13]

The foregoing explanation of how men are able to withhold satisfaction from instincts, and to exercise mastery over them in other ways, attributes an autonomy and causal efficacy to the human intellect which it could have only if the power of conceptual thought were an immaterial or non-physical power. Only if he possessed such a power would man be able to choose between diverse ways of gratifying his instincts; be able to decide whether to gratify them or not; and be able, in addition, to seek the gratification of desires that are not rooted in his instinctual urges at all, but arise from his capacities for knowing and for loving, as only an animal with the power of conceptual thought can know or love.[14] Thus, Freud's account of civilization and its discontents and his statement about the power of the human intellect in relation to man's animal instincts appear to lead to a conclusion that runs counter to his own commitments to determinism and to phylogenetic continuity; namely, the conclusion that man differs radically in kind from other animals by virtue of having a non-physical power that gives him freedom of choice and that has sufficient independence of instinctual energies to gain mastery or exercise control over them.

We find the same conclusion implicit in Konrad Lorenz' recent book on aggression; and there, as in Freud, its implicit presence is obscured and contradicted by many things that are explicitly said to the contrary. Lorenz acknowledges the uniqueness of man by virtue of his power of conceptual thought.[15] In addition, he attributes to man, because of his rationality, a "responsible morality" that is not possessed by other animals, and tells us, as Freud does, that "we all suffer to some extent from the necessity to control our natural inclinations by the exercise of moral responsibility."[16] In his discussion of the "behavioral analogies to morality," he clearly indicates that what morally responsible men do by reason, other animals do solely by instinctive mechanisms.[17] Nevertheless, he explicitly denies autonomy to reason; i.e., denies that man has in his constitution any power sufficiently independent of instinctual energies to exercise mastery over them.

By itself, reason can only devise means to achieve otherwise determined ends; it cannot set up goals or give us orders. Left to itself, reason is like a computer into which no relevant information conducive to an important answer has been fed; logically valid though all its operations may be, it is a wonderful system of wheels within wheels, without a motor to make them go round. The motive power that makes them do so stems from instinctive behavior mechanisms much older than reason and not directly accessible to rational self-observation.[18]

Reason, or the power of conceptual thought, has no driving energy of its own, and no causal efficacy of its own; all its commands or prohibitions draw their effective force "from some emotional, in other words, in-stinctive, source of energy supplying motivation. Like power steering in a modern car, responsible morality derives the energy which it needs to control human behavior from the same primal power which it was created to keep in rein."[19]

Because, like Freud, Lorenz is committed to determinism and to phylogenetic continuity, he leaves us with the puzzle of how reason and responsible morality can operate to thwart instinctual drives if they lack autonomy, i.e., if all their energy derives from instinctual sources. In

addition, there is the further puzzle of how man can have moral responsibility without having a freedom of choice that involves some measure
of independence of animal instincts. These puzzles vanish if one holds the
view that Freud and Lorenz cannot adopt, because it is irreconcilable with
their basic commitments—the view that man differs radically in kind
from other animals, and that he has the power of conceptual thought and
contra-causal freedom of choice by virtue of having a non-physical or
immaterial factor in his make-up, a factor that has a certain measure of
autonomy and causal efficacy.

(5)

The reader should not need to be reminded that, at this stage of our
inquiries, we do not *know* whether man's difference in kind is superficial
or radical; we do not *know* whether the materialist hypothesis or the immaterialist hypothesis is nearer the truth. Such arguments as can be
advanced in support of one or the other hypothesis have already been
examined [see the readings in Section I above, The Uniqueness of Man];
I have not, in the foregoing discussion of the role of instinct in human
life, offered any new arguments for the immaterialist hypothesis. My
sole purpose has been to see the alternative practical consequences
that would follow from a future decision in favor of one hypothesis or the
other. Let me summarize what has now become clear.

On the one hand, if man has an immaterial or non-physical factor
operative in his make-up and if, with that, he has freedom of choice and
some measure of independence of his animal instincts, then the resultant
discontinuity between man and other living organisms would require us
to desist from trying to explain human behavior by the theories or laws
that we apply to the behavior of subhuman animals. In spite of the fact
that the same instinctual drives are operative in man and other animals,
the radical difference in kind between them would mean that instinct does
not play the same role in human life that it plays in the lives of other
animals. Man would have a mastery over his instincts that no other
animal has; and he could have rational goals, ideals envisaged by reason,
beyond the satisfaction of his instinctual needs. We might then look upon
the future of man with the optimism that both Freud and Lorenz express,
but we would have grounds for that optimism which they cannot reconcile
with their scientific convictions.[20]

On the other hand, if determinism and the principle of phylogenetic
continuity hold true in the case of man, as they would if man differs only
superficially in kind from other animals, then the laws governing and the
theories explaining the behavior of subhuman animals would apply

without qualification to human behavior. In spite of the fact that man differs in kind by virtue of having the power of conceptual thought, instinct would play the same determining role in human life that it plays in the lives of other animals; and, in that case, we cannot be optimistic about the future of the human race, for so long as man is governed by his animal instincts, his behavior cannot be altered in its broad outlines and in its basic tendencies.

NOTES

[1] See J. Lilly, *Man and Dolphin* (Garden City, N.Y.: Doubleday, 1961), Chapters 1 and 12. Cf. Lilly's more recent book: *The Mind of the Dolphin: A Nonhuman Intelligence* (Garden City, N.Y.: Doubleday, 1967).

[2] *Ibid.*, pp. 211–212. Considering Dr. Lilly's prediction that, if dolphins and humans engage in conversation and thus appear to share a common intellectual power, some groups of men will probably advocate that we treat them as we treat human beings, a reader of this book in manuscript suggested to me that the opposite result might also occur. It is just as likely, he wrote, that some group of men "will take the view that a large part of the human race is *no better* than dolphins, and should therefore cease to have the rights currently accorded to human beings. In effect, this was the argument of the Germans with respect to Poles and Jews. Without denying that the Poles and Jews were biologically human, the Nazis maintained that they were in other respects sub-human, and more like animals or things. In short, the effect of the discovery that men do not differ in kind from animals is just as likely to promote malevolence toward some human groups as benevolence toward dolphins or other animals that are found not to differ in kind from men."

I have nothing to say about the relative probability of Dr. Lilly's prediction or my friend's prediction of the actual consequences that might follow from the discovery that men and dolphins do not differ in kind. I have no way of estimating which guess is shrewder or more likely to be true. When, in this chapter I consider the practical consequences of man's being different in kind from other animals—or, in the case of the dolphins, perhaps the same in kind —I am concerned *only with what ought or ought not to be the result of one or another state of facts, not with predictions of what might or might not actually result*. In other words, by *practical* consequences, I mean *normative consequences*—consequences in the form of the normative conclusions that we reach in the light of the facts as ascertained, not consequences in the form of actions taken, regardless of whether or not they can be justified in the light of the facts and sound normative principles [see *supra*, editors' introductory comments on this reading selection].

My friend obviously understood Dr. Lilly to be doing no more than making a prediction. I understood him to be considering the legal and ethical problem that the human race will have to face if and when dolphins show themselves to have the power of conceptual thought. I, therefore, read him as taking the position that if and when it is ascertained, as a matter of fact, that dolphins and men do not differ psychologically in kind, justice will require us to treat dolphins as persons and accord to them the same rights that we accord to men as persons. The action predicted by my friend would not be justified by the facts as ascertained, if they were subsumed under the normative principle that all persons (i.e., all living organisms that have the power of conceptual thought *in any degree*) ought to be treated in the same way—as persons, not as things. Nazi policies with regard to Poles and Jews, or similar policies with regard to Negroes, which have long prevailed and are still not eradicated, cannot be justified by the facts (that all human beings have the power of conceptual thought to some degree, and that every human being, even the least, therefore differs psychologically from the most intelligent animal) when those facts are subsumed under the correct normative principle that, as a matter of right or justice, all human beings ought to be treated in the same way (as persons rather than as things), and accorded the rights of persons. If the facts were otherwise—if men and other animals differ only in degree, and if some men are superior to other men in degree, as much as if not more than some men are to some animals—then Nazi policies in the treatment of Poles and Jews, or segregationist policies in the treatment of Negroes, would not be, *prima facie*, wrong as a matter of principle; the only question to be determined would be the question of fact about the inferiority of Jews or Poles to Germans, or Negroes to white men.

[2a] Editor's Note: "Turing's game" is a very specific performance designed to indicate sufficiently whether a given type of machine, one operating through a flexible and random network of connections around an "infant core" of fixed connections rather than on the basis of predetermined pathways laid down by programming (a *robot* as distinct from a *computer*), has the power of conceptual thought, by virtue of the fact that the robot could use propositional language conversationally. Since the fact that men have and animals (to say nothing of plants, minerals, etc.) lack propositional speech, alone justifies our inference that men have and other creatures lack the power of conceptual thought, we could be equally justified in *attributing* power of conceptual thought (or "intellect") to an unprogrammed robot that was able to engage in the flexible and unpredictable give-and-take of human conversation, and in *denying* that any machine failing this test possessed the power of conceptual thought—no matter what other intelligent or apparently "thinking" behavior the machine manifested. This is the point of Turing's Game, "a conversational affair using an ordinary language, such as English. It is derived from a game in which all the players are human beings. Two of the players are behind a screen; one of them is a male, the other female. The third player is the interrogator who asks the hidden participants questions in an effort to determine which one is male, and which one is female. The questions (unlimited as to content or variety) are submitted in typewritten form and answers return in typewritten form, so that tone of voice is eliminated as a clue. The hidden players are *not required to tell the truth* in answering. They can say anything that they think will serve to prevent their being detected. The Turing version of this game simply substitutes a robot for one of the human beings. All the rules of the game remain the same, but the problem becomes one of determining which hidden participant is a human being, and which a robot." Accordingly, *Turing's Machine* is "a mathematically conceivable robot of the future that will be able to play Turing's game as well—or almost as well—as men can play it." (Adler, p. 244). Cf. A. M. Turing, "Computing Machinery and Intelligence," in *Computers and Thought* ed. by E. A. Feigenbaum and J. Feldman (New York: McGraw-Hill, 1963), pp. 11–12, 20, 21–29, 30–35. Also Jeremy Bernstein, *The Analytical Engine* (New York: Random House, 1963), for a characterization of Turing and his work.

[3] *Op. cit.*, in *The Modeling of Mind*, p. 254.

[4] *Ibid.* Cf. Hilary Putnam, "Minds and Machines," in *Dimensions of Mind*, ed. Sidney Hook (New York: New York Univ. Press, 1960), pp. 175–176; Donald M. McKay, "From Mechanism to Mind," in *Brain and Mind* (New York: Humanities Press, 1965), pp. 180–190.

[5] Sellars, *Science, Perception and Reality* (New York: Humanities Press, 1963), pp. 39–40.

[6] For Professor Smart's views, see *Philosophy and Scientific Realism* (New York: Humanities Press, 1963), pp. 153–154; and cf. *ibid.*, pp. 93–105, 111–116, 119–125. For Professor Sellars' views, see *Science, Perception and Reality*, pp. 6, 15–17, 30–34.

[7] *Philosophy and Scientific Realism*, p. 155.

[8] In a brilliant essay, Jacques Maritain outlines the importance of the question of man's difference for the conception of human equality, showing how divergent conceptions of the equality of men stem from divergent views of man as a species and how they give rise to divergent normative recommendations (see "Human Equality," in *Ransoming the Time* [New York: Scribner's, 1941], pp. 1–32). Among its other current projects, the Institute for Philosophical Research is engaged in the dialectical clarification of the idea of equality in Western thought. Even at this early stage of the work, it has become clear that the central and controlling issue in the whole discussion is constituted by conflicting views about the specific equality of men as persons in relation to all the inequalities that arise from their individual differences, and that these views are resolvable into conflicting views of the difference of man.

[9] *Critique of Practical Reason and Other Works on Ethics*, trans. by T. K. Abbott, 6th ed., (New York: Longmans Green & Co., 1927), p. 180. Cf. *ibid.*, pp. 46–53; and *Critique of Teleological Judgment*, trans. by J. C. Meredith (Oxford: The Clarendon Press, 1928), pp. 99–100. The Christian conception of personality, like Kant's, involves an element of immateriality. The Christian dogma that man is made in the image of God, Who is pre-eminently a person, attributes personality to man as reflecting the divine being in this respect, i.e., immateriality.

[10] I have treated these problems at greater length in another book of much earlier date (see *A*

Dialectic of Morals [New York: Ungar, 1958], Chapter IV, esp. pp. 58–59). While I would revise what is there said in many respects were I to address myself anew to the problems of moral philosophy, as I hope to do in a book I am now working on, the points made there would remain essentially unchanged, at least so far as they bear on the relevance to morality of the way that man differs from other animals.

[11] See Robert Ardrey, *The Territorial Imperative* (New York: Atheneum, 1966); and also his *African Genesis* (New York: Atheneum, 1961). Both books are engaging popularizations of the findings of ethology, full of fascinating stories of animal behavior; but both are also flagrant examples of special pleading for the questionable thesis that instinct governs human life exactly as it does animal life. The truth of that thesis depends upon how man differs from other animals; if man differs even superficially in kind from other animals, it is in important respects false; if man differs radically, it is wholly false. In his zeal to explain human behavior and human life in terms of animal instincts, Ardrey does not pause to consider the facts bearing on the question of how man differs. He assumes the truth of the answer that suits his *ad hoc* rhetoric. The fact that books of this sort are dismissed for what they are by the scientific community does not prevent them from bemusing and misleading the laymen who read them for the enjoyable animal stories they contain and uncritically swallow the thesis along with the stories.

[12] *Op. cit.,* trans. by J. Riviere, 1930: p. 63. Cf. Chapter III, *passim.*

[13] *The Future of an Illusion,* trans. by W. D. Ronson-Scott (London: Hogarth Press, 1928), p. 93.

[14] Freud's attribution of an intellectual power to man that is not possessed to any degree by other animals is all of one piece with his theory of distinctively human erotic love, as contrasted with the sexuality of other animals, a sexuality that is devoid of love even when it involves the instinctive inhibition of aggressive behavior and so gives the *appearance* of tenderness and benevolence. Nevertheless, for Freud every form of human love is erotic, either overtly sexual or a sublimation of sexuality. But if the human intellect has the autonomy that it would have to have in order to control the instincts and to sublimate them, and if that, in turn, depends on the intellect's transcendence of physical causality, then, contrary to Freud's theory of love, the non-erotic forms of human love (such as the *amor intellectualis dei* of Spinoza, the appetitive character of which takes its special form from intellectual cognition rather than from sense-perception) would be explicable without reference to sex, sensuality, or sublimation.

[15] *On Aggression,* pp. 238 ff. Though Lorenz discusses free will in relation to "the laws of natural causation" governing human and animal behavior, he shows little or no understanding of free choice (see *ibid.,* Chapter 12, esp. pp. 225, 228–229, 231–232). An excellent critical review of the book by S. A. Barnett points out the illicit use that Lorenz makes of superficial analogies between human and animal behavior, and also the inconsistencies into which he falls by his effort to plead a case beyond what the acknowledged facts will support (see *Scientific American,* [CCXVI (1967)], pp. 135–137).

[16] *Ibid.,* p. 254. Cf. *ibid.,* pp. 240–254.

[17] See *ibid.,* Chapter 7, esp. p. 110.

[18] *Ibid.,* p. 248. In another place, Lorenz refers to "the functions of reason and moral responsibility which first came into the world with man and which, provided he does not blindly and arrogantly deny the existence of his animal inheritance, give him the power to control it" (*ibid.,* p. 215).

[19] *Ibid.,* p. 247.

[20] See especially Lorenz' concluding chapter, "Avowal of Optimism," in *ibid.,* pp. 275 ff.

Man in evolution:
A scientific statement
and some ethical implications

Francisco José Ayala

CONTEXTUALIZING COMMENTS

Dr. Ayala is former student and present co-worker of Theodosius Dobzhansky, the Dean of American geneticists, at the Rockefeller University in New York. His investigations into biological evolution and its implications have yielded contributions to such journals as *The American Naturalist, La Ciencia Tomista, Genetics,* and to books both in Spanish and English.

In the following reading selection, Dr. Ayala points out that evolutionary science has perhaps its most direct impact on the age-old problem of individual rights within the requirements of community. Its impact in this area is mediated by the development of *eugenics,* the science concerned with control of the factors determining improvement and degeneration of the genetic endowment of the human species.

The inescapability of the eugenic issue of course is rooted in the irreducibility of socio-cultural to biological heredity, even as the perplexity of the issue is rooted in the difference (as discussed in the previous essay on "Evolution and Ethics") between a human act and a cosmic event, or (more particularly) between the personal and the organismic

Reprinted from *The Thomist*, XXXI (January 1967), pp. 1–20, by permission of the Editor.

activities of men (every human act is an act of the human organism, but not every act of the human organism is a human act). For once the threshold of hominisation had been crossed with the consequent establishment of a cumulative hereditary endowment in principle distinct from that governed by the mechanism of gene transfer and biological parentage, evolutionary difficulties were bound to arise from the reciprocal impact of the two within the existential unity of personal life. As Ayala aptly points out, it is difficult, if not impossible, to think of any element of human culture that could not in one way or another have repercussions in the gene pool of the human species.[1]

Yet the fact remains that no known factor of correlation exists between individually desirable physical traits governed by genetic heredity (20-20 vision, for example, or a robust frame) and the spiritually desirable characteristics governed by cultural heredity and its individual base, self-determination (generosity, courage, and compassion, for example, or a capacity for sustained and penetrating research). Thus the two-levelled problem imposes itself: at the *first* level, "as far as possible, deficiencies already in evidence must be remedied and care must be taken that hereditary factors even of little value be not allowed to deteriorate still further. . . . On the other hand, it must be seen that positive characteristics at their full value join with those whose hereditary patrimony is similar";[2] and this moral obligation on a knowledgeable individual to take into account with respect to any given marriage opportunity the expectable consequences for physical and mental characteristics of potential descendants gives rise to the *second* level of the eugenic issue, for "eugenics applied to individuals leads to eugenics applied to society."[3]

---

[1] Mayr, for example, calls attention to the fact that "animal breeding has long abandoned all attempts to discover superior genes individually. . . . One could readily translate this into terms of desirable goals for human biological progress. Perhaps it is not unreasonable to assume that a person with a good record of achievement in certain areas of human endeavor has on the average a more desirable gene combination than a person whose achievements are less spectacular. In our present society, the superior person is punished by the government in numerous ways, by taxes and otherwise, which make it more difficult for him to raise a large family. Why, for instance, should tax exemption for children be a fixed sum rather than a percentage of earned income? Why should tuition in school be based, in large part, on the ability of the father to pay rather than inversely on the achievement of the student? Innumerable administrative rules and laws of the government discriminate inadvertently against the most gifted members of the community. Changing these laws so as to place a premium on performance (the "opportunity" of true democracy, rather than identicism) is entirely different from distributing privileges according to the artificial, arbitrary criteria of the racists, such as blond hair and blue eyes. I firmly believe that such positive measures would do far more toward the increase of desirable genes in the human gene pool than all the negative measures proposed by eugenicists of former generations." (*Animal Species and Evolution* [Cambridge, Mass: Harvard, 1963], pp. 661–662.)
[2] Address of Pope Pius XII to the *Primum Symposium Geneticae Medicae*, September 7, 1953, excerpt from translation prepared by the Vatican Press office, *Eugenical News*, XXXVIII (1953), 146–149.
[3] Teilhard de Chardin, *The Phenomenon of Man*, p. 282.

It would be more convenient, and we would incline to think it safe, to leave the contours of that great body made of all our bodies to take shape on their own, influenced only by the automatic play of individual urges and whims. 'Better not to interfere with the forces of the world!' Once more we are up against the mirage of instinct, the so-called infallibility of nature. But is it not precisely the world itself which, culminating in thought, expects us to think out again the instinctive impulses of nature so as to perfect them? Reflective substance requires reflective treatment.[4]

Thus Huxley points out that, for example, "the problem of avoiding nuclear war is more immediate, but that of overpopulation is, in the long run, more serious and more difficult to deal with because it is rooted in our nature."[5]

Alas, the progressive loss of valuable genes is not the only [biological] danger facing the human species. Indeed, overpopulation is a far more serious problem in the immediate future. I am not speaking of the material aspects such as the exhaustion of mineral and soil resources and the increasing difficulty of food supply for 6, 8, or 10 billion people. Human technology may find answers to all these difficulties. Yet I cannot see how all the best things in man can prosper—his spiritual life, his enjoyment of the beauty of nature, and whatever else distinguishes him from the animals—if there is "standing room only," as one writer on the subject has put it. It seems to me that long before that point has been reached man's struggle and preoccupation with social, economic, and engineering problems would become so great, and the undesirable by-products of crowded cities so deleterious, that little opportunity would be left for the cultivation of man's highest and most specifically human attributes. . . . Man may continue to prosper physically under these circumstances, but will he still be anywhere near the ideal of man?[6]

---

[4] *Ibid.*

[5] Sir Julian Huxley, *The Human Crisis* (Seattle: University of Washington Press, 1963), p. 43. See also Thomas Malthus, Julian Huxley, and Frederick Osborn, *Three Essays on Population* (New York: Mentor, 1960). *The Phenomenon of Man*, p. 282.

[6] E. Mayr, *Animal Species and Evolution*, p. 662. John R. Platt puts the issue more graphically still: ". . . world population has now shot up to more than three billion people, and the doubling time is now between thirty and forty years. In paleolithic times it is estimated to have been about thirty thousand years; so that our rate of increase is now about one thousand times greater than it was for prehistoric man.

"Obviously such a rate of increase cannot continue indefinitely. A doubling in 40 years means a fourfold increase in 80 years, eightfold in 120 years, and tenfold in 130 years. This would mean an increase from three billion people to thirty billion people by the year 2100, and to three hundred billion people by the year 2230—that is, in a time shorter than the time since the settling of New England. Yet this number would be far beyond the most optimistic estimates of what the world's food supply could support, even with the use of marginal land and farming the oceans. This consideration is quite aside from the question of whether life at one hundred times our present population density could still be called human. Such a level of crowding is no longer an affirmation of life but a denial of all that life might be.

"We see that the population of the planet, like the weight of a grown man, must sometime soon begin to level off to a 'steady state,' whether this is at some upper limit set by starvation, or at the low limit that would be set by nuclear annihilation, or at some intermediate level of well-being and decency set by sensible human choice. Some day all men will see that an excessively gross population is like an excessively obese man and shows a lack of control that damages its own humanity and its own potentialities. . . . If birth control can come to be

On every side then, be it from the side of improving the human stock, of preventing its material deterioration, or of striking a balance between the space and resources of our planet and the numbers of our species, the eugenic issue confronts us. And yet the tendency of the well-being of *homo biologicus* and that of *homo humanus* to fluctuate as relatively independent variables makes the task of coming to grips with the issue difficult in the extreme.

Still, the requirements of the issue are clear: "In the course of the coming centuries it is indispensable that a nobly human form of eugenics, on a standard worthy of our personalities, should be discovered and developed."[7] And if we meditate on the history of our species, on the general revulsion in the early days of anatomy to the idea of dissection, for example, or to the idea of organ transplants in the infancy of surgery, the contemporary opposition to the idea of eugenic control begins to take on a perspective.

If civilization survives, it is likely that we will be able to slow down and perhaps even to halt deterioration of the species. The methods that will be employed would probably not be palatable to many of us who are alive today. Nevertheless, the human animal is a flexible creature and has thus far been able to adjust his outlook to his needs with remarkable agility.[8]

As in most cases where the dire, tragic, or precarious aspects of the human condition are brought into an evolutionary focus, the attitude which seems to be demanded is the paradoxical one of short-range pessimism coupled with longe-range optimism, with the capacities of reason as the link between the two.

---

treated as a public health problem, like the control of disease, rather than an individual problem, we might achieve even within this generation the conscious worldwide control of population that all mankind must eventually have for the sake of its health and welfare. ("The New Biology and the Shaping of the Future," in *The Great Ideas Today 1968*, ed. by R. M. Hutchins and M. J. Adler [Chicago: Britannica, 1968], pp. 350–351.)

[7] Teilhard de Chardin, *The Phenomenon of Man*, p. 282.

[8] Harrison Brown, *The Challenge of Man's Future* (New York: Viking, 1965), p. 105. According to this author, "Precise control of population can never be made completely compatible with the concept of a free society; on the other hand, neither can the automobile, the machine gun, or the atomic bomb. Whenever several persons live together in a small area, rules of behavior are necessary. Just as we have rules designed to keep us from killing one another with our automobiles, so there must be rules that keep us from killing one another with our fluctuating breeding habits and our lack of attention to the soundness of our individual genetic stock." (*Ibid.*, pp. 263–264.) It may be, however, that if one defines freedom by reference to the good and distinguishes it at the same time from *license*—a deficient mode of freedom—Mr. Brown's conflict between the free society and the rational society can be arbitrated, if not entirely exorcised. See M. J. Adler, *The Idea of Freedom* (New York: Doubleday, 1958 and 1961), 2 volumes; see also Stephan Strasser's discussion of *Phenomenology and the Human Sciences* (Pittsburgh: Duquesne, 1963), esp. pp. 27–55.

"Man in evolution: a scientific statement and some ethical implications"

It is the main purpose of this paper to state that there exists, in twentieth century mankind, a great wealth of genetic variability, and that new environmental challenges are continuously arising. In other words, that man is evolving biologically. This fact has momentous implications for human life. Some of the philosophical and ethical questions arising from the biological fact of human evolution will be pointed out in the final part of the paper.

Mankind is engaged simultaneously in two kinds of evolutionary development, the biological and the cultural. Human evolution can be understood only as a result of the interaction of these two developments. They correspond to the two kinds of heredity existing in man, the genetic and the cultural, what Medawar has suggested calling *endosomatic* and *exosomatic* systems of heredity.[1] Genetic inheritance in man is very much like that of any other outbreeding, sexually reproducing species; it is based on transmission of genetic information in the form of deoxyribonucleic acid (DNA) from one generation to the next, via the sex cells. A fertilized human egg cell (technically, a zygote) contains two homologous sets of genetic information, which by interaction with the environment direct the development of the anatomical, physiological and psychological characteristics of the individual which will develop from it. The somatic cells of a human being contain also two homologous sets of genetic information; his sex cells, however, contain only one set. When a female sex cell, or ovum, is fertilized by a male sex cell, or spermatozoon, the double set of genetic information is restored. From that zygote a new human individual will develop carrying hereditary material coming in equal amounts from each one of his or her parents. By this process of sexual heredity the genetic endowment of the species is reshuffled every generation.

In addition to his biological system of inheritance, man transmits to other members of the species a cultural inheritance. Cultural inheritance is based on transmission of information by a teaching-learning process, which is, in principle, independent of biological parentage.

It was the appearance of culture, a superorganic form of adaptation, that made mankind the most successful living species. From his obscure beginnings in Africa man has become the most abundant species of mammal on earth. Numbers may not be an unmixed blessing, but they are one of the measures of biological success.

Nevertheless, the superorganic has not annulled the organic. I see no foundation to support the opinion that biological evolution stopped with human progress. Cultural and biological evolution interact and reinforce each other. The maintenance and development of human culture is possible only so long as the genetic basis of human culture is maintained or improved. There can be no human culture without human genotypes. At the same time, the development of culture is perhaps the most important source of environmental changes promoting the biological evolution of man. There exists a feedback cycle involving genetics and culture.

Biological evolution can be defined as change in gene frequencies in response to challenges from the environment. Heredity is basically self-copying of genes. Heredity is, therefore, a conservative force. Occasionally, however, genes produce imperfect copies of themselves. This is genetic mutation. Mutation is the ultimate source of all genetic variability. Without mutation there could be no evolution. All the genetic variability present in human populations has arisen by mutation in our immediate, remote, or very remote ancestors. The frequency of a particular genetic mutant in the population will increase or decrease depending on the selective value of the individuals carrying it. That is, if the average effect of a mutant gene in the environments in which the species lives is favorable, natural selection will tend to increase its frequency in the population. The selective value of a mutation is ultimately determined by the environment. Mutations favorable in a certain environment may not be so in a different one. The constellation of genes and the arrangements of these in present day populations are, therefore, determined by the environment to which the population has been exposed in the past. And today's environments condition the present and future genetic constitution of the population.

The limitation to the action of natural selection sorting out new genotypes comes from the availability of genetic variability. That human populations store a tremendous amount of genetic variability can hardly be doubted. No two human individuals, with the unimportant exception perhaps of identical twins, are genetically identical. Let me explain why it is so. Genetical studies involving a number of organisms, from bacteria through fruit flies to man, allow us to give 500 as a reasonable estimate of the minimum number of genes in a chromosome. With the 23 pairs of chromosomes existing in man, we can assume a minimum of 11,500 pairs of genes. Those genes carry the necessary information to direct the process of development from a zygote to an adult human being. The two members of each pair are not necessarily identical. Genes may exist in two or more alternative forms called "alleles." Different alleles arise by mutation. Let us assume, now, that mutations exist in only 200 of the 11,500 or more pairs of genes of man. Moreover, let us assume that there are only two alleles in each of those 200 loci. According to the simple rules of Mendelian heredity the number of possible human genotypes would be 3^{200}, or approximately 1 followed by 95 zeros. That number far exceeds the number of human individuals who lived or will ever live on earth, or to put it in a different way, that number is many billion times larger than the number of atoms existing on earth. And it is a gross underestimate. Mutations are actually known in man at more than 200 loci. Probably they exist or are possible at most if not all of the 11,500 loci that we assumed as a minimum number of gene pairs. Besides, at many loci there are more than two alleles; up to 15 different alleles are known to exist at certain loci. How a larger number of alleles increases the number of possible genotypes can easily be shown. For blood groups, there are in man at least 11 sets of alleles. If only two alleles existed at each locus the number of potential blood groups would be slightly larger than two thousand. More than two alleles exist at least at six of those loci, and the number of possible blood groups is considerably larger than two million.

All this amounts to saying that the existing genetic variability in the human species is essentially inexhaustible. Moreover, new variability is continuously arising in human populations by mutation. Genetic mutations occur in man as in any other living species. Due to the increased use of X-rays for diagnostic purposes, to the use of certain chemicals, and to exposure to atomic fall-out, mutation rates in man are probably higher today than they were in the past. And they are probably highest in technologically advanced countries. Rough estimates of the frequency of mutations in man are available. A majority of the mutation rates calculated for some drastic mutants, which produce fatal hereditary diseases and spectacular malformations, lie between one and ten mutations per 100,000 cells per generation. Small mutations with less dramatic effects are more difficult to observe, but they are believed to be several times as frequent as drastic ones. The average mutation rate per gene is then probably of the order of two per 100,000 (2×10^{-5}) or somewhat higher. If we assume that every human being contains only 10,000 pairs of genes, it will be a conservative estimate that $2 \times 10^{-5} \times 20,000$ or 40 per cent of all people will carry one or more mutant genes newly arisen in the sex cells of their parents. The supply of genetic raw materials for the operation of natural selection is, therefore, ample.

Most of the mutations studied in man have negative selective value, having deleterious effects which affect his welfare. Many other mutations affect characteristics such as color of the eyes, hair or skin, shape of the head, face or nose, tendency to be fat or slender, and many other aspects of the anatomy or physiology of a person which are not directly observable. Do these mutations have any effect on the survival and reproduction of the individual or are they adaptively neutral? One can speak of the usefulness or harmfulness of a genetic variant only in and with respect to a certain environment. A genotype which suffers in one environment may flourish in another. A well known case is the sickle-cell gene. The fact of being heterozygous for that gene makes a person at least relatively immune to certain forms of malaria. This is quite useful if one lives in a country where malarial infection is likely to occur, but not so where malaria is absent. The case is that almost every gene which has been adequately studied has proved to have some effect on the fitness of the individual, at least in certain environments.

Through the process of genetic recombination the harmful and beneficial genes present in the population become combined into new genotypes in every generation. Not all potential genotypes are realized in nature, but if genetic variants have different selective values in certain environments, natural selection will propagate them differentially in the following generations. Natural selection is the principal agent of evolutionary change. The mutation process provides the genetic variability, the raw materials upon which natural selection acts. Without hereditary variations natural selection would have nothing transmissible to work upon. Mutations, however, originate at random independently of their effects on the individuals in which they arise. Mutation is a chance process; without natural selection it would produce only chaos. The anti-chance force in evolution is natural selection. Natural selection is the essential directive force in evolution, as Darwin saw it. Natural selection, however, must not be understood in the metaphorical sense of "struggle for existence" that Darwin overemphasized. Natural selection is basically differential reproduction. To say that natural selection favors certain genotypes is simply to state that those genotypes are transmitted to the following generations more frequently than others. In many human environments, where early mortality, except for some drastic mutants, has been greatly eliminated, natural selection acts mainly through differential fertility. As Dobzhansky puts it:

The selectively fit, or, if you will, the fittest, is not necessarily a fellow with big muscles, or a lusty fighter, or a conqueror of all his competitors. He is, rather, a paterfamilias who has raised a large number of children who in turn become patresfamilias."[2]

Natural selection is continuously shifting its effects on man as a result of the endless variety and continuous change of the human environments. During the last fifty years or so the forces of selection acting on human beings have changed drastically, and will go on doing so in the future. Haldane expressed the opinion that for the last ten thousand years, in fact since man ceased to be a rare animal, selection has been mainly for immunity from infectious diseases.[3] During the nineteenth century, tuberculosis was the most important cause of death among young adults in Europe and North America. René Dubos has convincingly shown that the fall of death rate due to tuberculosis in recent times has been brought about not only by the improved living conditions and better medical services, but also by selection of more resistant genotypes.[4] Less susceptible genotypes were favorably selected, particularly during epidemic periods. Today, however, a large part of mankind lives in an environment where most people can go on without any serious infectious disease, except some virus diseases as common colds which we cannot yet control. Selection for resistence to infectious diseases is no longer operating, at least not with its past intensity.

An extreme example of how improved hygiene and medical services may change the selective value of a gene is the case of the gene for sickle-cell anemia mentioned above. Heterozygous individuals for that gene have an inborn resistance to subtertian malaria and enjoy higher selective values than the homozygotes for the "normal" gene in places where malaria is rife. But the price paid by the population to enjoy that inborn resistance in some of its members is high. In a population with 42 percent of its members resistant heterozygotes, 9 percent will be born homozygous for the gene and will suffer a severe form of anemia and die mostly before reproductive age. In countries where malaria has been eradicated by modern medicine, the sickle-cell gene loses its relative adaptive advantage and will gradually disappear. The population will not suffer from malarial infection, and the high price paid for the inborn resistance will no longer be justified. Reduction of the frequency of the gene has been observed in the Negro population of America at about the rate we should expect if malaria had ceased to be a scourge to it 200 or 300 years ago, i.e., when their ancestors came from malaria infected countries.

Improvements in human welfare may also have important genetic effects in the population. An example may be taken from the recently published work by Reed and Reed[5] on mental retardation. Nearly 50 per cent of a large sample of retardates in Minnesota proved to have one or both parents mentally retarded. If the mentally retarded would not reproduce, the frequency of mental retardation in the population is expected to be reduced by nearly half in a single generation. Institutionalized retardates have a considerably lower average reproductive rate than comparable retardates living outside the institutions. The provision of institutions for the mentally retarded has, then, a notable effect in reducing the frequency in the population of genes affecting mental retardation.

Some of the most radical changes in human environments have arisen from there being more humans. The world population has grown from 700 million to three billion during the last 200 years. If the present rate of population growth continues, there will be more than six billion humans by the end of this century. The genetic consequences of such "population explosion" are complex, but different attitudes towards birth control practices are creating new selective forces everywhere. Other associated phenomena, such as urbanism and mass migration, are of no less biological import.

Examples of natural selection in human populations could be multiplied at will. Natural selection continues to operate in man and is expected to continue doing so. To do away with natural selection a genetically identical mankind would have to be produced living in an environment where mutation does not exist. Or if genetic variability would exist, every pair of human beings would have to produce the same number of children, who would survive, marry and produce the same number of children. Other equally fantastic or practically not realizable methods of suppressing natural selection could be devised. So long as there is genetic variability and environmental change different genotypes will be differentially transmitted to the following generations. Human environments are changing faster than ever because of rapidly changing human culture. It is hard to think of any element of human culture that could not have, directly or indirectly, repercussion on the gene pool of mankind. Religious and ethical convictions, educational and fiscal laws, medical practices and social habits, and any act of legislation have genetic consequences because they modify the human environment. Environment instability presents challenges to the organism. These challenges are passed to the organism by natural selection which preserves the relatively fit and eliminates the relatively unfit. These changes in the hereditary endowment of the species create new environmental conditions, which in turn are responded to by new genetic changes. The process is expected to go on indefinitely. The only alternative to change is extinction if the organism is unable to cope with the environmental challenges.

Where is human evolution going? Biological change is guided by natural selection. However, natural selection is not a benevolent spirit guiding evolution towards sure success. It is an agent bringing about genetic changes that often appear purposeful because they are dictated by the requirements of the environment. Nevertheless, the end result may be extinction. The number of animal species which existed in the past and have disappeared without leaving descendants far exceeds that of living species. Natural selection has no purpose; man alone has purposes and he alone may introduce them in his evolution. The problem of directing human evolution raises a number of ethical questions of great import. Some of these questions I want to raise here.

Mutation rates in man have probably increased with the industrial civilization due to radiation exposures and by contact with and ingestion of some of the drugs and chemicals to which almost every body is exposed. The increase in mutation rates is, at best, a mixed blessing. Genetic variability must exist for evolution to occur, but the great majority of newly arising mutants are harmful. Because of the rapidly improving living conditions, in particular the increasing power of modern medicine, the elimination of some harmful mutants from human populations is no longer taking place as rapidly and effectively as it did in the past. Many kinds of hereditary disorders are cured today, and their carriers are able to survive and leave offspring, therefore transmitting their hereditary infirmities to the following generations. The more hereditary diseases and defects are "cured," the more of them will be there to be cured in the succeeding generation.

Some biologists have raised warning voices. H. J. Muller predicts a gloomy future for mankind if remedy is not provided. In a not very distant future, he writes:

Our descendants' natural biological organization would in fact have disintegrated and have been replaced by complete disorder. . . . It would in the end be far easier and more sensible to manufacture a complete man de novo, out of appropriately chosen raw materials, than to try to refashion into human form those pitiful relics which remained.[6]

It is undoubtedly true that with respect to some hereditary weaknesses and disorders natural selection has been tapered if not completely suppressed. But that is not the whole story. As Dobzhansky has noted: "With respect to some genetic variants natural selection in civilized societies has become more, not less, rigorous."[7] Emotional strains, for instance, and nervous tensions are far stronger in our modern cities than they were in the past. Overeating and food sophistication seem to be partially responsible for the high incidence of heart disease, peptic ulcer, diabetes, and other diseases common in abundant communities.[8] Many other examples could be added. Natural selection has not ceased to operate in the human species since differential reproduction of genetic variants exists.

The problem, however, remains that some undesirable hereditary conditions will spread more and more every generation if their carriers are cured and reproduce. The science and art of eugenics deal with the subject of the betterment of the hereditary endowment of mankind. Two forms of eugenics can be distinguished. Negative eugenics seeks methods to avoid the spread of undesirable genes, while positive eugenics encourages the spread of desirable ones. I cannot discuss in detail here the subject of eugenics, but I consider it worthwhile to point out the kind of ethical questions which are raised. Essentially, the basic problem here is that of the rights of the individual versus the rights of the community. A lot of thinking is needed here in the light of the new perspectives raised by modern science, and philosophers and theologians have a difficult task to face in meeting the problems which are being raised.

The main difficulty in the field of eugenics is the lack of adequate knowledge in many areas. But whatever knowledge we have we must use. The negative attitude of "hands off" or the simplistic attitude of "nature knows best" can hardly be acceptable. When we are afflicted with a disease or infirmity we do not let nature take care of it or wait for a miracle, but make use of the appropriate remedies to restore our health. We must act as rational beings in guiding our actions according to whatever knowledge we can master.

Programs of positive eugenics have been proposed in recent years. Preferential multiplication of the genotypes of well-gifted individuals can certainly be accomplished in man. The problem is, however, which criteria to use to decide what characters to multiply. There is another question, namely that we do not know what parts of many desirable traits are genetically and what parts are culturally determined. G. W. Beadle has made this point emphatically:

"Few of us would have advocated preferential multiplication of Hitler's genes through germinal selection. Yet who can say that in a different cultural context Hitler might not have been one of the truly great leaders of men or that Einstein might not have been a political villain." [9]

I fully subscribe to his opinion that at the present time, "in the absence of much more understanding than we now have, we will do best to preserve maximum genetic diversity." [10] He claims one exception only, that is those who are clearly and significantly genetically defective, to whom negative eugenics must be applied. This leaves still a number of unanswered questions. Should the cured carriers of fatal dominant genes be sterilized to prevent transmission of the gene? Should they be prohibited legally from having children, without sterilization? Should they merely be advised not to have children? What can be done if they do not comply, since half of their progeny is expected to have the same fatal defect? Could they, in that case, be considered as criminals and punished with sterilization or confinement? Another basic problem would be where to draw the line with respect to less severe disabilities equally transmitted as dominant genes. As for recessive deleterious mutants, how far can we go in legally prohibiting reproduction or should only marriage of two carriers be advised against, when one out of four children is expected to be homozygous for the deleterious gene?

Each case should perhaps be judged in its own merits, but some general principles ought to be considered. The problem is a serious one and it is being continuously aggravated with medical and biological progress. The collaboration of social scientists, legal experts and biologists will be needed here. The magnitude and social implications of certain eugenic problems is staggering. As a specific example, the studies on mental retardation quoted earlier could be considered. Reed and Reed found that 48.3 percent of an unselected sample of mental retardates had one or both parents retarded. [11] Similar results were obtained by Benda and his associates, who working with a smaller sample of mildly retarded persons found that 48.7 percent of them had one or both parents retarded.[12] There are five million mildly retarded persons in the United States. If the above results have general value, preventing the mentally retarded from having progeny would reduce the number of retardates in the population to half its present frequency in a single generation. This seems a socially desirable objective, but how to achieve it? As Reed and Reed have written:

It is unlikely that voluntary abstinence will be very effective with the retarded, and it would be unrealistic to expect the rhythm method to work with any appreciable proportion of them.[13]

Other alternatives are open, like institutionalization and voluntary or compulsory sterilization. Would the welfare of the community make ethically acceptable such measures? This is but a single case of many where our present and growing knowledge gives new dimensions to old problems and demands fresh and responsible thinking. The answer cannot come from biology alone. For, as Dobzhansky has written:

Man, if he so chooses, may introduce his purposes into his evolution. . . . The crux of the matter is evidently what purposes, aims or goals we should choose to strive for. Let us not delude ourselves with easy answers. One such answer is that a superior knowledge of biology would make it unmistakable which plan is the best and should be followed. Another is that biological evolution has implanted in man ethical ideas and inclinations favorable for this evolution's continued progress. Now, I would be among the last to doubt that biology sheds some light on human nature; but for planning even the biological evolution of mankind, let alone its cultural evolution, biology is palpably insufficient.[14]

NOTES

[1] P. B. Medawar, *The Future of Man* (New York: Basic Books, 1959), p. 92.

[2] T. Dobzhansky, *Heredity and the Nature of Man* (New York: Signet Science Library Books, 1966), p. 153–154.

[3] J. B. S. Haldane, "Biological Possibilities for the Human Species in the Next Ten Thousand Years" in *Man and His Future*, ed. G. Wolstenholme (Boston: Little, Brown and Co., 1963), pp. 337–361.

[4] R. J. Dubos and J. Dubos, *The White Plague: Tuberculosis, Man and Society* (Boston: Little, Brown and Co., 1952).

[5] E. W. Reed and S. C. Reed, *Mental Retardation: A Family Study* (Philadelphia: Saunders, 1965). See Tables 32 and 33, pp. 46–47.

[6] H. J. Muller, "Our Load of Mutations," *American Journal of Human Genetics*, II (1950, pp. 111–176.

[7] T. Dobzhansky, *Mankind Evolving* (New Haven: Yale Univ. Press, 1962), p. 306.

[8] J. F. Brock, "Sophisticated Diets and Man's Health" in *Man and His Future* (see footnote 3), pp. 36–56.

[9] "Genes, Culture and Man," *Columbia Univ. Forum*, VIII (1963), pp. 12–16.

[10] *Ibid.*

[11] *Op. cit.*, 46–47.
[12] C. E. Benda, N. D. Squires, M. J. Ogonik, and R. Wise, "Personality Factors in Mild Mental Retardation," *American Journal of Mental Deficiency*, LXVIII (1963), pp. 24–40.
[13] *Op. cit.*, p. 76.
[14] T. Dobzhansky, *Heredity and the Nature of Man*, p. 162.

4

The metaphysical issues

Rationale of this section

In ancient and medieval times, no distinct status was accorded to what we today call experimental science. Yet, the thinkers of these periods were generally of the opinion that a purely rational, a-mathematical analysis of certain pervasive natural phenomena (notably, changes of time, place, and identity) can yield an understanding of the essential framework of existence within which even the not always experienceable things and events studied by modern science must occur.

Such an understanding of being in its primarily intelligible aspect they called "natural philosophy"; but they assigned to that term a sense broad enough to cover equally the experimental labors of science and what we today distinguish from these labors as "metaphysics." [1] For, in present-

---

[1] This classical or generic notion of natural philosophy is clearly expressed, for example, in one of Thomas Aquinas' less well-known schema for the organization of human knowledge. "Since the human mind develops through disciplined exercise," wrote Aquinas in his *Commentary* on the *Nicomachean Ethics* of Aristotle (Bk. I, lect. 1, n. 2), "there are as many kinds of knowledge as there are respects in which reason can be methodically employed. This means in effect that there are only four basic areas for intellectual discipline, although each of these four areas admits of further subdivisions. The first respect in which reason can be methodically and systematically employed concerns the *universe of nature* [ad philosophiam naturalem pertinet], to which pertains all deliberation about the order which the mind discerns as obtaining or able to obtain in things independently of the activity of our thinking. ('Nature' here must be understood in such a way as to include whatever there is of being [ita quod sub naturali philosophia comprehendamus et metaphysicam].) The second respect in which reason can be

day circles of British and American philosophy, metaphysics is identified as that part of philosophical inquiry which is concerned with first-order questions—questions about that which is and happens in the world—to which laboratory and field studies are at most marginally relevant (and therefore to which science has in principle no answers), but to which tenable and reasoned answers are nonetheless supposedly able to be given.[2]

The classic usage of the term "natural philosophy" above indicated, therefore, enables us to employ the term "metaphysics" in this section in a sense that is common to the earlier periods of culture in which science did not exist in a state recognized as distinct from that of philosophy, and to the contemporary period in which the possibility of a philosophical knowledge of the world distinct in kind from the knowledge arrived at through science is challenged.[3] For in speaking of "metaphysical issues," we have in mind here precisely those questions that have arisen in the course of evolutionary and other scientific studies, questions which certainly concern our first-order understanding of that which is and happens in the world, but which equally certainly have not been resolved or significantly modified by any applications of experimental techniques or refinements in instrumentation.

Issues of this kind are not hard to come by in evolutionary studies. Take, for example, the dispute among biologists over reductionism. Eighteenth and nineteenth century controversies are usually left to historians of biology. But in this question such biologists as Ernst Mayr and Paul Weisz find themselves at the same point as their centuries-old predecessors. Weisz holds for reduction of life to chemical processes;[4] Mayr holds against it.[5] But what is disputed between these men is whether

---

methodically employed concerns the *universe of discourse* [pertinet ad rationalem philosophiam], to which pertains all deliberation about the order which the mind finds as obtaining only in and by virtue of its own activity, as, for example, the consideration of the relationships obtaining among the parts of speech, the consideration of the relations among principles, or of premises to conclusions. The third respect in which reason can be methodically employed concerns the *universe of morality*, to which pertains all deliberation about the proper ordering of human life so far as it is subject to our voluntary control. The remaining and fourth respect in which reason can be methodically employed concerns the *universe of productivity* or making, to which pertains all deliberation about the order which man can introduce into things through manipulation and control of his physical environment."

[2] A good discussion of this contemporary meaning and usage of the term "metaphysics," together with an indication of the recent literature in which this usage has been established, can be found in M. J. Adler, *The Conditions of Philosophy* (New York: Atheneum, 1965), pp. 42–48, esp. fns. 18, 19, and 21.

[3] E.g., see Richard Rorty, "Do Analysts and Metaphysicians Disagree?" *ACPA Proceedings*, XLI (1967), pp. 39–53; the several contributions to *The Linguistic Turn*, ed. by R. Rorty (Chicago: University of Chicago Press, 1967); and the essays by M. Schlick (pp. 53–59 and 209–277), R. Carnap (pp. 60–81), A. J. Ayer (pp. 228–243), F. P. Ramsey (pp. 321–326), G. Ryle (pp. 327–344), and F. Waismann (pp. 345–380), in *Logical Positivism*, ed. by A. J. Ayer (Glencoe: The Free Press, 1959).

[4] P. B. Weisz, *The Science of Biology* (2nd ed.; New York: McGraw-Hill, 1963), pp. 12–13.

[5] Ernst Mayr, *Animal Species and Evolution* (Cambridge, Mass.: Harvard, 1963), p. 6.

organism or molecule be the more or less unitary being in terms of which the other should be explained. Weisz wants to see the organism as a machine the functions of which are explained in terms of its parts, whereas Mayr denies that such a perspective is genuine.

Clearly, the disputed question is whether the relatively independent or "autonomous" unit be molecule or organism. And this dispute conceals an even deeper one: holding for organism or molecule as the absolute or "primary" unit leaves at issue whether cell, multicellular organizations, ecological community (species) or geographical community is the absolute unit of life—or whether life is not a continuum of all of these. Thus Julian Huxley takes the position that life is just such a continuum.[6] And in a later work, still claiming to speak scientifically, Huxley can claim that "the picture of the universe provided by modern science is of a single process of self-transformation, during which new possibilities can be realized."[7] Hence this most distinguished biologist opts for the total universe as the absolute unit in terms of which to understand the organism and the molecule.

These are of course not the only positions taken regarding the question of the fundamental structural units of the world. But we need not go deeply into the history of ideas in order to note that molecule, organism, and total universe differ notably in size. If the best biologists cannot agree in discerning differences so gross, it may be that biology as a science cannot deal with the type of evidence that indicates which are the more or less absolute unitary beings in the universe, in the sense that it cannot discern the fundamental structural features which the intelligibility afforded by the world postulates as its necessary condition. It is precisely this last type of evidence that metaphysical philosophy in its most formal reflections has concerned itself with discerning. Whether this task is proper and possible is the same as the question as to whether or not there is a metaphysical dimension to man's awareness of the world.

It may be, of course, that such issues as those just indicated, though indeed *raised* in the course of evolutionary and other scientific discussions, are strictly insoluble, as many modern philosophers contend, precisely because no "crucial test" can be devised as the means of resolution, i.e., precisely because they cannot be attacked through laboratory or field work. A variant of this view, which denies outright the possibility of metaphysics, is eloquently voiced by John Dewey in the first reading below. The second reading, by Benedict Ashley, argues for exactly the opposite position, contending that the evolutionary picture makes sense at bottom only in terms of a classically metaphysical analysis and scheme. The third reading, by C. H. Waddington, argues in effect that the great interest of evolutionary

---

[6] J. Huxley, *Evolution: The Modern Synthesis* (New York: Harper, 1942), p. 169.
[7] J. Huxley, *Religion Without Revelation* (New York: Mentor, 1957), p. 190.

biology for humanistic thought lies precisely in the fact that it renders inescapable many aspects of the classical philosophical issues concerning causality, development, and individual identity which are missed or not relevant in the mathematical schemes of the physical sciences.

The influence of Darwinism

John Dewey

CONTEXTUALIZING COMMENTS

Almost from his student days John Dewey was convinced that the recent discoveries in biology and psychology, especially Darwin's theory of evolution, demanded a radical change in the conception of nature and of philosophical thinking. Philosophy, in Dewey's assessment, was born and reared in the emotional and social life of man, and that is precisely where it must remain. Only a philosophy imperfectly aware of its nature and function would claim to be more than an intellectual expression of the aspirations and ideals of a particular culture. Thus "Metaphysics" is the name of philosophy so close to the Greeks in (cultural) aspiration and (cultural) ideal that it has not gotten around to meeting Darwin—or to realizing that inquiry into such problems as the existence of God, creation, direction or end in the cosmic unfolding, are futile and unrewarding, to say nothing of culturally obsolete. Our justification therefore for placing Dewey under the heading of "Metaphysical Issues" is not at all that we intend to force him to endorse a philosophical perspective which he thought best abandoned, but simply to focus attention on the most basic philosophical issue of all: is an intellectual understanding of the world at all possible which is not in every way bound to the cultural state of scientific progress, or, more generally, bound in every aspect to the level of historic conscious-ness attained by any given culture? Dewey's reply is in the negative.

The influence of Darwinism

That the publication of the "Origin of Species" marked an epoch in the development of the natural sciences is well known to the layman. That the combination of the very words origin and species embodied an intellectual revolt and introduced a new intellectual temper is easily overlooked by the expert. The conceptions that had reigned in the philosophy of nature and knowledge for two thousand years, the conceptions that had become the familiar furniture of the mind, rested on the assumption of the superiority of the fixed and final; they rested upon treating change and origin as signs of defect and unreality. In laying hands upon the sacred ark of absolute permanency, in treating the forms that had been regarded as types of fixity and perfection as originating and passing away, the "Origin of Species" introduced a mode of thinking that in the end was bound to transform the logic of knowledge, and hence the treatment of morals, politics, and religion.

No wonder, then, that the publication of Darwin's book, a half century ago, precipitated a crisis. The true nature of the controversy is easily con-cealed from us, however, by the theological clamor that attended it. The vivid and popular features of the anti-Darwinian row tended to leave the impression that the issue was between science on one side and theology on the other. Such was not the case—the issue lay primarily within science itself, as Darwin himself early recognized. The theological outcry he dis-counted from the start, hardly noticing it save as it bore upon the "feelings of his female relatives." But for two decades before final publication he contemplated the possibility of being put down by his scientific peers as a fool or as crazy; and he set, as the measure of his success, the degree in which he should affect three men of science: Lyell in geology, Hooker in botany, and Huxley in zoölogy.

Religious considerations lent fervor to the controversy, but they did not provoke it. Intellectually, religious emotions are not creative but con-servative. They attach themselves readily to the current view of the world and consecrate it. They steep and dye intellectual fabrics in the seething vat of emotions; they do not form their warp and woof. There is not, I think, an instance of any large idea about the world being independently generated by religion. Although the ideas that rose up like armed men against Dar-winism owed their intensity to religious associations, their origin and meaning are to be sought in science and philosophy, not in religion.

II

Few words in our language foreshorten intellectual history as much as does the word species. The Greeks, in initiating the intellectual life of Europe, were impressed by characteristic traits of the life of plants and animals; so impressed indeed that they made these traits the key to defining nature and to explaining mind and society. And truly, life is so wonderful that a seemingly successful reading of its mystery might well lead men to believe that the key to the secrets of heaven and earth was in their hands. The Greek rendering of this mystery, the Greek formulation of the aim and standard of knowledge, was in the course of time embodied in the word species, and it controlled philosophy for two thousand years. To understand the intellectual face-about expressed in the phrase "Origin of Species," we must, then, understand the long dominant idea against which it is a protest.

Consider how men were impressed by the facts of life. Their eyes fell upon certain things slight in bulk, and frail in structure. To every appearance, these perceived things were inert and passive. Suddenly, under certain circumstances, these things—henceforth known as seeds or eggs or germs—begin to change, to change rapidly in size, form, and qualities. Rapid and extensive changes occur, however, in many things—as when wood is touched by fire. But the changes in the living thing are orderly; they are cumulative; they tend constantly in one direction; they do not, like other changes, destroy or consume, or pass fruitless into wandering flux; they realize and fulfil. Each successive stage, no matter how unlike its predecessor, preserves its net effect and also prepares the way for a fuller activity on the part of its successor. In living beings, changes do not happen as they seem to happen elsewhere, any which way; the earlier changes are regulated in view of later results. This progressive organization does not cease till there is achieved a true final term, a *telos*, a completed, perfected end. This final form exercises in turn a plentitude of functions, not the least noteworthy of which is production of germs like those from which it took its own origin, germs capable of the same cycle of self-fulfilling activity.

But the whole miraculous tale is not yet told. The same drama is enacted to the same destiny in countless myriads of individuals so sundered in time, so severed in space, that they have no opportunity for mutual consultation and no means of interaction. As an old writer quaintly said, "things of the same kind go through the same formalities"—celebrate, as it were, the same ceremonial rites.

This formal activity which operates throughout a series of changes and holds them to a single course; which subordinates their aimless flux to its own perfect manifestation; which, leaping the boundaries of space and time, keeps individuals distant in space and remote in time to a uniform type of structure and function: this principle seemed to give insight into the very nature of reality itself. To it Aristotle gave the name, *eidos*. This term the scholastics translated as *species*.

The force of this term was deepened by its application to everything in the universe that observes order in flux and manifests constancy through change. From the casual drift of daily weather, through the uneven recurrence of seasons and unequal return of seed time and harvest, up to the majestic sweep of the heavens—the image of eternity in time—and from this to the unchanging pure and contemplative intelligence beyond nature lies one unbroken fulfilment of ends. Nature as a whole is a progressive realization of purpose strictly comparable to the realization of purpose in any single plant or animal.

The conception of *eidos*, species, a fixed form and final cause, was the central principle of knowledge as well as of nature. Upon it rested the logic of science. Change as change is mere flux and lapse; it insults intelligence. Genuinely to know is to grasp a permanent end that realizes itself through changes, holding them thereby within the metes and bounds of fixed truth. Completely to know is to relate all special forms to their one single end and good: pure contemplative intelligence. Since, however, the scene of nature which directly confronts us is in change, nature as directly and practically experienced does not satisfy the conditions of knowledge. Human experience is in flux, and hence the instrumentalities of sense-perception and of inference based upon observation are condemned in advance. Science is compelled to aim at realities lying behind and beyond the processes of nature, and to carry on its search for these realities by means of rational forms transcending ordinary modes of perception and inference.

There are, indeed, but two alternative courses. We must either find the appropriate objects and organs of knowledge in the mutual interactions of changing things; or else, to escape the infection of change, we *must* seek them in some transcendent and supernal region. The human mind, deliberately as it were, exhausted the logic of the changeless, the final, and the transcendent, before it essayed adventure on the pathless wastes of generation and transformation. We dispose all too easily of the efforts of the schoolmen to interpret nature and mind in terms of real essences, hidden forms, and occult faculties, forgetful of the seriousness and dignity of the ideas that lay behind. We dispose of them by laughing at the famous gentleman who accounted for the fact that opium put people to sleep on the ground it had a dormitive faculty. But the doctrine, held in our own day, that knowledge of the plant that yields the poppy consists in referring the peculiarities of an individual to a type, to a universal form, a doctrine so firmly established that any other method of knowing was conceived to be unphilosophical and unscientific, is a survival of precisely the same logic. This identity of conception in the scholastic and anti-Darwinian theory may well suggest greater sympathy for what has become unfamiliar as well as greater humility regarding the further unfamiliarities that history has in store.

Darwin was not, of course, the first to question the classic philosophy of nature and of knowledge. The beginnings of the revolution are in the physical science of the sixteenth and seventeenth centuries. When Galileo said: "It is my opinion that the earth is very noble and admirable by reason of so many and so different alterations and generations which are incessantly made therein," he expressed the changed temper that was coming over the world; the transfer of interest from the permanent to the changing. When Descartes said: "The nature of physical things is much more easily conceived when they are beheld coming gradually into existence, than when they are only considered as produced at once in a finished and perfect state," the modern world became self-conscious of the logic that was henceforth to control it, the logic of which Darwin's "Origin of Species" is the latest scientific achievement. Without the methods of Copernicus, Kepler, Galileo, and their successors in astronomy, physics, and chemistry, Darwin would have been helpless in the organic sciences. But prior to Darwin the impact of the new scientific method upon life, mind, and politics, had been arrested, because between these ideal or moral interests and the inorganic world intervened the kingdom of plants and animals. The gates of the garden of life were barred to the new ideas; and only through this garden was there access to mind and politics. The influence of Darwin upon philosophy resides in his having conquered the phenomena of life for the principle of transition, and thereby freed the new logic for application to mind and morals and life. When he said of species what Galileo had said of the earth, *epur se muove*,[1] he emancipated, once for all, genetic and experimental ideas as an organon of asking questions and looking for explanations.

III

The exact bearings upon philosophy of the new logical outlook are, of course, as yet, uncertain and inchoate. We live in the twilight of intellectual transition. One must add the rashness of the prophet to the stubbornness of the partizan to venture a systematic exposition of the influence upon philosophy of the Darwinian method. At best, we can but inquire as to its general bearing—the effect upon mental temper and complexion, upon that body of half-conscious, half-instinctive intellectual aversions and preferences which determine, after all, our more deliberate intellectual enterprises. In this vague inquiry there happens to exist as a kind of touchstone a problem of long historic currency that has also been much discussed in Darwinian literature. I refer to the old problem of design *versus* chance, mind *versus* matter, as the causal explanation, first or final, of things.

As we have already seen, the classic notion of species carried with it the idea of purpose. In all living forms, a specific type is present directing the earlier stages of growth to the realization of its own perfection. Since this purposive regulative principle is not visible to the senses, it follows that it must be an ideal or rational force. Since, however, the perfect form is gradually approximated through the sensible changes, it also follows that in and through a sensible realm a rational ideal force is working out its own ultimate manifestation. These inferences were extended to nature: (a) She does nothing in vain; but all for an ulterior purpose. (b) Within natural sensible events there is therefore contained a spiritual causal force, which as spiritual escapes perception, but is apprehended by an enlightened reason. (c) The manifestation of this principle brings about a subordination of matter and sense to its own realization, and this ultimate fulfilment is the goal of nature and man. The design argument thus operated in two directions. Purposefulness accounted for the intelligibility of nature and the possibility of science, while the absolute or cosmic character of this purposefulness gave sanction and worth to the moral and religious endeavors of man. Science was underpinned and morals authorized by one and the same principle, and their mutual agreement was eternally guaranteed.

This philosophy remained, in spite of sceptical and polemic outbursts, the official and the regnant philosophy of Europe for over two thousand years. The expulsion of fixed first and final causes from astronomy, physics, and chemistry had indeed given the doctrine something of a shock. But, on the other hand, increased acquaintance with the details of plant and animal life operated as a counterbalance and perhaps even strengthened the argument from design. The marvelous adaptations of organisms to their environment, of organs to the organism, of unlike parts of a complex organ—like the eye—to the organ itself; the foreshadowing by lower forms of the higher; the preparation in earlier stages of growth for organs that only later had their functioning—these things were increasingly recognized with the progress of botany, zoölogy, paleontology, and embryology. Together, they added such prestige to the design argument that by the late eighteenth century it was, as approved by the sciences of organic life, the central point of theistic and idealistic philosophy.

The Darwinian principle of natural selection cut straight under this philosophy. If all organic adaptations are due simply to constant variation and the elimination of those variations which are harmful in the struggle for existence that is brought about by excessive reproduction, there is no call for a prior intelligent causal force to plan and preordain them. Hostile critics charged Darwin with materialism and with making chance the cause of the universe.

Some naturalists, like Asa Gray, favored the Darwinian principle and attempted to reconcile it with design. Gray held to what may be called design on the installment plan. If we conceive the "stream of variations" to be itself intended, we may suppose that each successive variation was designed from the first to be selected. In that case, variation, struggle and selection simply define the mechanism of "secondary causes" through which the "first cause" acts; and the doctrine of design is none the worse off because we know more of its *modus operandi*.

Darwin could not accept this mediating proposal. He admits or rather he asserts that it is "impossible to conceive this immense and wonderful universe including man with his capacity of looking far backwards and far into futurity as the result of blind chance or necessity." [2] But nevertheless he holds that since variations are in useless as well as useful directions, and since the latter are sifted out simply by the stress of the conditions of struggle for existence, the design argument as applied to living beings is unjustifiable; and its lack of support there deprives it of scientific value as applied to nature in general. If the variations of the pigeon, which under artificial selection give the pouter pigeon, are not preordained for the sake of the breeder, by what logic do we argue that variations resulting in natural species are pre-designed? [3]

IV

So much for some of the more obvious facts of the discussion of design *versus* chance, as causal principles of nature and life as a whole. We brought up this discussion, you recall, as a crucial instance. What does our touchstone indicate as to the bearing of Darwinian ideas upon philosophy? In the first place, the new logic outlaws, flanks, dismisses—what you will —one type of problems and substitutes for it another type. Philosophy forswears inquiry after absolute origins and absolute finalities in order to explore specific values and the specific conditions that generate them.

Darwin concluded that the impossibility of assigning the world to chance as a whole and to design in its parts indicated the insolubility of the question. Two radically different reasons, however, may be given as to why a problem is insoluble. One reason is that the problem is too high for intelligence; the other is that the question in its very asking makes assumptions that render the question meaningless. The latter alternative is unerringly pointed to in the celebrated case of design *versus* chance. Once admit that the sole verifiable or fruitful object of knowledge is the particular set of changes that generate the object of study together with the consequences that then flow from it, and no intelligible question can be asked about what, by assumption, lies outside. To assert—as is often asserted—that specific values of particular truth, social bonds and forms of beauty, if they can be shown to be generated by concretely knowable conditions, are meaningless and in vain; to assert that they are justified only when they and their particular causes and effects have all at once been gathered up into some inclusive first cause and some exhaustive final goal, is intellectual atavism. Such argumentation is reversion to the logic that explained the extinction of fire by water through the formal essence of aqueousness and the quenching of thirst by water through the final cause of aqueousness. Whether used in the case of the special event or that of life as a whole, such logic only abstracts some aspect of the existing course of events in order to reduplicate it as a petrified eternal principle by which to explain the very changes of which it is the formalization.

When Henry Sidgwick casually remarked in a letter that as he grew older his interest in what or who made the world was altering into interest in what kind of a world it is anyway, his voicing of a common experience of our own day illustrates also the nature of that intellectual transformation effected by the Darwinian logic. Interest shifts from the wholesale essence back of special changes to the question of how special changes serve and defeat concrete purposes; shifts from an intelligence that shaped things once for all to the particular intelligences which things are even now shaping; shifts from an ultimate goal of good to the direct increments of justice and happiness that intelligent administration of existent conditions may beget and that present carelessness or stupidity will destroy or forego.

In the second place, the classic type of logic inevitably set philosophy upon proving that life *must* have certain qualities and values—no matter how experience presents the matter—because of some remote cause and eventual goal. The duty of wholesale justification inevitably accompanies all thinking that makes the meaning of special occurrences depend upon something that once and for all lies behind them. The habit of derogating from present meanings and uses prevents our looking the facts of experience in the face; it prevents serious acknowledgment of the evils they present and serious concern with the goods they promise but do not as yet fulfil. It turns thought to the business of finding a wholesale transcendent remedy for the one and guarantee for the other. One is reminded of the way many moralists and theologians greeted Herbert Spencer's recognition of an unknowable energy from which welled up the phenomenal physical processes without and the conscious operations within. Merely because Spencer labeled his unknowable energy "God," this faded piece of metaphysical goods was greeted as an important and grateful concession to the reality of the spiritual realm. Were it not for the deep hold of the habit of seeking justification for ideal values in the remote and transcendent, surely this reference of them to an unknowable absolute would be despised in comparison with the demonstrations of experience that knowable energies are daily generating about us precious values.

The displacing of this wholesale type of philosophy will doubtless not arrive by sheer logical disproof, but rather by growing recognition of its futility. Were it a thousand times true that opium produces sleep because of its dormitive energy, yet the inducing of sleep in the tired, and the recovery to waking life of the poisoned, would not be thereby one least step forwarded. And were it a thousand times dialectically demonstrated that life as a whole is regulated by a transcendent principle to a final inclusive goal, none the less truth and error, health and disease, good and evil, hope and fear in the concrete, would remain just what and where they now are. To improve our education, to ameliorate our manners, to advance our politics, we must have recourse to specific conditions of generation.

Finally, the new logic introduces responsibility into the intellectual life. To idealize and rationalize the universe at large is after all a confession of inability to master the courses of things that specifically concern us. As long as mankind suffered from this impotency, it naturally shifted a burden of responsibility that it could not carry over to the more competent shoulders of the transcendent cause. But if insight into specific conditions of value and into specific consequences of ideas is possible, philosophy must in time become a method of locating and interpreting the more serious of the conflicts that occur in life, and a method of projecting ways for dealing with them: a method of moral and political diagnosis and prognosis.

The claim to formulate *a priori* the legislative constitution of the universe is by its nature a claim that may lead to elaborate dialectic developments. But it is also one that removes these very conclusions from subjection to experimental test, for, by definition, these results make no differences in the detailed course of events. But a philosophy that humbles its pretensions to the work of projecting hypotheses for the education and conduct of mind, individual and social, is thereby subjected to test by the way in which the ideas it propounds work out in practice. In having modesty forced upon it, philosophy also acquires responsibility.

Doubtless I seem to have violated the implied promise of my earlier remarks and to have turned both prophet and partizan. But in anticipating the direction of the transformations in philosophy to be wrought by the Darwinian genetic and experimental logic, I do not profess to speak for any save those who yield themselves consciously or unconsciously to this logic. No one can fairly deny that at present there are two effects of the Darwinian mode of thinking. On the one hand they are making many sincere and vital efforts to revise our traditional philosophic conceptions in accordance with its demands. On the other hand, there is as definitely a recrudescence of absolutistic philosophies; an assertion of a type of philosophic knowing distinct from that of the sciences, one which opens to us another kind of reality from that to which the sciences give access; an appeal through experience to something that essentially goes beyond experience. This reaction affects popular creeds and religious movements as well as technical philosophies. The very conquest of the biological sciences by the new ideas has led many to proclaim an explicit and rigid separation of philosophy from science.

Old ideas give way slowly; for they are more than abstract logical forms and categories. They are habits, predispositions, deeply engrained attitudes of aversion and preference. Moreover, the conviction persists— though history shows it to be a hallucination—that all the questions that the human mind has asked are questions that can be answered in terms of the alternatives that the questions themselves present. But in fact intellectual progress usually occurs through sheer abandonment of questions together with both of the alternatives they assume—an abandonment that results from their decreasing vitality and a change of urgent interest. We do not solve them: we get over them. Old questions are solved by disappearing, evaporating, while new questions corresponding to the changed attitude of endeavor and preference take their place. Doubtless the greatest dissolvent in contemporary thought of old questions, the greatest precipitant of new methods, new intentions, new problems, is the one effected by the scientific revolution that found its climax in the "Origin of Species."

NOTES

[1] *Epur si muove!* : And yet it moves! (Ed.)

[2] "Life and Letters," (New York: D. Appleton & Co., 1888), Vol. I, p. 282; cf. p. 285. (Ed.)

[3] "Life and Letters," Vol. II, pp. 146, 170, 245; Vol. I, pp. 283–284. See also the closing portion of his "Variations of Animals and Plants under Domestication." (Ed.)

Change and process

Benedict M. Ashley

CONTEXTUALIZING COMMENTS

Benedict Ashley's philosophical reflections have led him to convictions the very opposite to those we have just seen Dewey reach. Like Dewey, he as a young man felt the immense pressure of recent discoveries in science, especially of the refinements on Darwin's theory of evolution with their possible implications for the social order. Thus in his student years at the University of Chicago he was powerfully drawn by the dialectical thought of Marxism.

But in the course of his intellectual odyssey, and while continuing to be deeply in sympathy with the insistence of Marx and Dewey on the intermediate relation of theory and practice, Ashley found himself situated by his assessments of more and more features of the scientific world-picture exactly within the conceptual tradition stemming from the classical Aristotelian way of posing the central question of philosophy:

The question which was raised of old and is raised now and always, and is ever the subject of doubt, viz. what being is, is just the question, what is substance? For it is this that some assert to be one, others more than one, and that some assert to be limited in number, others unlimited. And so we also must consider chiefly and primarily and almost exclusively what that is which *is* in *this* sense.[1]

This essay was specially adapted by Dr. Ashley for this volume from an unpublished manuscript by the same author on man in the scientific age. The headings and subdivisions are supplied by the editor.

[1] Aristotle, *Metaphysics*, Bk. VII, ch. 1, 1028 b1–7.

On the one hand, Ashley came to consider that the logical and experientially consistent alternative to a fixist world of eternal substances (exactly what the ancients and medievals essayed) is not a world of reified processes, but rather a world of relatively stable fundamental natural units (which was from the outset the notion alone basic to the contingent substances of Aristotle's sublunary world) which come into existence through processes of generation, exist in an interaction process through activities and passivities, to be in the end transformed into other natural units or substances by continuing processes which in the course of interaction overcome the original units' central stabilizing tendencies and so yield their passing away or destruction.

On the other hand, Ashley became equally convinced that the fundamental objection to a "pure" process philosophy is that, in the end, it would require us to eliminate from the world picture our knowledge of ourselves as observers. Denial of the natural unit, of the contingent substantial nature, vitiates the consistency of process philosophies with the data of cognitive transcendence, or compels the introduction of "eternal objects" à la Whitehead—in either case evades the real issue of change, which is to understand the sustainment of temporal identity in a world where nothing is exempt from process. Since, moreover, in discerning a plurality of natural units as the fundamental structural features which the intelligibility afforded by the world postulates as its necessary and adequate condition, all actualization comes by interaction, at one stroke monism is ruled out of consideration and the conception of causality as expressible in the form of conservation laws is laid bare in its ultimate ground.

The ancient and medieval cosmos was divided into two fundamentally different regions, the sublunary region in which alone radical change and process could be found, and the celestial region which was subject to no change save that of circular, mechanical movements in place, movements susceptible moreover of perfect mathematical expression. With the relegation of that dichotomized cosmos to the cultural ashcan through the demonstration that *all* of nature exists in and through process, modern science, far from requiring a radical change in the conception of nature and of philosophical thinking, has precisely freed the original thrust of Aristotelian naturalism from the alien elements of mathematicism and mechanism which deflected it from the first and led to its complete turning aside in the seventeenth century: that is Ashley's key conclusion.

And in the dialectical framework of this book, it is the conclusion which holds the particular interest for us in the light of the previous selection. For Dewey is not the first nor will he be the last to proclaim in the name of "evolutionary science" a radical revolution and clean break in philosophy; and yet, in view of the question of being as posed by Aristotle, it is incumbent on those who would relativize philosophy in terms of the cultural state of scientific progress and who would see in philosophy accordingly no more than an effort to draw out the ultimate implications of scientific theories in terms of "world-view," to *demonstrate* and not merely *proclaim* that the data of evolutionary science render all substantialist interpretation of nature radically inept. Only then would their position be a reasonable and not authoritarian or dogmatic one; for it can hardly be claimed that evolution renders metaphysics in the classical sense "impossible" and outdated by time if an empirically sound assessment of the materials on which evolutionary thought is primarily based can be shown to be in accord with the basic insights of an act/potency analysis of substance.

Change and Process

I. THE PROBLEM OF NATURE

In trying to understand a changing universe, we must begin by a fundamental option with regard to "the one and the many." That the universe is one being appears to me not only a dangerous assumption, but clearly contrary to fact. You and I are not each other. This radical plurality of things is not so evident among non-persons, yet can we really doubt the plurality of animals or of trees, or of two stones? As we move away from persons to animals, to plants, to minerals, the distinctness of units becomes more and more obscure, and in many cases, becomes quite doubtful; yet, if I am convinced that my body is mine and not yours, I can hardly doubt that two pebbles awash in a stream are just as distinct from one another and from the liquid in which they jostle. Of course, these units exist tied together by a network of external relations, but there seems no evidence that their external relations are more fundamental than their inner unity. The monist who sees the universe as a continuous whole, rather than as an aggregation of distinct natural units, can appeal to scientific theories which picture the cosmos as a *field* in which atoms or sub-atomic particles are only disturbances, but such theories, valuable as mathematical models, do not invalidate or replace our observation of the reality of gross units like a man or tree, or minute units like the molecule or atom.

These units are not themselves mere artifacts of observation, nor are they themselves chance aggregates. They are *natural* units, in that sense of "nature" which has been fundamental in all western scientific thinking. A natural unit is one which develops from some *inner* characteristic; it is not put together from the *outside* as is an artifact or a chance aggregate. Naturalness admits of degrees. The natural units like a man, a dog, a tree (or even a crystal, a molecule, an atom when these exist separately and not in a chemical combination with other units) are the primary natural beings. All their combinations and collections and interactions are consequent on this basic existence as primary units. As we go up the evolutionary scale, things become more natural, less mechanical. The behavior of a seed as it unfolds into a plant is more natural than is the ebb and tide of the ocean. As things become more distinctly individual and inner-determined in their behavior, as they rise above the chaotic and random, their naturalness appears more clearly. It is only in this sense that the term "natural law" makes sense. A natural law is not a disembodied plan or decree, it is simply the expression of the fact that each natural unit has a behavior primarily determined by its own inner constitution and not by imposition from without.

Nature is thus something which is always in *process*. The word "process" indicates "to go forward," to go toward order out of chaos, toward unity out of plurality, toward regularity out of chanciness. In the term "process" (made so popular by Whitehead in his attempt to give a philosophical analysis of the modern world-view), there is an inevitable *teleological* implication.

There is, however, a broader and more fundamental concept than that of process. It is *change*, a term which does not necessarily imply teleology, and which is thus more neutral. It was the great contribution of Heraclitus to focus western thought on this concept of change, and his own view of it seems to have been anti-teleological.

At first sight, we are impressed by the stability of the earth and the heavens. Deeper investigation, constantly reinforced by the development of the physical sciences, shows that everything changes. In the nineteenth century, it appeared that once the atomic level is reached, matter is essentially inert. Now we know that the atom is in perpetual internal change, and is composed of "particles" constantly transformed into one another in everlasting metamorphosis. Most particles decay in very brief periods, but even those that seem relatively stable are constantly being "annihilated" and "created," absorbed and emitted. There is literally nothing in the universe which is permanent; all things constantly pass into other things and these into others.

Most of the apparently stable bodies, the planets, the stars, are mere collections of many materials which are loosely joined together and constantly in turmoil. Even living things, which strangely are in a way the most stable of substances, are found to have only the stability of *homeostasis*, the stability that results from an equality of input and output in an on-going process. The human body itself is completely renewed in its actual material content in a period of a few years. Thus the scholastic view that the universe is *contingent* has been borne out with a vividness that the medievals could never have anticipated. Thus the world is in perpetual flux and transformation.

This universality of change raises a paradox for our personal self-knowledge. On the one hand, I am vividly aware of myself as a "stream of consciousness." I experience my own existence as a perpetual flow of images, impressions, ideas, desires, fears, enterprises which can never be fixed. I can appreciate the immense discipline required of the yogi who seeks to quiet the flux of his interior life, and I suspect that he does so only at the expense of psychic vacuity.

Even as I become aware of my own body, I find that I exist in a perpetual physiological flux of hunger and tensions, of growth and of decay. My contact with the world through my body is always a flow of events and happenings. Lights impinge on my eyes, sounds on my ears, objects press against my body, or give way before me. I more and more realize that as a human being I am a transient, a cloud that passes.

On the other hand, the fact of my unity as a person brings to me an equally profound conviction of my permanence and stability. Looking back in memory, I find myself emerging from nothing (I have no awareness of the previous existences claimed by some), and I fear death as an inevitable dissolution of at least my present personal bodily wholeness. Yet I am firmly convinced that I am the same person as in my youth. My memory ties my life together. More than that, I am aware that when I make a decision, I am committing myself in the future, and that I am guilty or deserving of reward for the decisions of the past. I see my life as somehow unified in plan by my decisions, however much my plans have gone awry. What is still more striking is that I am convinced that through thought and decision, I somehow transcend time. I can see the past, the present, and the future as exhibiting a meaningful order, some aspects of which are of my own making. If this were not so, all scientific research or creative planning would be impossible. The very notion of natural law and order in a changing and temporal universe implies that the scientist himself or the moral man rises above the flux of change. The very ideal of scientific objectivity, as well as the ideal of moral responsibility, imply that within the singular events of history there are also aspects of universal order.

Parmenides first posed this dilemma for us. Either we admit the universal flux of Heraclitus and deny the possibility of universal truth and certitude, or we stand for the ideal of universal truth and certitude, and deny the empirical reality of change. The thinkers of the East seem to have accepted the latter position. For them the empirical world of flux is Maya, an illusion. But the main line of western thought has sought to reconcile these apparently irreconcilable poles of experience. In my opinion, however, not a few of these western thinkers have in fact adopted crypto-Parmenidean solutions, by minimizing the radical reality of change. Two of these minimizations are *mechanism* and *mathematicism*.

A. Mechanism

The way of mechanism, originally proposed by Democritus, remains even today very strong in the thinking of many scientists. In its purest form, it attempts to explain nature and man by reducing all the complex objects of which we are aware to simple particles, "atoms." These particles are the only primary reality, and are to be considered genuinely atomic, that is, irreducible to other more fundamental particles. These atoms are incapable of any *internal* change. They have quantitative properties, but not qualitative properties, except *motion* and resistance to motion. These particles exist in a void. This void cannot be considered as an attenuated bodily reality, a gas or a fluid (since this would be to introduce a non-atomic, non-mechanical, second type of matter) but is absolutely empty.

All qualitative characteristics such as color, temperature, etc. which are found in bodies larger than the atoms can be explained by the arrangement or the motion of the atoms in the void. Except for such rearrangement, the atoms do not change in any way. All change is purely *external*.

Few thinkers after Democritus had the courage to propose this pure form of mechanism. Inconsistently, from a logical point of view, they usually introduced some non-atomic fluid to fill the interstices between the atoms, and admitted that such a fluid was subject to *internal* change since it had to expand or contract sufficiently to permit the atoms to move through it, or at least to flow by them. Thus Descartes admitted an inter-atomic fluid, while the physicists of the nineteenth century admitted an ether which actually interpenetrated the atoms. With the twentieth century, however, and the discovery of the electron and the photon, it began to appear that mechanism of the purest sort might be re-instituted, and Robert Millikan, the discoverer of the electron, did not hesitate to say that Democritus had been proved right.

It must be admitted that the mechanical theory of nature has been an extremely fruitful working hypothesis. It led scientists to explain every quality in terms of quantity and in this way led to greater and greater precision and objectivity. Furthermore, it led scientists to resolve every natural whole into its parts, and to seek to explain the whole in terms of the organization of its parts. There is, thus, in mechanism a basic dualism: the *dualism of the whole and the part*. The dualism of the atom and the void only contributes to the dualism of the whole and the part, since each of the actual bodies of experience is a whole made up of two kinds of parts, atoms and empty spaces.

In each of these wholes, there is on the one hand a *matter*, namely the parts, the atoms and the voids, and on the other a *form*, namely the arrangement or order of these filled and empty bits of space. From this matter and form together arise qualities, and activities, the phenomena of our actual experience. The success of mechanism is to be found in this dualism, since it opens the way to an analysis of every natural phenomenon in terms of something perduring and something variable. Does not my own subjective understanding of myself exhibit this duality of experience? I exist only in the flux of experience, yet I continue to be myself in this flux.

However, in spite of its successes, mechanism has been repeatedly exposed as an inadequate account of change. It avoids the issue of change by positing the basic reality, the atoms, as unchanging. Change is transferred from real bodies to empty space. This concept of "empty space," moreover, turns out to be self-contradictory.

Either empty space is literally *nothing* real, or it is somehow real. If it is literally nothing, how can it explain the reality of change? If two atoms were located in an absolute nothingness, then it would be true that these two bodies would be in contact, since there would be "nothing" between them. If a body moved through absolute nothingness, it would be moving through nothing, and hence would not be progressing at all, since its starting point and its goal would be indistinguishable.

Some reality, therefore, would have to be attributed to the void to make it capable of playing an explanatory or constitutive role. What the mechanist has in mind, obviously, is that it is pure "space," that is, that it has dimensions, distances, like the pure space of the geometer. But how can reality be attributed to this geometrical abstraction? The geometer is able to imagine pure space precisely by removing from his conception every trace of physical substantiality and reality. We would think it absurd to suppose that the Number 1 or the Number 2 could exist in the physical world as physical entities. It is just as absurd to suppose that pure space could so exist. Yet if we add to it the least touch of concrete, physical reality, we run into the position of Descartes or the ether-theorists who introduced a continuous matter filling the interstices of the atoms. But this interstitial matter, since it is a positive reality, must itself be made up of atoms, or be another non-atomic type of matter. If it is made up of atoms, these are either packed tightly (and then no motion is possible) or able to move about (and then we must again introduce an absolute void, or get into an infinite regress). If it is another type of matter which is able to flow past the atoms which move within it, then we have introduced a matter capable of internal, qualitative change (expansion, contraction), and this is to destroy the very essence of the mechanistic position. The one other possibility seems to be an ether which perfectly compenetrates the atoms so that they move through it without in any way affecting it. This, however, involves the notion of two material entities occupying the same place, which seems contradictory to the quantified character of matter.

Thus mechanism really evades the problem of change, since it ascribes change to the void, without being able to give a consistent definition of the void. To resolve the paradox of change by something still more paradoxical is not much help. We can conclude, therefore, that while mechanistic explanations have been and will continue to be useful as an imaginative way of understanding wholes in terms of the arrangement and motion of their parts, mechanism is inadequate as a basic theory of nature.

This has been exemplified repeatedly in scientific experience, but at no time so vividly as in this century. At the very moment when it looked as if the quantum theory had finally come down to explaining all natural phenomena, even light and gravitation, in terms of quanta, i.e. of ultimate particles, the wave-particle complementarity was discovered. It became apparent that no particle in nature can be imagined simply as a discrete "atom," unchanging in itself, but moving in a void. The principle of uncertainty now appears to be not merely a subjective condition of our knowledge, but a basic fact about nature, namely, that "particles" cannot be said to have at the same time a determinate position and a determinate velocity. The physical event in which the electron can be detected in a definite position is one in which it is no longer in motion, and a physical event in which the electron appears as moving is one in which it lacks a precise position, but seems like a wave out in space. We cannot conceive what the particle is like "in between" these events, because in fact the particle exists only in one type of situation or the other, it has existence only in a particle-like or a wave-like situation, not in some intermediate situation in which it has determinate simultaneous position and momentum. We must conclude, therefore, that the "fundamental particles" now known are not the "atoms" of the mechanist. If mechanism is to be applied here, then these particles must be resolved into something more fundamental. Probably a good many scientists hope for this, since mechanism seems to them a very comfortable way of thinking; but as yet their hopes are receiving little consolation from the facts. It now appears that these fundamental particles *all change into each other*. This at once eliminates the whole mechanistic mode of thought, since it shows that in nature there are no unchanging fundamental entities, but that all things are radically changeable. Today, physics seems to have passed definitely beyond mechanism.

B. Mathematicism

Another mode of approaching change which has been closely associated with mechanism, but which is conceptually quite distinct from it, is that initiated in ancient times by Pythagoras and given definitive shape by

Plato and his school, namely *mathematicism*. We have seen that mechanism has as one of its features the elimination of the qualitative aspects of nature by reduction to largely quantitative ones. Mathematicism carries this procedure to its logical conclusion. Mechanism is embarrassed by its adherence to the idea of the unchanging atom, and its puzzling counter-concept of the void. The pure mathematicist overcomes this difficulty by eliminating the atom altogether, and retaining the void, still equipping it, however, with dimensions.

Thus nature becomes essentially simply empty space. For Pythagoras and Plato this space was Euclidean and three-dimensional. With advances in mathematics, it becomes possible to form a more general notion of space by which it can have a desired number of dimensions and any possible curvature. Space can even be conceived without determinate dimensions (a topological space). In such spaces, it is possible to describe any physical phenomenon, whether a static or a dynamic one. A static one is described by locating any body in shape and position within the space. A dynamic phenomenon can be described by considering *time* as a dimension, so that local motion becomes a world-line through space-time.

A difficulty which mathematicism, like mechanism, had to meet at an early stage of its development was the description of qualitative differences like heat or temperature. It was soon found, however, that these qualities could be considered as *intensities* and measured on a scale. Thus heat becomes measurable as a temperature of so many degrees, and this is reduced to the length of a column of mercury; or electric charge can be measured by the movement of the hand on the scale of an instrument. Thus qualities became as quantified as spatial dimensions.

There still seems to remain a qualitative element in geometric definitions, since the term "curvature," for example, is not itself a quantity, but rather a mode or quality of a quantity. The Cartesian analytic geometry, however, and the development of calculus which followed in due time, made it possible to arithmetize geometry, so that all mathematical relations can, to any accuracy desired, be reduced to relations between numbers; then the notion of number can itself be generalized so as to leave only the most tenuous connection with any mechanical concept.

Now it is true that at the present time, the program of mathematization is far from complete in science. Many of the features of the world-view, particularly at the biological and psychological level, have not been developed in the mathematical mode and cannot at the present be reduced to it. There are still some psychologists and biologists who remain frankly sceptical as to whether the phenomena they study can ever be so reduced, without over-simplifying reality. It is quite true that it is possible for mathematization to be disastrously reductive in cases where the scientist considers only those aspects of observed reality for which he has the technique of easy mathematical representation. However, the history of science shows that mathematization need not be reductive in this way. By a refinement of mathematical technique and of methods and instruments of measurement, the mathematical description can be more and more refined, so as to take into account in a wonderfully delicate way the qualitative aspects of reality. Before Pythagoras, who would have imagined that the wonderful qualitative harmonies of music could be expressed numerically? Then for a long time such mathematical descriptions could account for the pitch of notes, but not for their special qualities of timbre. Finally it was discovered that the difference between the sound of a violin and a horn and the human voice could also be exactly expressed mathematically by the proportions of overtones. Similarly it is possible to take every nuance of color and shading in a painting and to reduce it to numbers and then to reproduce this picture exactly on a television screen out of the numerical recipe. What can be done for sound and vision can in principle be done for the other senses. Thus it appears that while mathematical description is undoubtedly a reductive simplification of reality, in principle it can be refined to approximate the complex reality as closely as one desires.

There remain, of course, the phenomena of the subjective realm. Can we mathematicize feelings or thoughts? The non-scientist is likely to reject this notion at once, but I doubt that we should be so hasty. If color can be mathematically described, why may it not be possible to find a mathematical description of feelings? After all, feelings are not totally unanalyzable. The poet and the psychologist find it possible to indicate a mood or a feeling as a particular blend of certain basic feelings in different intensities, just as a shade of color is a blend of certain primary colors in certain intensities. Primary colors can be resolved by the scientist into a wave phenomenon of quantifiable intensities. I see no reason in principle that this cannot be done for feelings.

As for thoughts, it is obvious enough that modern mathematical or symbolic logic has proceeded very efficiently to reduce the *forms* of thought to relations which are mathematically expressible. There remain only the *contents* of thought, but this content again is derived from our experience of the natural world, generalized, refined, analogized, etc. If our experiences of the natural world are mathematically expressible, then it would seem that all of our thinking can be expressed mathematically.

The ambition of mathematicism, therefore, is to translate all that we know of nature and of ourselves as part of nature, into the pure language of mathematics, where every symbol has a precise meaning, and where ultimately all concepts seem to be *relational*. Thus the world will appear as a network of relations, and these relations will be those of identity, difference, equality and inequality, etc. One does not know which to admire most, the boldness of this Pythagorean ambition, or its actual success. The modern world-view (and the technical achievements that have followed on it) is above all rooted in the mathematical method; and its greatest builders, Galileo, Newton, and Einstein, have been men of mathematical insight.

It would seem therefore that what we are really finding out about man is that in essence man is a mathematician. It is by being able to *count* that man has become man, capable of intelligent thought and creativity.

Nevertheless, there is a fundamental difficulty about mathematicism. The Pythagorean and Platonic mathematicism based on the idea of "saving the phenomena" which produced the Ptolemaic system of astronomy had a marvelous success but ultimately proved to be an inadequate method. Galileo and Newton belonged to this tradition but they added to it an important element, namely *empiricism*. By this I do not mean to deny that Ptolemy based his theory on empirical facts; indeed, his whole method was "to save the phenomena," i.e. the facts. But with Galileo and Newton, the search shifted from the mere saving of phenomena, to what may be called a *"creative empiricism."* It is no longer sufficient that a theory save facts, it must be *fertile*, it must enable the researcher to look for new facts. It thus becomes a kind of dialogue with nature, in which nature is coaxed into revealing herself. This is why Newton said "non fingo hypotheses," because for him an explanation must not rest on the mere "saving of phenomena," but must somehow arise from a progressive exploration of nature.

This means that a mathematical theory, no matter how brilliantly clear, taken merely in itself is not a scientific explanation. This mathematical theory must somehow be given a *physical* meaning for it to satisfy the human mind. But what is a "physical" meaning, other than the mathematical relations which are verified as holding in observation? Let us cite a significant instance. From a mathematical point of view, time can be treated as a dimension in a manifold, just like a spatial distance. This is possible because there is an analogy between a length and a time. Yet in human experience it is obvious that a distance and a time are wholly different realities. A mathematical representation which does not distinguish between them is simply not an adequate description. The same holds for qualities. The reduction of color, or sound, or, perhaps, thought or feeling, to quantity can be carried through very well by a process of analogy, but it is a reductive description. A sound or a color are not numbers, or in any univocal way like a number. No matter how refined the mathematical description may be, it gives us a substitute for what we experience.

Now the scientist is apt to say that our experience is something subjective and very vague, while the mathematical description is objective and precise. This is indeed the case, but the point is that the objectivity and precision have been achieved by an abstraction which renders our description of the thing essentially incomplete. The evidence of this is that in order to use a mathematical prediction in actual experimental work, it is necessary to interpret the mathematics in physical terms. For example, a given quantity must be interpreted as *weight*, as *heat*, as a difference of *color*, as distance, or motion, or time. The experimenter knows what these physical terms mean by reference to his own sense experiences in the laboratory. He knows how to distinguish between a thermometer and a clock and a measuring stick, and he does this because he knows what weight, heat, color, distance, motion, or time "mean" in qualitative experience. Without this mediating step where numbers are translated into qualitative experiences, the whole mathematical theory would be empirically unverifiable.

This is why not a few scientists have spoken of mathematics primarily as the "language" of science, and have admitted that the basic questions of physics are deeper than the mathematization. Thus Einstein's theory is fundamentally a reflection on the meaning of the terms "time" and "space" as realities of physical experience, not on the mathematics in which these are expressed.

The apparent ability of the scientist, therefore, to dissolve the universe of experience into mathematical relations depends, as Eddington well said, on "pointer readings," but pointer-readings are not mere mathematical entities, they are a black pointer moving across a black scale on a white dial, and this scale stands itself for other qualitative sensible realities which are not reducible to the quantitative.

This is why, for a long time, in order to actually operate in science, mathematicists made a supplementary use of mechanism, since mechanism provided a way of giving to a mathematical theory a physical, imaginable model. In the twentieth century, beginning especially with the historical writings of Pierre Duhem, it again became fashionable to say that this mechanical model was a mere heuristic device, that in fact the scientist never really needed it, since pure mathematics was sufficient. Today it is often said that the scientist must get used to thinking in merely mathematical terms, without any attempt to imagine or visualize the physical reality which the mathematics describes. However, as I have indicated, this cannot ultimately succeed, since the empirical verification of the theory must come back to a physical experience which is not homogeneous with the mathematical theory.

This criticism is confirmed by the history of modern mathematics. The claims of Bertrand Russell, and, from a different point of view, of Hilbert, that mathematics is a purely logical system without reference to the physical world, now appear untenable. Goedel has shown that it is not possible to prove that any mathematical system is consistent, without ultimately testing it by application to a model, and even if this model be only the system of natural numbers, these natural numbers cannot be shown to be consistent except by reference to some human, non-mathematical experience. The Intuitionists have tried to find this experience in something purely internal and subjective, namely, our awareness of the instances of subjective time. It would be much more satisfactory to admit that the numbers are derived from our objective experience of countable objects. Since we have already established that in the world there are distinct natural units, this objectivity of natural numbers is verified. The fact that you and I are two bodies, and that my body is a continuous whole with parts exterior to one another, is a sufficient basis for all of mathematics.

The matter-form dualism found in mechanism appears in a more abstract form in mathematicism, as is apparent from Plato's own dualism between Necessity and Reason. The formal factor is given by the mathematical order, the material element is space, or (in arithmetized geometry) the numbers or elements. It is obvious that the more abstract the mathematics becomes the more and more the material factor is reduced, yet there always remains present the notion of discrete elements or sets. These cannot be illustrated or defined except by reference to space, and this space in turn cannot be imagined without reference to the Euclidean space of our local experience. Thus we might characterize the mathematical approach as one which seeks to reduce the material element in the world picture to the ghostly minimum, but which is never able entirely to formalize the world picture. When the stage of physical interpretation needed for empirical verification is reached by the scientist, this material factor has to be re-introduced in a concrete fashion.

It is just here that the contemporary emphasis on "operational definitions" comes in. The philosophers of science have increasingly stressed the necessity of interpreting every mathematical theory in terms of specific experimental operations. When we refer to "space," or "time," or "weight," or "mass," if we are to preserve scientific objectivity, we must be able to define these in terms of definite procedures of observations and of measurement. "Time" for example can be defined as that which is measured by a clock, but we must then be able to describe just what a "clock" is and how we use it to tell time.

But what if someone were to ask "How do you define a clock? How do you define 'reading a clock'?," etc. Clearly it would be circular for us to give a mathematical description of a clock, or of the operation called "reading a clock." These instruments and operations must be defined in a qualitative way, in terms of ordinary, unmathematicized experience. Thus scientific operationalism is a way of seeking maximum objectivity in the material half of the scientific duality, just as mathematization is the technique of objectifying the formal half. We have to conclude, therefore, that mathematicism as such is an inadequate theory of change.

The greatest defect of mathematicism, however, is similar to that of mechanism. Both minimize the basic significance of change and hence are Parmenidean, rather than Heraclitean. In a mathematical theory, *change* as such disappears, and is replaced by static relations, namely by the notion of mathematical *function*. Some mathematicians have the delusion that there is change in the mathematical world, because such expressions as "the rate of change" are used, but a moment of reflection will show that such expressions are metaphorical. Actually the geometrical lines or surfaces which graph a function are static entities, all parts of which exist simultaneously and extra-temporally. When reduced to algebra, these functions are seen to be static relations existing between sets which also are extra-temporal objects. Indeed it is precisely the triumph of modern mathematics that it was able by the calculus to find a way to reduce motion and change to absolutely static descriptions. In the modern theory of relativity, a marvelous objectivization and fixation of the world of change is achieved. We are furnished a cosmic picture where space and time are conceived as a single manifold in which the past, the present, and the future are all fixed simultaneously and every change in the world is reduced to a world line, as permanent as the chart of a voyage on a globe. The line that represents the motion of a ship is not itself in motion, but is something frozen forever. In such a universe, *novelty* has no meaning.

C. Process and life philosophies

Heraclitus and all his later followers have always risen in protest against mathematicism, although many have been distinguished mathematicians and quite aware of the enormous value of mathematization as a technique. A recent outstanding example of this protest was supplied by Alfred North Whitehead. A mathematical physicist and one of the chief authors of modern mathematical logic, Whitehead nevertheless developed what remains one of the most striking efforts to find philosophical meaning in the modern world picture. Whitehead vigorously accepted the Heraclitean view and developed a philosophy of *process*, in which the accent is on change. However, his reconciliation of mathematicism with Heraclite-anism was strongly Platonic. While he saw the concrete universe as made up of processes and actual events, rather than of substances, nevertheless, he conceived these events as the concretization (so to speak) of "eternal objects" which have their reality only as they "ingress" in actual concrete events.

Another Heraclitean school is that of Marx, Engels, Lenin and the whole Communist orthodoxy. Lenin in his "Empirio-Criticism" specifi-cally opposes mechanism and mathematicism. In the latter, he sees an inevitable collapse into idealism, and quite rightly (I think) demonstrates that the trend of thought introduced by Ernst Mach and leading on to logical positivism and to the positivistic and empiricist "philosophy of science" schools of Reichenbach, Feigl, etc. of today are essentially of the same type. In mechanism he sees a non-idealistic, but essentially static conception of the world.

Marxism accepts the fact of change as the primary fact about the cosmos. Everything exists in a state of change. Every apparent stability is in fact a state in which a hidden contradictory is already present and beginning to develop. The appearance of a new stability heralds at the same time the hidden positing of still another contradiction. This process is not circular but progressive. The progression, however, is not a mere evolution, or advance in degree of complexity. Quantitative intensification and evolution always lead to a crisis or revolution in which quantitative change passes over into qualitative change, i.e., a change in kind. Hence, the universal flux does produce genuine novelty, which contains the past within its own synthesis, yet which is not simply predictable from the past. Determinancy, therefore, can be found in what has already happened, but looking forward it does not exclude freedom and novelty.

This emphasis on the importance of the Heraclitean position accounts for the rather striking similarity between certain of the Marxist positions and that of the Aristotelian tradition which opposed Plato and the mathematicists so vehemently. Aristotle also criticized Democritus for much the same reasons that Marx and Lenin used, namely that mechanism is static, denying the intrinsic character of change, and reducing it to merely external rearrangement of unchanging parts. The essence of Aristotle's position is that the matter-form dualism (which I have tried to show has its analogues in all these systems) has been misconceived by the mechanists. They have thought of the form (the arrangement of the whole) as changing, and the matter (the atoms) as static, stable, and actual. Aristotle argued that the material factor in the natural dualism is something *purely potential*, a capacity for any actualization. It is not, therefore, a Necessity resistant to Reason (form) and rendering form imperfect and unstable (as for Plato and the mathematicists) but is an appetite or tendency to actuality and perfection, which exists only in and through the process of actualization. The Thomists who today are the main proponents of this position often seem to express it in a manner which tends back toward the Platonic view that Aristotle opposed.

In spite of this commitment to a view of nature which emphasized its dynamism, Aristotle and the medieval Aristotelians failed to follow out its full implications. They were restrained both by the lingering influence of Hellenic religious attitudes and by the actual state of astronomy in their day which had been developed under Pythagoreans and Platonists. This astronomy accepted the mythological view that the cosmos is divided into two fundamentally different regions, the sublunar region in which alone radical change can take place, and the heavenly spheres in which no change can take place, except pure mechanical change (the frictionless motion of the unalterable spheres, motion which is absolutely uniform and capable of perfect mathematical expression). Thus mechanism and mathematicism were realized in the principal parts of the universe, while dynamic naturalism applied only to a restricted and inferior region. This drastic restriction of dynamic naturalism in Aristotle's world view survived through the middle ages. St. Thomas Aquinas recognized it as hypothetical rather than definitive, but Thomists abandoned it only reluctantly as modern science made it completely untenable. Contemporary Thomists have made only a feeble effort to rethink the consequences of the Aristotelian view of change for the modern world-picture, and have been largely content to argue that changes in empirical science cannot affect the metaphysical principles of Thomism, which, it is claimed, rest on a superior ground.

Another stream of thought, which also has Aristotelian roots, but which stems from Aristotle's biological rather than his physical thought, is that of the "life philosophies." Bergson, Teilhard de Chardin, Huxley and others (it should not be forgotten that Whitehead called his own system "the philosophy of organism") react to mechanism and mathematicism in a way which links them with this tradition. They insist that mechanism and mathematicism are abstract and reductionistic, that they destroy the dynamic character of nature which exists only in process and development. These thinkers find in the biological theory of evolution a strong confirmation of the dynamism of nature which indicates an increasing *interiority* found in beings higher up the evolutionary scale, an interiority that cannot be adequately described in terms of merely mathematical relations.

D. Outstanding difficulties

This Heraclitean and Aristotelian trend of thought, emphasizing process and change, has thus furnished a persistent critique of mechanism and mathematicism. Nevertheless, its proponents have not been able to meet the pragmatic test by showing striking success in scientific discovery, while the mechanists and mathematicists have been notably successful, especially in the physical sciences.

Nevertheless, mathematicism, for all its power of scientific creativity, is showing itself incapable of resolving many questions. Books on modern physics are filled with the complaint that it has become increasingly difficult to make modern science intelligible to the layman. Its mathematics are beyond him, and when the results of science are "popularized" in imaginable terms, they are distorted into a gross mechanism, or even to a mythology. The result is that a mythological attitude toward science continues to dominate the popular mind. Many students of our culture also bewail the growth of the "two cultures"—the opposition between the sciences and the humanities—and even deplore scientific advancement as a de-humanization of our culture. This romantic attitude is reflected in the growth of existentialist and phenomenological philosophies which, while not denying the value of natural science in the technological sphere, tend to minimize its capability of giving human "meaning" to life.

Perhaps the source of these criticisms, which certainly have merit, lies in the reliance of contemporary science on the mathematical and mechanical modes of thought. If so, perhaps the gap between the "two cultures" can be closed if science itself makes more use of a "process philosophy" in interpreting its own results. This was certainly the hope of Whitehead and of Teilhard de Chardin. We might wonder, however, whether their approach was sufficiently thorough-going. Whitehead with his "eternal objects" which "ingress into actual occasions" has often been accused of platonizing, and Teilhard, with his all-embracing evolutionary unfolding of a predetermined cosmic plan, has not escaped the same accusation.

II. THE ELEMENTS OF SOLUTION

What would such a thorough-going process philosophy be built upon? Obviously it could be easily devised in an idealist perspective, if by "process" we mean a flux of phenomena, like the oriental Maya. Thus we would posit either the individual mind or the World-Mind as eternally existing, in which the flux of the world appears like a movie on a screen. Whatever the ultimate value of idealism, such a philosophy does not admit real process in an objective world, and this is one of the fundamental assumptions of modern science.

A. The datum of natural continuities

If we speak of objective extramental processes, we must account for their continuity and order. If process is simply a series of disconnected "events," then once again scientific research becomes impossible. Of course, some comfort is given this view by current theories that physical processes are completely quantized, so that even the trajectory of a subatomic particle is a series of discrete "jumps." But such a theory still calls for a physical basis for the continuity of the series. And what makes a process continuous? It cannot be the substrate of the process, since this would be to reintroduce substantiality, which is forbidden in an extreme process philosophy. It must, therefore, be something in the process itself. But if we conceive process like a continuous line or surface in which continuity is given by the simultaneous existence of parts, one joined to the next by a point, which point has its reality only as a term of a part, then we have mathematicized process and given it static fixity. Or the reality of process is like that of time, nothing other than the reality of a point joining a past which no longer exists, and a future which does not yet exist. But in the point itself, there is obviously no principle of continuity; the continuity of time arises from the perdurance of a substance undergoing process. Hence we are back to a process made up of points, without continuity. Actually, I believe, an analysis of the attempts that have been made at applying process philosophy to science will show that they fall either into idealism, or mathematical fixism. Thus, the pure process philosophies are really a hidden fixism.

To admit the perduring existence of natural units as that in which process takes place, is not, however, a fixism, since (1) these units are themselves produced and destroyed, their endurance and stability being only relative, (2) while they endure, their endurance is not static, but is always an unfolding tendency to full development. They exist only in process and through process. As soon as the process by which they are is checked by counter processes, then they begin to disintegrate and are transformed into other units. There are therefore in them no ingression of eternal objects, no eternal natures. Their natures are nothing other than a relatively perduring tendency to characteristic processes. Thus the nature of a rabbit is nothing other than the tendency of the fertilized rabbit ovum to develop by characteristic embryological processes into an adult rabbit which grows, moves, eats, mates, sleeps, etc. The rabbit has a nature only in the unity and continuity of such processes. If the continuity of process is broken, the nature is destroyed.

If, therefore, a process philosophy is to be useful, it cannot be "pure" in the sense that it asserts that it denies the existence of primary units, or "substances," which exist in process. A "pure" dynamism which asserts that the world is nothing but events, activities, movements, without *things* or *persons* which encounter each other or act on each other is a needless extravagance that runs counter to daily experience and common language. In order to accept in an honest and thorough-going fashion the reality of universal change and process, we need not accept the fantasy of the smile of the Cheshire Cat which lingered after the Cat herself had disappeared, or of the school of Japanese Buddhism which played its trick of one-upmanship by countering to a rival school which asserted that "the world is an imaginary wave on the ocean," the more subtle view that "the world is an imaginary wave on an imaginary ocean." The logical alternative to a fixist world of eternal substances is not a world of reified processes, but rather a world of relatively stable substances which come into existence through process, exist in process through activities and passivities, and are transformed into other substances by continuing processes.

B. The datum of observed events

The fundamental objection to a "pure" process philosophy is that it would require us to eliminate from the world-picture our knowledge of ourselves as observers. The "events" and "processes" for which science and philosophy must account are *observed* events and processes. While it might make sense to say "I observe an event," it would hardly make sense to say "an event is observed," without implying that there is an observer, a person, who is not merely a process, but a primary unit which exists in and through processes. Someone might say that the "event" is precisely the "observation," so that it is better to say "an observed event takes place." But how then could we distinguish between an "observed event" and simply an "event," without introducing an observer? Is the observer himself a pure event, or something to which events occur? If he himself is a pure event, then we have the notion of an "event which observes another event," which seems absurd.

It does not seem possible for science to accept a view of reality which cannot account for at least the presence of the scientist himself, the observer, within the world of events which he observes. In trying to develop a thorough-going process philosophy, therefore, let us not begin by assuming that the world is *nothing but* processes. Rather let us begin with the fact of experience, more and more strongly confirmed by the progress of science, *that nothing in the world of nature is exempt from process, that all of nature exists in and through process.*

The best point to start is with an observed event, since certainly this is primary in all scientific understanding of the world. All scientific theories rest on observations, and these observations must be operationally specifiable, i.e., there must be a real physical interaction between observer and observed. Such an interaction is an observed event. Science knows of no way in which we can ascertain a "fact" (and every theory must be verified by facts) except by some kind of physical operation. Perhaps the least physical way of observing is when we see something with the unaided eye, yet even in this case we know that the object must have a physical effect, however minimal, upon the eye through the mediation of light, and that the eye sees by an interaction with the light.

It might be objected that psychologists (except for the extreme behaviorists) admit events observed by introspection, and that such "psychic events" are of a wholly different order than "physical events."

Without here entering into this question, it suffices to say that just as the very notion of "observation" implies a psychic aspect to any "observed event," so also the *typical* psychic event (we can leave open the question as to whether this is true of *all* psychic events) involves some physical process. Even introspectively we are aware that even in highly abstract thinking there are involved images and words and feelings that are somehow derived from sensory experiences of physical contact with the world, and that, whatever the differences, there remains a genuine continuity with physical experience.

In the observed event, we are, at least by reflection, able to distinguish between the event and the observation of the event, between the object which the act of the observer intends, and the act of the observer. If this were not so, we would have to deny that there are unobserved events. Yet that unobserved events are real is assumed by every scientific law. The notion of a natural law or general order of processes in the universe would be meaningless if we have no grounds for extrapolating from the small number of events actually observed by the scientist. Whatever theory of induction we adopt, we must attribute to it some element of validity, and hence imply the reality of unobserved events.

Whenever we attempt to distinguish observer and observed at the extreme limits of observation, whether of the "elementary particles" or of the "total universe," the attempt breaks down. At this level, objectivity itself demands that we recognize that the event which we seek to observe is in fact nothing other than the very act of *contact* between observer (or his instrument of observation) and the "observed." When we probe the atom with a beam of light and by this fact produce an *elementary* event, obviously this elementary event cannot be further divided into the object event and the subject event; it is one simple event. Similarly when we attempt to make an assertion about the universe as a whole, we cannot abstract from the fact that we are a part of that whole, and that if the relation between universe and man were omitted, it would not be the same universe. Again the event in question, namely the mere fact that the universe is "aware of itself" only in an observer, is something simple and indivisible. Short of these limiting cases, however, the separation of the object and the subject is possible, and is the very aim of the scientific method.

Let us, therefore, bracket the contribution of the observer to the event by assuming that he is able to reduce his interference in the event to a minimum and so recognize it that he can distinguish it from the event. For example, the physicist watching the collision of two iron balls is aware that the effect on this collision of his seeing the collision, although real, is minimal and insignificant. In every such controlled experiment, irrelevant factors are reduced to a minimum, and we can improve the objectivity by proper techniques, up to the limiting cases, the one macrocosmic, the other microcosmic.

C. The datum of natural tendencies

What then is an event? We can distinguish at the outset two types of events: (1) unique events, (2) reoccurring events. Of course, in the concrete, all events are unique. There is never perfect repetition of an event, unless we admit that time is somehow circular and that history can literally repeat itself. We can call events taken in this concrete sense, in their uniqueness, *historical events*. However, it is obvious that similar events do occur and reoccur. While the historian is interested in the concrete unique event as such, the scientist is not. He abstracts and generalizes, seeking to find uniformities, regularities, "laws" in nature. Experience has convinced him that this is possible, and that most events, at least at the sub-human level, are largely classifiable under abstract categories. He no longer claims, however, as did scientists in the last century, that every concrete event is rigidly determined and deducible from general laws. The laws he looks for are statistical, i.e., they show a generalizable order, but one which admits of exceptions. We will call these *reoccurring* events *natural* events, something which behaves in a regular manner from some internal principle.

But why the regularity? Are we to say that the universe is an endless chain of events in which certain events occur again and again for no reason at all? Is science simply a description of regularities without questions about the source of regularity? To admit this would be to make the scientist contradict himself. On the one hand, he would desire to find order and unity (otherwise he would not look for regularity or natural laws) and on the other hand he would not be interested in unity and order, since the existence of a plurality of instances of one thing suggests to the mind an inquiry as to a source of this unity.

The obvious resolution to this dilemma is to admit that the event occurs *in* a natural unit of the sort already discussed, an atom, a molecule, an organism, a person. The event happens regularly because there continues to exist a natural unit (or a number of similar units) which tend to produce this act over and over again. This explanation occurs to us by the analogy with our experience of ourselves. We are aware of ourselves as persons, natural units, and we are aware that we tend to perform the same acts again and again, even without deliberation or choice. We find in ourselves recurrent events of hunger, eating, evacuation, sleeping, etc. We say that it is "natural" to men to do these things. This behavior keeps reoccurring in us because we have an inner nature which is manifested in exterior action. This experience of ourselves is easily transferred to the things we encounter: food regularly nourishes us, the ground regularly supports our body, etc. We see animals having similar behavior to our own. Finally, we observe inanimate objects repeatedly acting in a certain way, stones falling, water rippling, etc. This is all quite intelligible to us if the events are seen as residing in natural units like ourselves.

Now if we ask ourselves again more exactly what we mean by a natural unit, we will see that such a unit is nothing other than a unity of recurring behaviors. We know ourselves to be a unit precisely because we are alive and our life manifests a regularity and continuity, very complex in some respects, but also resting on simple recurrences of physiology and sensation. Thus the natural unit is not something *occult*, lying underneath events and impervious to them. It is not a static *substance* as this term is often understood by philosophers today, which remains as unchanging substrate over which changes pass like shadows. A natural unit exists precisely in activities, yet it is not identical with any single activity (since it has many recurring activities), nor is it merely a collection of activities (since these activities reveal a unity), but the principle of their unity and reocurrence.

III. RESOLUTION OF DIFFICULTIES

Perhaps this last point should be expanded. Why cannot we say that a natural unit is simply a collection of activities or events? If this collection of activities were dissimilar or unrelated, then it would suffice to say that they were just a collection. But what are we to say if (at the least) they are regularly occurring activities? Or if (at the maximum) they are an inter-related hierarchy of reoccurrent activities? Obviously, this unity in the collectivity must have a principle, and this principle is not any instance of the activity, but a principle of all, i.e., not an activity but a source of many activities. This is what we mean by a natural unit, a concept derived from our own awareness of our unity as persons underlying the multiplicity of our actions, and most clearly manifest when those activities are regular.

Since this explanation is such a simple and common-sense one, why have so many philosophers rejected it, and gone so far as to posit processes without any subject? I think the difficulties should be divided into two sorts. (1) Hume's difficulty: How can we know anything at all except the direct stream of perceptive events? (this difficulty passes beyond the realm of scientific thought, since clearly science is quite willing to admit entities that are not directly observed, provided that they are justified on the basis of observation and logic); and (2) the fear of introducing into science *occult* entities. This is why Einstein advised that the concept of "ether" be dropped from science, when observational evidence for its existence could not be found. If there is no reason for positing natural units as the source of empirical events, then they are occult entities.

This latter objection would be valid if natural units are conceived as static realities beneath observable events. But as we have seen they are the principles of unity in observed events. Thus the atom is nothing other than the principle of unity posited for the collection of reoccurring and inter-related events which we call atomic behavior. The same holds for the molecule, for the living organism, for the human person. The natural unit exists in process, as the source of processes, as having no other being except that unity which is expressed in these processes. Hence if we were to attempt to define a particular unit, we should do so by listing its observed behavior or activities, and then seek for the unity which these activities exhibit. That unity is definitive of the natural unit, and can be called its "nature."

Since a natural unit exists only in its natural or recurrent activities, in events, we have still to make clear what we mean by an event, and what a unit will be which exists eventfully.

An event is a change, and what is a change? We use the word "change" to indicate the appearance of something which was not previously there. It is a coming to be from not being or vice versa. Yet this is not enough to describe a natural change. In nature, we do not observe merely a series of isolated events, appearances and disappearances of phenomena. An event appears in the midst of (or, as it were, against the background of) other more enduring phenomena (which are themselves *relatively* stable events). When we watch an event in a Wilson cloud chamber, the track of mist appears in a steady field, otherwise it could not be observed. For this reason, we do not conclude that what appears, appears from nothing; rather it appears from something already existing, namely the background. Thus in every change or event we must analytically note:

1. The new phenomenon (appearance).
2. The previous absence of this phenomenon.
3. The perduring background.

As we have seen already, the perduring background is made up of natural units, so that the absence of a phenomenon and its appearance are features of a natural unit in which this change is seen to regularly occur. Thus in our own person we feel our heart constantly beating, and we call this a natural change, or process.

A. The natural duality

We are now back to the matter-form duality, which as we have seen is a feature of all efforts to explain natural phenomena. Even the process philosophy does not really escape this duality, since process philosophers either have recourse to eternal objects which ingress in process, or to the notion of space or time *in which* processes occur, or to some other equivalent for the background against which change takes place and which gives unity to events which otherwise would lack all coherence.

The problem is, how to conceive this duality so as to escape the various difficulties to be found in the theories I have already described. Obviously the aspect of *form* or *order* is to be found:

a. In each of the phenomena which appear, and which exhibit definite qualitative or quantitative character. They are the actualities which are observed directly or indirectly, they are what happens in an event.
b. In the collectivity of phenomena as unified in a natural unit, i.e., in its nature, as an internal principle from which these events arise.

The problem is with regard to the "background." This background is also the natural unit (or better the natural units which act upon each other, since every event turns out to be an encounter of units). We have seen that the natural unit has form as its *nature*, the principle of its regular behavior. It would seem, therefore, that duality is sufficiently cared for by saying that there are two levels of form:

1. The nature of units as source of activity.
2. Their particular activities.

Not a few philosophers have opted for this position. If we do, then we will find ourselves back in a version of mechanism, since we will have to say that change affects the activities of units, but it never produces or destroys a unit. Thus the natural units would be eternal things. We have seen that this is contrary to modern physics which discovers no eternal units. Aristotle and the scholastics were willing to accept it for the heavenly bodies which were eternal natural units, subject only to superficial change. However, both Democritus and Aristotle pointed out an interesting consequence of this possible position. If a natural unit is to remain in permanent existence, the only kind of change it can undergo will be change of place, since any internal qualitative or quantitative change, or even any irregularity of motion, would eventually result in its destruction, since we observe that no body is able to maintain itself under indefinite variation. Since it is very clear that no entity now known to science is exempt from qualitative and quantitative variation, this ancient argument suffices to exclude this possible position.

How then are we to conceive this aspect of natural units which makes it possible for one natural unit to be converted into another natural unit? We now know that this goes on at every level of the universe: not only do living things die and turn into inanimate molecules, but molecules are broken up into atoms, and atoms as the simplest stable natural units are constantly converted into unstable particles which pass into each other and finally into new atoms. Thus no natural unit is permanent. There is, therefore, no primordial stuff, capable of independent existence, out of which natural units are made, i.e., no natural unit which is basic to the others. We have already seen that this is evident from the very existence of different kinds of natural units. From the fact that we as persons are natural units, i.e., not merely an aggregate of more basic natural units, and that we are not permanent, the same consequence follows.

It is necessary therefore to posit, as Heisenberg and others have made clear, that while the different natural entities have specific characteristics that distinguish them (their forms), they have in common the capacity to be converted one into the other.

This material aspect, therefore, must be conceived as of itself totally "formless." If we were to attribute a form to it, then we would have to posit its existence as the basic natural unit, and other natural units would not be possible. It is "pure potentiality" as the scholastics said. As such it is not independently observable, but is observed simply as the capacity of natural units to be transformed into other natural units. It is not imaginable, or even conceivable as an absolute entity, but only relationally, i.e., as a capacity. Yet it is logically entailed once we admit that a natural unit is transformed into another natural unit, and does not merely appear from nothing.

The explanatory value of this type of matter-form duality is that only in this way can we avoid the evasion of the fact of change found in mechanism, mathematicism and other similar philosophies of nature, or the denial of the natural unit which vitiates process philosophies or compels them to introduce eternal objects.

B. The subject-object distinction

We have seen earlier the paradox of our own knowledge of ourselves as a natural unit, and the dualism which it involves. On the one hand, I am conscious of the flux of experiences, on the other of the relative stability of myself as subject unifying these experiences by memory and by synthetic interpretation. This unification, however, is not merely subjective, as Kantians say, but takes account of an order which is in the experiences themselves. This order or intelligibility is only in a very small degree immediately evident, although from the outset something is orderly in experience. My creative effort is gradually to untangle the various levels of intelligibility in experience, not to create these levels. I do this by a process of hypothesis, if you like, "trying out an order for size," to see what will fit, using analogy from already discovered order to try to find the pattern that will fit given data. This is exactly the method which science has proved pragmatically useful. This requires of course that I can *recognize* when a pattern does fit, which implies a genuine power of insight on my part, and not a mere positivistic correlation of data. I can recognize when a pattern fits because I have some insight into my own basic experiences, my own immediate encounters with the world.

How does this extension of my encounter with reality take place? Do I first know myself as thinking subject, and then try somehow to break through this enclosure into the outside world? This has always been the dilemma of idealism, which begins with the assumption that what I know first is that I am a thinking being. But is the awareness of myself as a thinking subject my basic awareness? Is it not rather the experience of *encounter* in which the subject and object are not yet distinguished? It would seem that in any philosophy where *change* is the primary fact, the primary datum must be that awareness of *changing being* which comes when I encounter some object in process, not a merely mental event, since psychologically it is doubtful that merely mental events that do not arise out of physical experience are possible.

Any human being encounters reality as a manifold of experiences, sights, sounds, colors, etc. But what do these mean to him? What he is first aware of is simply that *something exists*, that *being is*. This being includes all that he experiences in a confused complexity. This awareness of being is often most vivid to us at that moment of awakening from sleep, when the succession of dream images suddenly yields to the sense of reality. This being of which we are aware is real, without qualification, it simply is, i.e., it has substantiality. It does not appear to us as a *phenomenon*, either as an illusion, or as the *outward appearance* of something hidden; but simply and unqualifiedly as that which *is* in a primary and irreducible sense. There just *is* something.

This something, however, is not simple, but confused, since it contains many different experiences of color, sound, etc. But at the first, this confusion is not troubling, since we are not aware of any other possibility. What is important is that this something *changes*. All of our experience is in flux, and this initial something does not enter our awareness as something static or permanent, but as something which changes in our very awareness. Hence what was present ceases to be present. Yet this change is not all; rather it is a ceasing of some aspect of the complexus of experience, against a background, which does not immediately change. We thus become aware of the contradictory of being, namely not-being. This not-being is not some general thing, but is simply the absence of some aspect of being previously present; it is not an absolute (as being was an absolute), but is relative to the being.

With the awareness of being and not-being, we become aware of distinction and identity. The presence and non-presence of some aspect of being gives us this notion of *distinction*, while the continuity of the background shows us what sameness is. The primary point here is that being as we meet it in experience is always *changing* being—not absolutely changing being, but relatively, partially changing being.

At this point also comes the formation of the first human judgment, the principle of contradiction, "A thing cannot be and not be in the same sense." It simply asserts the distinction between being and not being found in change. Those who claim that the first principle of our thinking is the principle of identity, since we must first assert that "something is," do not see that we must be aware of "something existing" in an intuitive non-judgmental manner *before* we are able to make any judgments of anything.

Up to this point, we are not aware of the object-subject distinction. The Cartesian assumption that what we first know is the thinking subject (*cogito, ergo sum*) is the sandy foundation of much of modern thought. Undoubtedly included in our first awareness that something is must also be a certain experience of our subjectivity, since the experience of the world includes objective and subjective elements; but as we have already seen, this experience is manifold, and at first all its elements are com-mingled in confusion. Our analysis must show how these elements of experience emerge as distinguishable.

We are also at this point able to form the notion of *unity*, since this is the negation of distinction. We perceive in experience as a background both unity and plurality—unity by reason of sameness, plurality by reason of distinction.

It is at this point that the subject-object difference emerges. We become aware that in experience there are two realms which to some extent coincide, and to some extent are distinct. The first realm is that of being in the unqualified way. This is the realm of the *real*. On the other hand, I become aware also that there are in experience elements which do not have this unqualified being, namely images, mental fictions, mental relations (such as the mental relations formed in making a judgment, the relation of predicate and subject). These now appear as less real than reality, because they can in some measure violate the principle of contradiction; I can imagine that something which is no longer present is actually present. I can now label these two realms as "real" or "objective being," and non-real or "purely subjective being," with the principle of contradiction as the criterion of distinction. I thus become aware of myself as a knowing subject. I am a *unity* (the idea of unity as the negation of distinction in being has already been achieved) which perdures through the experience of change, and I am a knowing subject because to me appear both real being and fictional being. I find that I have a control over fictional being which I do not have over real being, so that I recognize fictional being as dependent on me, as in some measure my creation, while objective being is not my creation. I become aware of myself as real being, since I find that my power to know is not something simply under my control, but is a part of the objective world. Thus idealism is refuted, for the man who is interested in exploring the objective world, by the fact that the knowing subject is discovered only within a world of change subject to the law of contradiction.

C. Causality as the factors of process

It is important at this point to face the difficulty raised by Marxism. Marxism is uncompromisingly realistic, but it does argue that contradictions exist in the real world, since *change* is something contradictory. In saying this, Marx carried over from Hegel an idea that goes back to Parmenides, and which seems to be the very soul of Eastern philosophy. I have already tried to show, however, that change does not imply *contradiction* in what changes, but *potentiality*. It is the refusal of Parmenides to admit that being can be potential, relative rather than absolute and actual, that is the root of this fundamental difficulty. This in turn is basically a refusal to admit that change is a reality, external to our way of thinking about it.

It will be noted, against many neo-Thomists, that thought does not begin with some perception of *universal being*, but simply with the being of experience which is in its initial contact, *changing being*. Whatever analysis may later show us, we do not initially know of any real being except changing being, and whatever we are able to know as human intelligences must be derived from changing being.

As we have seen, this changing being is seen upon analysis to be dualistic, a potentiality perpetually tending toward actuality, but never achieving total actualization, because each actualization is limited, and excludes other contrary actualizations.

It is important, here, however, to consider whether this concept of tendency to actualization is to be conceived after the manner of Marxism and of many other vitalists. Is this tendency active or passive? If our conception of the universe is monistic, then it will be necessary to conceive it as a single natural unit which develops from within. Hence there is in its potentiality a tendency to self-actualization. The fact that Marx actually accepts this alternative seems to betray the fact that he did not succeed in wholly overcoming the idealistic monism of Hegel. In the system of Teilhard de Chardin there seems to be a similar danger. As a theist, Teilhard introduces God, extrinsic to the universe, as the ultimate actualizer of the potentialities of the universe; but this is hardly a naturalistic solution.

The way out of this dilemma is provided once we admit a pluralism in nature. If there are many natural units, then actualization comes by interaction. One natural unit, being itself in some manner actual, can actualize another natural unit in that respect, and be actualized by it in another. Hence the potentialities of things are tendencies to actualization, not in the sense that they actualize themselves, but in the sense that they are open to actualization by mutual interaction. We have thus the conception of *causality*, i.e., a thing cannot actualize itself, but is actualized by another with respect to some characteristic by which the other is actual. This finds its expression in the universe in the form of the *conservation laws* which some physicists believe are the only laws of nature. A conservation law is simply a statement "that a thing cannot give what it does not have," i.e., one thing is actualized by another only with respect to what that thing actually is. Hence a thing can energize another only to the degree that it itself is first energized.

There is another aspect of the problem, however, and a very difficult one. This is once more the problem of *teleology*. Teleology is a condemned term in science today for two reasons. It is believed that teleological explanations are anthropomorphic, attributing to nature thoughts, feelings, etc. Indeed, there are some scientists, like Julian Huxley, who accept the idea that the first traces of psychic life are to be found at the lowest levels of the universe. It is not necessary, however, to go this far, if we accept the method of comparing other natural units to the human person. Anthropomorphism is not only not forbidden, but, as we have seen, is necessary if we are to make sense out of nature, and is rooted in man's goal-seeking tendency to posit teleology in non-conscious, non-psychic natural units. It is only necessary to make this analogy clear, and not to obscure the basic differences between knowing and non-knowing units.

The second, and a very important reason for distrusting teleology is that it appears to close the way to *causal* explanation. If we see evolution, for example, as due to a tendency of animals to evolve into higher forms, then we will cease to look for the causes which produced this evolution. But if we know the causes, then the teleology seems redundant. Furthermore, teleology seems to introduce an occult, unobservable factor.

However, in spite of these formidable objections, it would be a mistake to omit from our explanations of nature the element of *natural tendency*. Since the fundamental fact of nature is change, and since some changes are regular and recurrent, while others are unique or random, we must notice the precise difference between the two types of changes. In regular change, we *observe* a process which issues in a determined and uniform effect. We are able, therefore, whenever we observe this effect, to infer the process by which it came about. This is what is *properly* called a teleological explanation: it is one which reads backward from effect to cause. The whole theory of evolution is such a teleological theory. Obviously, understood in this way, teleology is not occult, since it begins with observation (the observed effect); nor does it eliminate the search for causes, since it is precisely an effort to discover causes. Nor does it imply conscious effort in the cause, but merely that it acts with regularity and determinance.

It will be noted that teleology also is linked closely with the idea of the natural unit. We can recognize regular occurrences, and separate them from the background of randomness and chanciness, because they produce, preserve, and develop a natural unit. For example, the physiological processes in the human body are seen as preserving and developing the human body as a unit. When these processes are swamped by chance events, disease, or accidents, then the natural unit is destroyed. Hence, teleology can be seen as the tendency in natural units to self-preservation and development in their characteristic activities.

We are thus equipped with four aspects of every natural change or process which correspond to the classical Aristotelian four "causes," only one of which is a cause in the modern sense of the term. The natural unit has (1) some *organization* or order (formal cause) and (2) at the same time has *potentiality* (material "cause") for becoming other than it is. (3) This potentiality is actualized from outside by *another natural unit* (efficient cause), and this actualization is either destructive of the unit, or actualizes it in its own line of stability and activity, and hence is (4) *teleological* (final "cause").

IV. CONCLUSION

Has my lengthy analysis only resulted in a return to the old hylomorphism of Aristotle and St. Thomas Aquinas? This objection would not be mere prejudice, since historically the Aristotelian tradition proved a formidable barrier against the development of modern science, and only when hylomorphism was put aside by Galileo did the modern development get under way.

But what was it in Aristotelianism that had this disastrous effect? It was precisely the conception of the unalterable heavenly bodies with their perfect circular motions which proved the great barrier. These were the features of Aristotelianism which, as I have explained, were the survival of mechanism and mathematicism in his view of the world. Making use of Aristotle's very trenchant criticism of other philosophies of nature, and indicating that all of them in fact introduce a matter-form duality, I have tried to show what is still viable in Aristotle's insight in view of the fundamental discovery of modern science that all natural units are transformable, and in view of modern philosophy's emphasis on the central significance for all knowledge of man's discovery of himself as a person distinct from and related to other persons.

The shape of biological thought

C. H. Waddington

CONTEXTUALIZING COMMENTS

Dr. Waddington, Professor of Animal Genetics at the University of Edinburgh, is one of the prominent figures of contemporary evolutionary science. In addition to numerous books and articles of a technical nature, Dr. Waddington has also published such books of general interest as *The Scientific Attitude* (1941), *The Nature of Life* (1961), and has edited a volume on *Science and Ethics* (1942). His familiarity with the doctrines of Logical Positivism and Linguistic Analysis and with many of the key personalities expounding those doctrines (he was a personal friend of Wittgenstein, for example[1]) makes his conclusions from evolutionary biology all the more interesting for the philosopher.

In the present reading, Dr. Waddington suggests three currents of thought arising from evolutionary biology which require a shift in philosophy to a broader base of understanding than can be provided by reflection upon the mathematical texture of modern physics which has for years been almost the exclusive preoccupation of philosophers of science.

The third or epistemological current which Waddington claims to discern concerns the concept of causality in the world of nature, and he thinks that the sequences of simple causal or logical relations exemplified

From *The Ethical Animal* by C. H. Waddington, copyright © 1960 by George Allen and Unwin, Ltd., Publishers. Reprinted by permission of George Allen, and Unwin.
[1] See C. H. Waddington, *The Ethical Animal* (New York: Atheneum, 1961), ch. 3, "Squaring the Vienna Circle," pp. 34–45. More generally, see *ibid.*, chs. 3–7, pp. 34–71.

in theoretical physics must be replaced or supplemented by a concept of causal structure as circular.

It was just this which we had in mind when referring in our discussion of "The Aristotelian Explanatory Mode" in the Introductory Essay to the "quite general sense which the ancients gave to the term 'cause' " (see "The Impact of Evolution on Scientific Method," at fn. 55). Modern or mathematical physics deals with a small number of universal equations and with about one hundred fundamental particles. Compared to biology, as John Platt has put it, it is a relatively "low-information" type subject, concerned always with a reductive analysis. In biology, on the other hand, as in ancient or philosophical physics, the properties of the subject matter vary in variety in proportion to the complexity of the organization of the individuals which make it up (no longer fundamental particles, but now organisms and specific populations).

Here understanding depends on a factorial rather than reductive analysis, for the objects of interest to the evolutionist are not as much invariant relationships (though these are involved, as expressed for example, in the Hardy-Weinberg equation which underlies modern population genetics[2]) as subjects of processes, organisms taken as undergoing constant change; and in such analysis it is necessary to take account both of structure, with its correlative composition and organization (what Aristotle and the ancients designated material and formal "causes"), and of function, with its correlative agent and product (efficient and "final" or teleonomic "causes").[3] Just as we cannot describe an organism except by telling what its parts are made out of and indicating how these parts are put together to form the whole, so we cannot fully understand the organism unless we grasp why each step in its development was necessary if maturity was to be reached; and this in turn requires a grasp of the forces and processes involved and of the agency which gives rise to them. Just as, in

---

[2] More generally: "To consider the organism apart from its mechanistic components and functions is patently absurd. This is such a fundamental fact that one must be quite sure not to mistake it for an implicit guarantee that mechanistic biology will be successful by itself. . . . Our contention is that the understanding of the mechanistic processes is a prerequisite for a broader analysis of the functioning of the organism that arises, not through the operation of some separate agency but as a result of the radical inhomogeneity of the microscopic state of the system combined with its intrinsic indeterminacy." (Walter M. Elsasser, *Atom and Organism* [Princeton: The University Press, 1966], p. 105.) "The fact that the relationship of the mechanistic to the autonomous process components is dependent upon, and therefore largely patterned after, the physical relationship of macrovariables and microstates has been the origin of many of the difficulties and misconceptions which have arisen in biological theorizing. The tendency of so many investigators who are concerned with specific macroscopic mechanisms to make short shrift of any organismic concepts, and to generalize mechanistic views too readily beyond their original limits, can no doubt be traced to this source." (*Ibid.*, p. 106.)

[3] Like Elsasser, "We are equally far removed from a pat mechanism as from an intrinsically dualistic vitalism." (*Atom and Organism*, p. 60. Cf. J. Maritain, *The Degrees of Knowledge*, pp. 192-199).

brief, composition and organization are correlative aspects of the natural unit which cannot be described separately, so the forces which produce a thing and the thing itself as end product cannot be described separately.

These are four kinds of reason, four kinds of answer, four necessary conditions— necessary for understanding the process: we need to know all four if we are to find it intelligible. Only one of the four, the By What, the agent, the efficient cause, is a "cause" in the popular sense today—if "cause" have any clear meaning in our ordinary language. The unfortunate neglect of the other three has been due to the dominance of mechanical thinking [and mathematicist explanation of nature] since the day of Newton, complicated by the popular heritage of Hume and John Stuart Mill.[4]

The essential nature of this factorial or "process" analysis was summed up by the ancients in a famous axiom, *causae sunt ad invicem causae* ("causes are causes one to another," or "causes are reciprocally active"). In this way the ancients expressed their accord with Waddington in seeking to replace sequences of simple causal or logical relations required in absolutizing Forms, by organized causal or logical frameworks, which ideally and when complete should be self-contained in the sense of necessitating no reference to anything outside the system. Similarly, the ancients reached an advance agreement with Waddington in realizing that factorial process analysis has a wider reference than the quantitative analysis of various material systems to which cybernetic or "feedback" ideas are applied, and they did this by a careful analysis of the ontological conditions for reciprocal activation of the causes. Just as their meta-physical analysis showed that the common condition of beings was such that they were both actual and potential under different formalities, so their analysis of physical interactions showed that the same reality can be both cause and effect under certain circumstances. Following Aristotle, Aquinas set the matter forth thus:

It must be recognized that as there are four causes, two of them correspond to each other because the efficient cause is the *principle* of motion and the final cause the *termination* of motion; and likewise the other two correspond to one another, for the form *gives* being and the matter *secures* being. The agent thus is the cause of the end, and the end, cause of the agent: the former is true as regards being, because in initiating motion the agent continues to the attainment of a term; while the latter is true not as regards being, but as regards the formality or intelligible character of causality, since the agent is cause insofar as it acts, but it acts only in a determinate fashion. Thus the agent has its very causality from the term. Form and matter are causes in relation to each other and with regard to being: the form is the cause of matter by giving it existence, but matter is a cause of form in sustaining it.[5]

[4] J. H. Randall, *Aristotle* (New York: Columbia, 1962), p. 124: "It is worth noting, incident-ally, that the empiricist notion of causation as constant succession, of 'cause' as the invariable antecedent of its effect, is wholly lacking in Aristotle."
[5] *In V Met.*, lect. 2, n. 775.

Thus, the ancients explained the circularity of natural causes by distinguishing. Causes are reciprocally active, are causes one to another, either according to being (as matter and form), or according to becoming (agent in respect to determinate productions), or according to causality in its intelligible ground (as the effect or end product in respect to the efficient cause). But not every combination of causes exhibits this circularity of reciprocal activation: only the end and agent (final cause by bounding the agent's efficacy, the agent by exercise) and matter and form (matter by receiving form, form by actuating matter so that in mutually communicating they have being as partial principles in the whole itself). In other combinations, causes need not be reciprocally active.

Waddington is therefore quite correct in observing that thinking in terms of circular causal systems has a much longer history in biology than the special terminology "referred to by the term 'feedback' and the word cybernetics" and a much wider reference than the quantitative analysis of cybernetic material systems. If he is equally correct in considering that the further developments of evolutionary theory require incorporating a circular concept of causality (consisting, as he explains elsewhere,[6] of four basic systems: the genetic system, corresponding analogously to Aristotle's sense of material cause; the epigenetic system, corresponding to formal cause; the natural selective system, analogous to efficient causality; and the exploitive system, corresponding to Aristotle's final cause[7]), then it may well be that in reflecting on the exigencies implicit to the causal view of modern physics, on the one hand, and on those implicit to the causal

---

[6] See Waddington's essay, "Evolutionary Adaptation," in *The Evolution of Life*, Vol. I of *Evolution After Darwin*, ed. by Sol Tax (Chicago: University of Chicago Press, 1960), pp. 381–402; and ch. 9 of *The Ethical Animal*, "The Biological Evolutionary System," pp. 84–100. This point of Waddington's is taken up and developed in terms of its philosophical implications in John N. Deely's essay on "The Philosophical Dimensions of the Origin of Species," *The Thomist*, XXXIII (January and April, 1969), in Part I, Section IV, pp. 102–130, esp. pp. 105–108, and in Part II, Section VII, pp. 290–304, esp. pp. 299–301.

[7] Cf. Randall, *Aristotle*, p. 229: "'final causes,' as they were developed during the predominance of the religious traditions, tended to become a way of showing how under the ministrations of God's providence everything in the universe conduces to the self-centered purposes of man. In sharp contrast, Aristotle's natural teleology is, in the technical sense, wholly 'immanent.' No kind of thing, no species, is subordinated to the purposes and interests of any other kind. In biological theory, the end served by the structure of any specific kind of living thing is the good—ultimately, the 'survival'—of that kind of thing. Hence Aristotle's concern is always to examine how the structure, the way of acting, the 'nature,' of any species conduces toward the preservation of that species, and enables it to survive, to exist and to continue to function in its own distinctive way. This Aristotelian emphasis on the way in which kinds of living things are adapted to their environment brings Aristotle's thought very close to the functional explanations advanced by evolutionary thinkers: in both cases the emphasis is placed on the survival value of the arrangement in question.

"It might be well to add, that such functional and teleological conceptions are just the notions that modern biologists, no matter how 'mechanistic' their explanatory theory, actually have to employ in describing the subject matter they are attempting to explain. Teleological relations, the relations between means and ends, or 'functional structures,' are an encountered fact. Like all facts, they have to be explained in terms of certain mechanisms that are involved."

view of modern biology, on the other hand, we have a clear illustration and justification, not this time historical, but properly philosophical, of our introductory thesis concerning the divergence implicit in the explanations of nature worked out in what we have called respectively the Platonic and Aristotelian Explanatory Modes.[8] As Elsasser has so carefully pointed out in what Etienne Gilson rightly considers his epoch-making study of *Atom and Organism*, "to introduce irreducible logical complexity on a purely abstract basis . . . brings us much closer to the preoccupations of the naturalist than we are in the absence of this idea, and it separates us to some extent from the more rigorous methods [because more formally reducible to mathematics and the style of causation mathematics allows for] of the physical scientist."[9]

The shape of biological thought:
or the virtues of vicious circles

The disagreement or even distaste and scorn which many modern philosophers evince towards theories such as I am putting forward here[1] probably have their origin in rather deep-lying disagreements about what constitutes a convincing argument. Philosophical thinkers have, in the last few decades, been profoundly influenced by many advances in modern science. The advances which have made most impression on them have been those in the physical sciences. Open any book of the present day dealing with epistemology or the general problems of philosophy, and you will find a discussion of pointer readings, the theory of relativity, the quantum theory, the indeterminacy principle, and so on. These are

---

[8] Cf. *Ibid.*, pp. 126–127: "Aristotle's viewpoint and approach are, as we often say, biological, rather than 'merely' mechanical. They spring out of the experience of the biologist that Aristotle was. He takes biological examples, living processes, as revealing most fully and clearly what natural processes are like. He analyzes the behavior of eggs, not of billiard balls. He seems to have spent much time with the chickens, while the seventeenth-century founders of modern dynamics seem to have spent their lives, like Pascal, at the billiard and the gaming table.

"But Aristotle expands his essentially biological approach into a generalized functional conception and analysis for understanding any natural process. He takes motion in place— the billiard ball behavior from which modern dynamics started—as a limiting instance of more complex 'motions' or processes. In this respect, his procedure is not without analogy to that of our own physics, which has likewise passed beyond billiard balls and the motions of masses to the more complex processes of the field of radiation. Aristotle puts the emphasis, not on beginnings, but on outcomes, not on initiating efficient causes, or By Whats, but on results achieved, ends, For Whats. For him, real understanding of processes comes not primarily from past efficient causes, but from present and future activities and operations. Our own physics is likewise subordinating efficient causes to formal or mathematical causes and to operations."

[9] Elsasser, *Atom and Organism*, p. 137.

undoubtedly exceedingly important matters, but one would have thought them somewhat remote from the general activities of human beings, except in the very special field of the quantitative analysis of the behaviour of material bodies. Man is, after all, a biological entity. It is only in his most generalized characteristics, which he shares with sticks and stones, that he is a part of the subject matter of physics or chemistry. In his full being—or at least if we do not wish to beg the question, over a much wider range of his being—he falls within the province of biology. There are three currents of thought arising within biology which seem to me much more relevant to epistemology and general philosophy than anything that physics has or could discover about the behaviour of the ultimate units into which matter can be analysed. Two of these points relate to the subject matter of thought; the third to its logical structure.

The first unavoidable biological fact is that of evolution. For at least the last hundred years, since Darwin wrote, biologists have had to consider all living things, including man, as being produced by a process of evolution which operates in such a way as to bring its results into adjustment to the circumstances surrounding it. It is by now absolutely conventional and a matter of first principles to consider the whole physiological and sensory apparatus of any living thing as the result of a process which tailors it into conformity with the situations with which the organism will have to deal. The same principle undoubtedly applies to behavioural characteristics, and there is no obvious reason to deny it out of hand in relation to intellectual and even moral characteristics in those organisms which exhibit them. Within the professional field of biology there is a very general recognition of the occurrence of evolution of types of thinking, and of behaviour patterns, including behaviour to which we attribute a moral value. For instance, Roe and Simpson have recently edited a large symposium on *Evolution and Behaviour*; and we find such eminent biologists as J. B. S. Haldane and H. J. Muller discussing the technical genetical problems involved in the evolution of moral quantities.

The second major contribution of modern biology to the way in which we envisage living things derives from its emphasis on the importance of individual development. A few decades ago the growing point of biological thought was the analysis of the operations of the living machine. The most advanced biology dealt with problems of metabolism, of the intake of oxygen, foodstuffs, etc., the changes they undergo in the body, and their final excretion, together with problems of co-ordination between activities going on in different regions of the body. The fundamental problem of biology was seen as the understanding of the nature of enzyme action. More recently we have seen an increasing importance attached to questions concerning the mechanisms by which the functioning apparatus becomes gradually transformed as the individual develops from the fertilized egg onwards. This movement of thought, which has its origins in the work of such men as His, Roux, Driesch and Spemann, eventually and inevitably became linked with concepts derived from genetics. Its full depth and profundity then became apparent. The dominant position it now holds within the technical field of biology may be recognized in the fact that almost any biologist nowadays would admit that the crucial problem for theoretical biology is an understanding of the way in which genes control the characters of the organisms which develop from newly fertilized zygotes. The biology of metabolic systems can in some ways be compared to physics and chemistry in the days of the Daltonian irreducible atoms. The serious introduction of developmental thought introduces a new dimension, in much the same way as the Bohr-Rutherford atomic model opened new vistas for the physical sciences.

Both these new types of biological subject matter—the evolutionary and the developmental—have called for a new type of thinking, a type for which the most convincing causal structure is a circular one, in which A influences B and B again influences A.

I will return to the point of logical structure later, but, prior to and independent of any conclusions we may reach about it, I should like to emphasize that, however they may be thought about, the facts of evolution and development simply cannot be omitted from any discussion of the human condition which hopes to carry conviction at the present time; and yet, in spite of the publicity which is given to Darwin, they are very generally neglected by thinkers who are not professional biologists. H. J. Muller has recently published a paper with a somewhat angry title 'A Hundred Years Without Darwin Are Enough.'[2] He was not thinking of epistemologists and philosophers, but he well might have been. Remarkably few professional philosophers of the present day so much as mention the fact that the human sensory and intellectual apparatus has been brought into being by an evolutionary process whose observed effects in all other instances are to produce operative systems conformable to the situations with which they will have to deal. Take two examples more or less at random: the word evolution does not occur in the index of either Gilbert Ryle's *The Concept of Mind* or A. J. Ayer's *The Problem of Knowledge*. To the biologist, I think it is bound to remain almost inconceivable that one can talk much sense about the relation between man and the external world if one leaves out of account the fact that man has been brought into being by evolution in relation to the external world. This is not to deny, of course, that many different interpretations of the situation might be possible in place of the particular interpretation that I am offering here. But even if one were to finish up by concluding, for instance, that evolution had not occurred in the human epistemological apparatus—even if one were to retreat to the pre-scientific concept of man as a special creation, springing fully-armed from the head of his creator—the presupposition that evolution has affected him is so strong that it needs special arguments and not mere silence for its rejection.

This does not imply, of course, that evolutionary processes have already supplied mankind with an intellectual apparatus which is perfectly efficient in dealing with the external world. As we shall see later,[3] our stock has evolved a sociogenetic mechanism for transmitting information down the generations which, although much more effective than anything which preceded it, exhibits many obvious defects. It is clear also that our sensory apparatus also offers many opportunities for improvement, for instance, in discriminating scents, in visual accuity at small dimensions, or in responding to a broader spectrum of electro-magnetic vibrations. There can be no reason to doubt that our conceptualizing and logical faculties might also be susceptible of betterment. Indeed it seems probable that the progress of experimental physics has brought us in contact with phenomena to which our mental apparatus is not at all well suited, so that we find it exceedingly difficult to formulate a structure of concepts which corresponds to the facts. The point is, however, not that evolution produces perfect instruments for coping with all conceivable aspects of the world, but that the active systems it brings into being must have at least sufficient effectiveness to "get by" in the circumstances of life in which they are used. There must be some correspondence, of a degree which requires notice and not neglect, between the structure of our mental activities and the structure of our environment.

Here is one example typical of many in modern philosophical works in which a train of thought seems to call out for the invocation of evolutionary ideas but in which they fail to appear. Consider the following lines from Hannah Arendt's recent, most stimulating and in many ways profound book *The Human Condition*: "In other words, the world of the experiment seems always capable of becoming a man-made reality, and this, while it may increase man's power of making and acting, even of creating a world, far beyond what any previous age dared to imagine in dream and fantasy, unfortunately puts man back once more—and now even more forcefully—into the prison of his own mind, into the limitations of patterns he himself created. The moment he wants what all ages before him were capable of achieving, that is, to experience the reality of what he himself is not, he will find that nature and the universe 'escape him' and that a universe construed according to the behaviour of nature in the experiment and in accordance with the very principles which man can translate technically into a working reality lacks all possible representation. . . . It is therefore not surprising that the universe is not only 'practically inaccessible but not even thinkable,' for 'however we think it,' it is wrong; not perhaps quite as meaningless as 'triangular circle' but much more so than a 'winged lion'."[4] Here the phrases such as "the prison of his own mind," "the limitations of patterns he himself created," "to experience the reality of what he himself is not," all imply that one can conceive of a radical distinction between man and the rest of nature. They presuppose the possibility of considering man as a being created independently of the rest of the world, which he observes and acts on as something essentially foreign to himself. They leave out of account altogether the fact of evolution, the fact that the faculty by which man creates patterns to fit the natural world has itself been brought into being by a process which has moulded that faculty in such a way that it has the ability to form patterns which are in some way appropriate to what it has to deal with.

In the last sentence quoted from Miss Arendt above, she is herself quoting from the eminent physicist Erwin Schroedinger.[5] It is true indeed that eminent physicists have proved no less unaware of the consequences of accepting an evolutionary origin for man than have the philosophers. Schroedinger closes another book of his[6] with the words "For the purpose of constructing the picture of the external world, we have used the greatly simplifying device of cutting our own personality out, removing it. . . . This is the reason why the scientific world view contains of itself no ethical values, no aesthetic values, not a word about our own ultimate scope or destination, and no God, if you please." This puts very succinctly one side of the paradox into which non-evolutionary thinking has fallen. The other side is equally clearly stated by another eminent theoretical physicist, also quoted by Hannah Arendt, namely, Werner Heisenberg (pp. 22–23. Arendt quotes the same passage on her p. 261). Heisenberg writes: "If, starting from the conditions of modern science, we try to find out where the bases have started to shift, we get the impression that it would not be too crude an over-simplification to say that *for the first time in the course of history, modern man on this earth now confronts himself alone, and that he no longer has partners or opponents.* . . . Thus even in science, the object of research is no longer nature itself, but man's investigation of nature. Here, again, man confronts himself alone."[7]

Thus, while Schroedinger suggests that we have left man wholly out of our picture of nature, Heisenberg argues that we have nothing but man included in it. Surely it is clear that the paradox arises as a consequence of the attempt to draw a distinction between man and nature of a more ultimate kind than the facts warrant. Man is a part of nature, he forms a certain picture of what we may crudely call the external world, not as an outside observer of it but just because the forces of the external world have moulded his evolution into a being capable of reflecting it in a way adequate for carrying on the activities of life. We can, I think, be quite confident of this statement in relation to the physiological, sensory and intellectual capacities of man for handling his environment. Theologians might wish to reserve a small but crucial element in the human constitution outside the sphere of relevance of evolution. Such a thesis cannot be rejected out of hand, but it requires special arguments to support it. The great bulk of human nature, and the part that is most easy to observe, has undoubtedly been produced by evolution and has been moulded by the necessity to interact reasonably successfully with the non-human components of the universe.

Such a point of view has of course many implications for questions of general epistemology. It implies, for instance, that we have a mind capable of grasping logical structures because the universe exhibits regularities which make logical thinking a useful activity. It has implications also for the theory of perception. It argues that we experience tables and chairs, and not only (if at all) mere sense data, because it is evolutionarily useful to perceive, as Whitehead put it,[8] in the mode of causal efficacy as well as in that of presentational immediacy. Dewey, of course, explored some of these ideas, not always very convincingly. But I do not wish, even if I were capable of doing so, to pursue here the complex ramifications of this line of thought into these fields.

The point I wish to make is that also in the context of the evaluation of ethical systems it is inadmissible to try to erect a firm dualism between man and nature. When, for instance, Ewing writes: "There is nothing logically absurd about the supposition that the whole evolutionary process was harmful, and that it would have been better if life had never developed beyond its first stage or any intermediate stage we might happen to fancy,"[9] his contention cannot in my opinion be accepted. One is bound to ask "harmful to whom?" If we could conceive of man outside the evolutionary process, and in possession of a logic unrelated to the rest of the universe, Ewing's conclusion might follow; but if harmfulness is to be assessed entirely from the point of view of the products of evolution, and without bringing in any exterior, non-evolutionary point of reference, it is I think illogical (in the sense of being a contradiction in terms) to suppose that the whole evolutionary process is harmful.

Again, when Emmet writes: "It will not do to call a world outlook 'scientific' because it can bring natural and ethical phenomena under the same concepts, if those concepts were in fact derived from analogies with human actions and purposes in the first place and read back into the natural world. The concept of 'evolutionary progress' may be a case in point";[10] she is once again basing her thought on the implicit assumption that human actions and purposes could be something completely external to, and independent of, the natural world. The same assumption again underlies this quotation from Arendt: "The conviction that objective truth is not given to man but that he can know only what he makes himself is not the result of scepticism but of a demonstrable discovery,[11] and therefore does not lead to resignation but either to redoubled activity or to despair. The world loss of modern philosophy, whose introspection discovered consciousness as the inner sense with which one senses his senses and found it to be the only guarantee of reality, is different not only in degree from the age-old suspicion of the philosophers towards the world and towards the others with whom they shared the world; the philosopher no longer turns from the world of deceptive perishability to another world of eternal truth, but turns away from both and withdraws into himself."[12] Again, what the biologist notices is that the philosopher is not supposed to turn to himself as one constituent of the world, someone who not only makes it himself but is himself made by it. Arendt, who has just been discussing Cartesian doubt, is still caught by what Whitehead called the Cartesian dualism. Her philosopher is a would-be quite independent observer set over against the world. To any evolutionist he must appear as merely one part of it.

An outlook which sees the human observer as something quite separate and distinct from the external world which he perceives finds a sympathetic logical form in relations between two clearly separable terms, and a congruous type of causal analysis in statements such as that A causes B. For a more biological outlook, which sees man as simultaneously an observer, and a product, of the remainder of the world, such simple logical and causal structures are not adequate. For those who feel at home in the climate of present-day biology, a statement such as that A causes B inevitably has an air of incompleteness. What causes A, and what effects does B have? The system is not complete but leaves loose ends, which can only be tidied away by inventing something outside the system.

We feel more confident in an analysis which arrives at a conclusion of the form that A has a causal influence on B and B has a causal influence on A. In recent years, such circular causal systems have been referred to by the fashionable term "feedback," borrowed from engineering, and the word cybernetics has been introduced as a general term to cover the study of properties of systems organized in this manner; but this type of thinking has a much longer history in biology than this special terminology, and a much wider reference than the quantitative analysis of the operation of various material systems to which cybernetic ideas are usually applied. The general point is the replacement of sequences of simple causal or logical relations by organized causal or logical networks, which ideally and when complete should be self-contained, in the sense that they necessitate no reference to anything outside the system.

Theories which involve such self-contained and organized causal networks may easily be taken to be no more than vicious circles. For instance, Arendt writes[13] that it seems that science "has fallen into a vicious circle, which can be formulated as follows: scientists formulate their hypotheses to arrange their experiments and then use these experiments to verify their hypotheses; during this whole enterprise, they obviously deal with a hypothetical nature." But, one asks, if the experiments *do* verify the hypotheses, what more could one want? To call the nature with which the scientists deal "hypothetical"—is this any more than to make a depreciatory emotional noise about it? If science can produce—and this is its aim, never of course finally to be attained—a closed but consistent causal network in which the scientist himself is included, can any meaning be attached to a demand for something more?

Although some types of circularity in causal organization appear more convincing than mere open-ended causal sequences, it cannot of course be denied that certain other types of circular argument merit being dismissed as vicious circles. If, for instance, one says, as Herbert Spencer did and probably Julian Huxley also, that evolutionary progress is good and therefore the good can be defined by means of evolutionary progress, the argument does not escape from the imputation of being a mere vicious circle. The second clause, that good can be defined from evolutionary progress, adds nothing whatever to the first, which states that evolutionary progress is good.

It is very necessary to distinguish a circular argument which is "vicious" in the usual sense from one which validly expresses the structure of a causally organized system. The basic distinction can perhaps be expressed as follows. Consider two terms A and B. A vicious circle arises if we attribute some property to A, or claim that some sub-systems within A produce an attribute which belongs to A as a whole, and from these premises attempt to deduce some character of B. The viciousness arises from the fact that B has not been essentially referred to in the premises. For instance, if we say that goodness is an attribute of evolutionary progress (which would correspond to A), we cannot use this to make deductions about anything other than evolutionary progress, for instance, about human life (which would correspond to B). On the other hand, a valid circular statement of causal organization is of the form that A conditions the appearance of an attribute of B, and that B produces an attribute of A. Thus one does not commit any logical fallacy, but instead exhibits the real structure of the situation, if one says that nature through the processes of evolution has produced in the human race a certain perceptive-intellectual apparatus (influence of A on B), that this apparatus sees nature in a particular way (influence of B on A). Moreover, such an organized causal system is not necessarily, or even usually, stable in time. If at any moment A influences B, that is quite likely to change the manner in which B is acting on A, so that the system as a whole will become modified as time passes. The scientist, for instance, on the basis of the way in which he now sees nature will formulate hypotheses as to how he might see it more fully; and when he carries out experiments he offers nature the opportunity to influence him and his world view. It is this type of argument which has here been applied to ethics. The processes of evolution have produced the phenomenon that the human race entertains ethical beliefs. Man can then, not so much through experiment but rather by taking account of its results, use evolution to guide the way in which those beliefs will develop in the future.

The temporal instability of organized causal systems is one of their most important properties and it will be advisable to discuss it somewhat further. Much of the thought devoted to cybernetics has been in the context of various mechanical devices designed by man. In these, circular causal systems (systems involving feedback) have been designed for the specific purpose of changing in time in such a way as to attain some pre-set goal. The theory of their operation usually involves the concept of "negative feedback"; that is to say, the input into some part of the machine of an influence (for instance, an electric or mechanical force) whose value is a function of the difference between the present state of the machine and its pre-designed end. The invocation of a pre-designed end-point is, however, not at all a necessary part of an analysis into a closed causal organization. For instance, if we mix together definite quantities of a certain number of chemical substances, all of which are capable of interacting with one another, the mixture as a whole will form a causal system which will change in time until it reaches a certain equilibrium constitution. In the biological field, progressive changes of this kind form the major subject matter of the study of embryonic development.

A fertilized egg is provided with a certain number of genes in its nucleus and an organized structure of different regions of cytoplasm. The genes influence the cytoplasm by controlling the specificity of the substances which are synthesized within it; and the different regions of cytoplasm affect these gene-directed processes by controlling the relative rates at which they proceed. The system as a whole changes along a defined and recognizable course as time passes. Such a time-trajectory of developmental change arises from the characteristics of the closed circular causal organization of the system of genes and cytoplasm, but its mechanism does not involve anything strictly comparable to a "negative feedback," dependent on a predetermined end-point—a concept which biologists would consider teleological and therefore inadmissible. I have proposed[14] the word "creode," derived from the Greek words χρη "necessity" and ὁδοσ "a path," as a name for such time-trajectories of progressive developmental change, which arise from the nature of the causal organization of their starting-point.

Although the concept was first derived in an embryological context, creodes are a type of phenomena which occur in many other fields also. For instance, although the course of any particular evolutionary lineage cannot usually be considered a creode, since it is dependent perhaps on chance and on external environmental circumstances which may not be fully determined at the beginning of the evolutionary process, yet certain aspects of evolution as a whole do probably show the essential features of creodes. They are found, for instance, in the evolution of genetic systems as proposed by Darlington.[15] If there is as I suggest[16] an evolution of evolutionary systems, this also would be a creodic phenomenon.

The importance of developmental considerations was the third of the biological currents of thought to which attention was drawn at the beginning of this chapter. Any characteristic of a living organism must be thought of not only as something which has a functional role in the life of the organism, and not only as something which has been evolved through a considerable course of history, but also as something which undergoes a process of development during the individual lifetime of the organism. The concepts which have been derived from the scientific study of development are no less relevant to philosophical thought than those of evolutionary theory; and no consideration of the attributes or faculties of mankind can be accepted as satisfactory which neglects these two routes by which biology approaches the problems presented by living things. The inadequacy of the fashionable method of linguistic analysis in this respect has already been pointed out in Chapter 4.[17]

NOTES

[1] Ed. note: Waddington is referring to the general argument of his book which attempts to relate the development of moral codes and religions to the general trends of evolution.

[2] See H. J. Muller, "A Hundred Years Without Darwin Are Enough," *School Science and Mathematics*, LIX (1959), 304–316. Cf. George Gaylord Simpson, "One Hundred Years Without Darwin Are Enough," Ch. II of *This View of Life* (New York: Harcourt, 1963), pp. 26–41. (Ed.)

[3] Waddington is referring to Ch. 13 of *The Ethical Animal*, "Human Evolution and the Fall of Man," pp. 155–174. (Ed.)

[4] H. Arendt, *The Human Condition* (Chicago: Chicago University Press, 1959), p. 288.

[5] E. Schroedinger, *Science and Humanism* (New York: Cambridge University Press, 1952).

[6] E. Schroedinger, *Nature and the Greeks* (New York: Cambridge University Press, 1954).

[7] W. Heisenberg, *The Physicist's Conception of Nature* (New York: Harcourt, 1958).

[8] See A. N. Whitehead, *Modes of Thought* (New York: Macmillan, 1938).

[9] A. C. Ewing, *The Definition of Good* (New York: Macmillan, 1947), p. 74.

[10] D. M. Emmet, "The Choice of a World Outlook," *Philosophy*, XXIII (1948), 217.

[11] She was referring to Galileo's telescope.

[12] Arendt, *The Human Condition*, p. 293.

[13] *Ibid.*, p. 287.

[14] See C. H. Waddington, "The Cell Physiology of Early Development," in *Recent Developments in Cell Physiology* (New York: Academic Press, 1954).

[15] See C. D. Darlington, *The Evolution of Genetic Systems* (New York: Macmillan, 1939). Waddington takes up the discussion of this point in Ch. 10 of *The Ethical Animal*, "The Human Evolutionary System," pp. 101–124. (Ed.)

[16] On p. 102 of *The Ethical Animal*.

[17] "The Relevance of Developmental Facts," pp. 46–49.

5

The impact of evolution
on Christian thought

Rationale of this section

By far the most serious difficulties which evolution has created for man in the past one hundred years have been in the area of his religious sensibilities. In the Judaeo-Christian tradition, evolution has been seen as a threat to the central religious *datum* of the revelation of God through the Bible. Evolutionary thinking has been openly condemned on occasion by religious-minded persons, on the grounds that it undermines the whole superstructure of Christian theology and belief, inasmuch as it requires a view of the origin and nature of man sharply at variance with the picture of mankind's creation by God as painted by the book of Genesis and the theological traditions, and hence challenges—or seems to challenge—the credibility of the very core of Christian faith, namely, the requirement that one accept the Scripture as the word of God.

The biblical account of human history seems to require a view of human nature today as something intrinsically inferior to what it was in the beginning. Man is presented as a fallen creature, in the literal sense of having lost the high and balanced condition of mastery over himself and the world originally established for him by the grace of God. Not only has he fallen from this original state, but his nature has been wounded in the fall, wounded by an ignorance, malice, weakness, and concupiscence that cannot be healed by any human effort.

The evolutionary account of human history, on the contrary, requires a view of human nature as better integrated today than it could possibly have been in the primitive days before the development of culture and civilization. Man has not fallen, he has risen, risen from a state of relative isolation and helplessness before the forces of nature to a position of relative autonomy and control of nature. By comparison with his original state, man today is incomparably less ignorant and impotent. Moreover, his rise to mastery of the world inside and around him has only begun. The potential for further growth and better integration of human needs with natural forces is virtually limitless. Moreover, it is no longer this earth alone that contains man. Already the first tentative steps into the larger universe have been taken. Man has risen far, but he will rise farther still. And for this rise he has only his own initiatives to thank, and only further consolidations of knowledge to pin his hopes for the future on.

Hence the question seems fairly and inevitably posed: For post-Darwinian man, is a biblical or Christian faith still possible?

It has taken all of the hundred years of tedious research and discipline to mend the breakdown of communication between the world of evolutionary views of Darwin, Spencer and Thomas H. Huxley, and the world of Christian thought. Unquestionably the advances in professional evolutionary sciences such as prehistory, genetics, ecology, and physical anthropology have made the inferences of evolutionary theory more sure, its methods more rigorous and better documented, and its conclusions more available to the educated public. Since 1900, equally unprecedented advances in the sophistication of methods and clarification of principles of biblical exegesis have made it unmistakably clear that the bible is a religious and moral document, not a manual of scientific origins. In still another perspective, it must be noted that much of the philosophical dust has settled about the relations of order and chance in an evolving universe. Professional competence has cut away much of the opposition among the theological, philosophical, and scientific disciplines affected by the facts of evolution by the simple process of deepening insight and clearer formulation of issues.

But there is still great failure of professional competence, especially on the part of the scientist and the theologian. The matters at issue require a specialized knowledge on several levels which only a few men have had the opportunity or shown the willingness to acquire. On the side of theology, how a theologian can think and teach and write today about creation, about the nature of man, and about the relationship between man and God without bothering to establish for himself a competence in metaphysics, cosmology, psychlogy, sociology, anthropology, and comparative religion, is beyond the comprehension of contemporary enlightenment. And with good reason: it simply cannot be done. On the side of science, the scientist is not always so careful about neutrality in areas of moral value and religious affirmation.

Still, on the whole, this second half of our century has opened with a remarkable respect manifested by professional scientists and theologians for each other's efforts in probing the mystery of human origins. We realize now that the old, acrimonious exchanges among these professionals were in the order of constructs, products of what was referred to in the introductory essay as reason's second explanatory mode. The basic oppositions were only superficially on the level of reality; they were fundamentally caused by a clash of world-pictures, and the theologian and the scientist each had his own into which, before all else, he was concerned that the discussed realities would have to fit. Such an attitude was possible only so long as and to the extent that those entertaining it remained unaware of the dependence of the human mind upon traditions, upon cultural environment, and often upon, consequently, even the deadweight of history.

The distance is still very great between theoretical communication of common issues and the free, personal, creative expression about these realities, thus, because there is a deep, basic, preconscious impediment to human understanding and sympathy in these matters which stems not as much from the issues themselves as from the cultural traditions in which man's speaking and thinking are formed. Scientists and theologians are professional, and they are carefully raised in distinctive traditions necessarily and carefully constructed upon rigorous canons of judgment, rules of procedure, method. Most of them do not claim to be men of genius, but simply men of competence, bound, for the most part, to think and feel and act each within the walls of his tradition, within the security of the professional guidelines. The *creative* scientist or theologian is rare—even as the creative philosopher is rare—for tradition generates a condition in which the *status quo* is always preferred. Man is the historical animal. Most scientists and theologians—like most philosophers—tend to a hopeless *impersonality*, objectivity's half-sister; they place a premium upon communicating their disciplines in the appropriate language, without expressing themselves personally; they become victims of the "package deal," inasmuch as their interdisciplinary "discussions" tend not toward problem solving and sharing of insights, but toward the airing of clichés of the trade (each one of his own trade, of course), toward defending a system or attitude. The issue supposedly up for "free discussion" is examined in terms of the consequences to system (if it is the theologian) or in terms of restrictions of method and prior presuppositions never themselves brought out (if it is the scientist), rather than on the level of independent and detached research assuming responsibility for an issue in *all* its implications *and as a problem*.[1] But this "package dealing"

---

[1] In another context, but touching formally on this very point, Noam Chomsky comments: "The moral is not to abandon useful tools" (i.e., techniques which a given profession has forged in the solution of certain problems); "rather, it is, first, that one should maintain enough perspective to be able to detect the arrival of that inevitable day when the research that can be

is an ideological attitude, no more befitting the theologian or scientist than the philosopher.[2] It is a special case of wishful thinking.

Meanwhile, in the drama of living, the tension between concern for the timeless and concern for the timely remains at the heart of contemporary cognitive anxiety. Man's entire world-view is in the process of reformation, just as the churches are at this moment re-defining their relation to the world and its contingencies. But this much seems clear. By disengaging erroneous ideological interpretations from established scientific facts, it is by no means certain that one cannot view evolution in accord with the wisdom of traditional Christian thought and the wisdom of the Christian revelation. Evolution is indeed a kind of wisdom in its own right, the wisdom of the timely; but it leaves open the question as to the meaning of it all.

And if the Christian Revelation presents itself as a discourse on ultimate meaning, there is also the problem of religion in the cultural context: what role does *homo religiosus* have to play in the transformation of *homo sapiens* (*homo biologicus*) into *homo humanus,* in the shaping of mankind evolving and in the architecture of the human city?

Our readings in this section, accordingly, focus on these two key issues: the challenge of evolutionary thought to the very tenability of the Christian revelation on origins (Dubarle), and the reconcilability of a transhistorical belief with the insistence of evolutionism, historicism, and existentialism that history has no value if it has not an intrinsic one (Teilhard and Nogar).

---

conducted with these tools is no longer important; and, second, that one should value ideas and insights that are to the point, though perhaps premature and vague and not productive of research at a particular stage of technique and understanding." (*Language and Mind,* New York: Harcourt, Brace & World, 1968, p. 19). A concrete example of this attitude of independent and detached research not shirking responsibility for the full complexity of a matter at issue is given by C. H. Waddington in his discussion of man as *The Ethical Animal,* pp. 31–32. There are, of course, very many other examples that could be indicated.

[2] An example of this ideological "package dealing" on the part of a philosopher seems to be provided by Jacques Maritain, in pp. 154–159 of his recent book, *The Peasant of the Garonne;* or again on p. 142 of J. Messner's *Social Ethics.*

Original sin in the light of modern science and biblical studies

André-Marie Dubarle

CONTEXTUALIZING COMMENTS

A leading authority on the Old Testament, Père Dubarle is Professor of Sacred Scripture at the Dominican House of Studies of Le Saulchoir, France. His book, though not definitive, is perhaps the most profound and thorough study yet essayed of the question of an orginal sin of the human race. The question is so acute today precisely because it is chiefly in this area that scientific inferences seem to run counter to a moral and religious truth as traditionally traced to the biblical account of origins. The confiict has arisen over the knotty issue of the origin of "original sin": *monogenism* vs. *polygenism*. Evolutionary science, dealing principally with populations in its accounts of species development, thinks of man's physical origins among the Australopithecines in terms of a group (which for the theologian means polygenism). Christian theologians, interpreting *Genesis* and the account of the Fall through the eyes of Saint Paul and the Council of Trent, think instinctively in terms of a unique couple or single pair (the theological meaning of "monogenism"), Adam and Eve, at the origin of the human species. This issue has long since seemed to bring evolutionary science and the Bible into direct confrontation. But of late, many scholars have searched the biblical texts, the teaching of St. Paul, and the writings of the Council of Trent contextually, and it may well be that the moral and religious teaching of the Bible (and Trent) are neutral on the question of physical numbers and therefore neutral too on this evolutionary question. Patient research and discussion will tell. In this selection, Père Dubarle considers directly the bearing of modern theories of evolution on the idea of original sin.

Original sin in the light of modern science and biblical studies

I. WAS THERE A SINGLE OR A MULTIPLE SIN AT THE BEGINNING?

. . . Does the narrative of the fault of the first couple describe the clearly defined failure of two individuals, who alone are the ancestors of all men, or does it give a general representation of the law of spiritual heritage dominating the life of mankind?[1] Parables like that of the wicked rich man or of the pharisee and the publican give us a living and concrete description of a religious and moral way of life that numerous individuals can realize, independently of each other, in very different circumstances. But there is more than that here: the essential aim of the Genesis narrator is to show that the free conduct of the ancestor foreshadows and conditions the situation of his descendants. The passing on of a state of separation from God and the various misfortunes that are its consequences forms the basic teaching of this passage. In consequence all are sinners, as St. Paul was to say: sinners, i.e., not necessarily responsible because of any personal act of guilt, but separated from God, incapable of behaving as his sons and prone to act against his will.

All have need of redemption, for all are sinners by reason of the corrupting influences passed on from generation to generation. But do these ascending lines of filiation in evil, which accompany natural filiation and embrace the whole human race like a fan, converge at the beginning in one, absolutely unique point? Must we affirm the existence of a first couple as Genesis at first sight seems to say? Or may we think of a more complex network of sin and contagion? Are we faced here with a dia-grammatic representation, true in the way of such representations, but not as a photograph is true with the truth of an individual document?

In Genesis the characteristics and the destiny of different peoples or tribes are traced back to the action of one ancestor giving to each his name and his psychological characteristics. There is some artificial simplification here, but to speak of error would be to fail to understand that truth can be expressed in many and varied ways. The biblical authors were very conscious of the connection between the successive links of a line. In their eyes the divine choice was conditioned by membership of a particular race. And conversely the faults of an individual were visited on his children and their children to the third and fourth generation (Exod. 20:5; 34:7; Deut. 5:9; Num. 14:18; Jer. 32:18). So they were able to portray in one single ancestor and the sentence pronounced on his descendants the common effect of multiple sins, the consequence of disturbances which go on diminishing. It is then possible that the whole of mankind with the constant factors of its condition, was consciously represented in the story of Adam, whose name means "man".

If this interpretation is admitted, original sin is not a unique catastrophe at the birth of our species; it is the continually perpetuated perversion of mankind, in which new sins are conditioned more or less by the preceding sins and carry on the existing disorder. Instead of a disturbance that would die away in three or four generations, there is a generalized and anonymous corruption, with everyone its victim and many its authors, but in such a way that more often than not it is impossible to pinpoint any individual responsibility. In the story of the Tower of Babel (Gen. 11: 1–9) one of these collective faults is shown bringing in its train dire consequences, which remain for later generations. It is easy to see that this passage involves a great deal of generalization and simplification, when it reduces to one localized event a number of facts leading to the divergence of language and opposition between peoples. The question is whether it is unfaithful to the intention of the inspired author, or whether the teaching of St. Paul is gravely compromised, if we admit an analogous literary genre in Chapter 3 of the same book.

St. Augustine was struck by the Bible texts that told of ancestors being punished in their posterity and he put forward a theory that theologians have sometimes considered strange, but that could still be very stimulating.[2] In his eyes original sin is modified in each generation by the merits or demerits of individuals; parents increase or lessen for their children the burden of penalty and sin, which is passed on by the act of generation. Compared with the catastrophic corruption brought on by the fault of our first father these additions or subtractions remain relatively unimportant and original sin is not gradually obliterated in a long line of just men. Only the grace of Christ can succeed in remitting it entirely. These ideas, when compared with the questions being asked by modern literary criticism, can help in forming the concept of a universal state of original sin, but one that proceeds from multiple sources. This leaves room for certain individual variability, although the main common features are present everywhere.

What we suffer from is not only the fault of a distant ancestor but as much and above all far closer sins, which in their turn were provoked by earlier sins. We could speak today of a chain reaction and such an idea seems to us to fit in with what we see elsewhere of mankind's condition. Each of us, because he is born into a world and a race contaminated by sin, is born a sinner. "That which is born of the flesh is flesh" (John 3:6). He cannot enter the kingdom of God without first being cleansed.

Many modern exegetes refuse to see in the story of the Fall the story of an individual event. In their eyes it is only a kind of parable, illustrating the universal fact of sin in mankind. From the point of view of literary criticism such an interpretation is far from unfounded. But what it ignores or overlooks, what catholic dogma proclaims with insistence and teaches us not to add to the text but to recognize in it, is the fact that sin like all other elements of man's destiny is not a strictly individual matter. Its consequences weigh heavy on the posterity of the culprits. In the case of man and his wife it is the whole of mankind that is now in the grip of fear at God's approach, banished far from the garden of happiness, condemned to death and the penalty of work by the sweat of the brow. Underlying the text is the idea of a heritage of sin.

Between the interpretation of many modern exegetes, who see in Chapter 3 of Genesis only the stylized outline of individual sin, and the early and usual interpretation, which sees in it the account of a particular sin at the beginning which had consequences for the whole of mankind, there is room for an intermediate position, admitting the schematic and universal nature of the narrative but not missing the main point, a sin handed on by inheritance: what the text describes is the effect of a countless multitude of individual sins.

Perhaps a faithful catholic is not today in a position to provide unanswerable arguments in support of this interpretation. Perhaps the positive science of human origins, which at the moment is unable to affirm the existence of a single couple, the ancestors of all men, will one day succeed in finally rejecting this or in incontrovertibly proving it. Perhaps the theologians will agree to accept the idea offered them by biblical exegesis: the idea of a symbolic and schematic account, intending to describe not a strictly individual event but a universal condition passed on by inheritance. Perhaps on the other hand they will succeed in showing that strict unity of physical origin is so necessarily bound up with the universality of original sin that the first cannot be denied without the second being abandoned. It is not always easy to discover at the outset the remote consequences of a new idea, nor is it easy to form a complete picture of all the internal connections of revealed truths and so estimate the repercussions of a denial that may at first seem unobjectionable. . . .

II. THE ORIGINAL STATE OF MAN

Modern theories of evolution have already been mentioned in the discussion on the unity or plurality of the source of original sin. We can now consider them more directly, not in order to expose or discuss them in themselves, but to consider them in their bearing on the idea of original sin.

The evidence of modern science leads us to think of mankind as issuing from an animal stock. Gradually the human organism, as we see it today, was formed by multiple modifications, which we can observe or reconstruct thanks to the discoveries of fossil remains. Parallel to this the human psyche came from an animal psyche, which does not imply stupidity or brutishness: we should rather think of the industrious spirit, provident and co-operative, that can be shown by an animal society. Human industry, recognized as such by the manufacture of tools on a large scale, however primitive they may be, is associated with skeletons that seem to us a far cry from our ideal of plastic beauty.

Our Christian faith assures us that at a given moment a qualitative threshold was passed. God created man in his own image, as a spiritual being, using and dominating from then on the corporal and psychic organism that was the result of animal evolution. Arguments from analogy show us that this threshold is not necessarily perceptible to external observation. If, and this experiment has been made, the reactions of a human baby and those of a young monkey are simultaneously followed and compared, it is not possible to say at what moment truly human intelligence begins to appear. But it is certain that in three or four years the differentiation is complete. In the same way, if *per impossibile* we had sufficient palaeontological documentation at our disposal to follow closely over the course of time the transformations in body and technical ability of a group of hominids, we admit that we should still be unable to pinpoint the generation in which human nature appeared. For our positive sciences hominisation is a fact that extends over a period of time, but this does not contradict the religious or philosophical idea that maintains that a creature either is or is not a man, whatever may be the actual appearances of human spirituality perceptible to external observation.

These extremely modest beginnings suggested by the modern theory of evolution form a strange contrast with the descriptions in classical theological treatises, brought to life for us in the great Christian works of art: an Adam clothed in splendour, endowed with perfect physical beauty and wide knowledge, and of such moral integrity that he not only dominates his passions but has the same control of will over them as over the movements of his muscles, an Adam blessed with sanctifying grace and preserved from suffering and death. We are told that all this is the effect of a privilege of grace, which could be lost without nature being affected or mutilated thereby. If it is admitted that this nature was the result of a slow process of evolution and not instantaneous creation, this privilege determines a new state that can really be nothing but a "marvellous parenthesis,"[3] since the Fall caused by sin soon intervenes. The laws of evolution, suspended or surpassed for a moment, come into play again, and we cannot hope to be able to look back and discover definite traces of these brief moments when an emerging humanity was crowned with grace.

Such a reconciliation between the modern theory of evolution and the classical representation of the original state of man is not absurd, considered in itself. But there is still the danger that it will not entirely remove the uneasiness felt by many people who are accustomed to the disciplines of science and who see the happiness of paradise as a poetic legend, in which we can, of course, as in many legends, pick out a more or less profound moral truth but need not recognize a real happening.

This is the point at which we may well ask ourselves what our Christian faith teaches us in its most authentic documents. In the first place we can say this. The Council of Trent passed over a doctrinal draft in which Adam's exceptional gifts were described in the classical way,[4] and simply promulgated a canon stating that by his sin the first man lost "the holiness and righteousness in which he had been constituted."[5] These words correspond to those defining man's role on earth in the book of Wisdom (9:2-3). They do not imply that this holiness and righteousness were the equal of what the theologians call sanctifying grace, making men "partakers of the divine nature" (2 Peter 1:4). In the following session, which was devoted to justification, the Council went on to teach that authentic Christian righteousness replaces the original righteousness lost by Adam, without making clear whether it surpasses it or not.[6]

The Genesis narrative shows us man in a divine garden, a sort of luxurious oasis (cf. Gen. 2:19), giving names to the animals, thus exercising over them his powers of intellect and will, but there is no need to see all this as a sign of any exceptional infused knowledge. He recognizes the value of the gift made to him in the person of the woman: he is not a child. The first couple are naked and feel no embarrassment, for the physical climate is clement and there is not yet any dissimulation between individuals, nor any of those artificial social distinctions symbolized by clothing. Peace also reigns between God and man.

The author of this passage wanted not so much to give a detailed description of, as to suggest, a state in which the harsh facts of the human condition that we experience did not make themselves felt (death, mutual distrust, slavery, the difficulty of work, the anxieties of motherhood, the hostility of surrounding nature and the moral struggles that men now experience). He had no intention of describing any magic state, supposing other physical or psychological laws than those now in force. Clearly he had no idea of evolution and the ideal picture that he drew was a retouched picture of the life of the small Palestinian peasant seeing in his home and the fruits of the earth the marks of divine favour (cf. Psalm 128).

Then sin appeared to disturb this happy harmony and to spoil the excellence of the Creator's work. And the essential point in the Garden of Eden story, as also in St. Paul's teaching, is that this perversion has disastrous consequences for the sinner and also for his posterity. Man hides from God and this foreshadows the future reaction of even the most favoured persons; individuals experience embarrassment before one another. Man must work hard to live; woman must put up with subjection to her husband and the pains of childbirth; all must die; and finally all must struggle against the serpent. But while the narrator described the main features of the state of fallen humanity so precisely, he only sketched, and then with great restraint, the better conditions that went before. The Wisdom authors laid particular emphasis on all that is normal in the limitations and difficulties of a creature whose freedom is being tested.

In the mental reconstruction that we may be inclined to make of the state preceding the Fall, we should be warned by these examples against increasing the divergence between this state and our present condition and imagining exemption from the laws governing life or the physical world. This can be applied even to death, the wages of sin according to St. Paul (Romans 6:23), who brings together in this passage many Old Testament texts, beginning with Gen. 2:17 and 3:22. On this point we should remember what a wealth of meaning the Bible has put into this word "death." At the beginning of its semantic development there was certainly the idea of corporal decease; but around this basic meaning crystallized the idea of misfortune, shame and separation from God. Death is a return to the earth and from this point of view it is normal for man, who was taken from the earth; but, considered in the light of all the secondary meanings attached to the word, it is also the downfall of the sinner. In this sense it did not exist before sin: "God did not make death" (Wisdom 1:13).

To be faithful to the statements of Genesis and St. Paul we do not need to postulate a corporal existence immune from decease. The inspired authors saw a consequence of sin in these tragic experiences surrounding decease (physical sufferings, family separations and a feeling of hopeless failure). They did not speculate explicitly on what a state of innocence would have been like or on the possible dissociation between physical decease and death as the sum of human ills.[7]

The magisterium of the Church has made no irrevocable pronouncement on this point. One provincial Council, that of Carthage in 418, did directly condemn the denial of Adam's corporal immortality.[8] But this canon was not specifically approved by Pope Sosimus in the letter that he addressed to the whole Church on the subject of grace and original sin.[9] The ecumenical Council of Trent did not take up this canon of Carthage, while it did reproduce almost the exact text of another canon of the same council. A draft condemnation was left in the archives, and it was worded in this way: "if anyone says that Adam was bound to die in any way, even if he had not sinned, let him be anathema."[10]

So without contradicting any irrevocable doctrinal authority, scriptural or ecclesiastical, we can, in conformity with the suggestions of the evolution theory, admit that mankind emerged from the animal world. Leaving aside the existence of a spiritual soul, the transition may have been very gradual. In the case of the soul, which either does or does not exist, there can have been no gradual transition, although all its rich potential would not have been immediately manifest to external observation. Man began gradually to diverge from the animals in his way of life, while sin began to form that heritage of perversion that was handed on at the same time as the heritage of technical culture.

Such a theory, attempting to do justice to the evidence of our modern knowledge of human origins as well as to the biblical data taken as a whole and not just the two solitary texts of Genesis 3 and Romans 5, does not compromise the gravity of original sin. Instead of concentrating on the loss of wonderful and gratuitous gifts, whose disappearance does not really injure our nature in itself, this theory fixes its attention on quite concrete troubles that we find in our actual experience and that the Genesis narrative represented in a stylized way: a poisoning of our trusting relationship with God and various sufferings. The doctrine of original sin consists in stating that not everything that worries us can be explained by the still incomplete development of man's spiritual powers or by the failures of an evolutive system that would leave room for mistakes or a process of trial and error on man's own level as well as on the level of the formation of the species. In the present state of humanity there is a disorder (not just something missing) on the religious as well as the human level, and this is the result of deliberate sin. Individuals are embroiled in this disorder whether they like it or not: and it is of small importance whether the point of departure was close to the animal state, as modern evolution theory thinks, or raised far above it, as was thought for a long time by theologians who did not have the information that we possess today. The essential point is that the present state of mankind, with the baneful influence that it exercises on newcomers to existence, is the result of deliberate faults and that even the initial religious state of young children is vitiated by it.

NOTES

[1] The conviction that the first sin did not become known through historical tradition or immediate revelation, but as the result of a mental reconstruction, is expressed by H. Renckens, *Urgeschichte und Heilsgeschichte*, (1959), p. 40, (translated from the Dutch, *Israëls visie op het verleden*, (1957); and by K. Rahner, in the article on 'Atiologie, in *Lexikon für Th. u. K.*, I, (1957), col. 1011–1012; and by many others.

[2] On this Augustinian theory cf. A. -M. Dubarle, 'La pluralité des péchés héréditaires dans la tradition augustinienne', in *Rev. Et. Aug.*, II, (1957), pp. 113–136. This theory found a certain echo among the theologians of the Middle Ages and at the Council of Trent. At the end of the article the discussion that this theory provoked at that time is studied.

[3] This expression comes from M. Labourdette, *Le péché originel et les origines de l'homme* (1953), p. 178. However the author already introduces some restrictions into the pictures of the original perfection drawn by classical theology, particularly as regards the extent of their knowledge.

[4] *Concilium Tridentinum edidit societas Goerresiana*, vol. XII, pp. 566–569: Decreti de peccato originali minuta (ineunte mense iunio 1546) . . . '. . . "Fecit Deus hominem rectum et inexterminabilem" (Eccles. 7:29; Sap. 2:23). Et hoc secundum corpus et secundum animam et secundum mentem ac spiritum. Secundum enim corpus fecit eum non subjectum corruptioni et morti, non laboribus aut doloribus subditum et infirmitatibus, sed sanum, integrum . . . Secundum animam vero ita bene compositum reddidit ac temperatum rectitudine et justitia, ut omnino corpus animae subiceretur et pars inferior animae, ubi passiones gignuntur, superiori parti, in qua ratio viget, sicut ordo ipse bene institutae naturae poscebat, miro modo consentiret, minime contumax, superior autem pars id est ratio et ipsa mens, Deo creatori suo, ut parerat, obedienter obtemperaret. Quare cum nudi essent, "non erubescebant" (Gen. 2:25), nulla scilicet existente erubescentiae causa, cum nihil in eis contra decorem et rationem pugnaret. Denique secundum mentem . . .' (p. 567, 1.10-21).

[5] 'Sanctitatem et iustitiam in qua *constitutus* fuerat" (Denz. 788 or 1511). This formula was definitively sanctioned and replaced one in an earlier draft: 'sanctitatem et iustitiam in qua *creatus* fuerat', which was inspired by Eph. 4:24: 'novum hominem qui . . . creatus est in iustitia et sanctitate veritatis'. The similarity of words, which was quite deliberate, might suggest an equality between the condition of the first man and the sanctifying grace given to the Christian. The formula that was actually adopted by the council reproduced, but apparently without realizing it, the words of Wisdom 9:2–3: 'constituisti hominem . . . in aequitate et iustitia', where 'sanctitate' would be a better translation of the Greek *'hosioteti'* than the 'aequitate' of the Vulgate.

[6] Session VI, Decree on justification, ch. 7; Denz. 800 or 1531.

[7] We do not attribute any real theological authority to these lines of C. Péguy, but they are quite apposite.

> Ce qui depuis ce jour est devenu la mort
> N'était qu'un naturel et tranquille départ.
> Le bonheur écrasait l'homme de toute part.
> Le jour de s'en aller était comme un beau port. *Eve*, st. 26.

[8] 'Whoever says that Adam, the first man, was made mortal, so that, whether he sinned or not, he was going to die in his body, that is he was going to leave his body, not through the merit of sin but by the necessity of nature, let him be anathema' Denz. 101 or 222.

[9] The letter (Tractoria or Tractatoria) of Sosimus on the council of Carthage has been lost. Only a few extracts have been preserved by St Augustine (cf. *PL*, 20, 693–695), who would certainly have reproduced with care any passage concerning the immortality of the first man. On this and related points cf. P. M. de Contenson in *RSPT*, XXXIX, (1955), p. 58, n. 40, in a review of the work of M. Labourdette, *Le péché originel*.

[10] 'Qui ergo dixerit Adam omnino moriturum etiam si non peccasset, anathema sit'. This draft, which was not discussed, can be found in *Concilium Tridentinum . . .*, vol. XII, p. 567, 1.47. The Council of Trent did take up a canon of the provincial Council of Orange (529), but with a modification that has made it lose its indubitable bearing on the present question. The Council of Orange declared that Adam had handed on to his posterity bodily death, which is the penalty of sin, and sin, which is the death of the soul (Denz. 175 or 372). The Council of

Trent states that Adam passed on to his posterity death and bodily sufferings and sin, which is the death of the soul (Denz. 789 or 1512). After this rearrangement it is no longer clear that the text is speaking of bodily death to the exclusion of 'death' in the very full biblical sense of the word, all the more so as the definitive text left out a clause of the draft, which spoke of sin 'to which is due both the death of the body and of the soul as a penalty' (*Conc. Trid.*, vol. V, p. 196).

The heart of the problem

Pierre Teilhard de Chardin

CONTEXTUALIZING COMMENTS

In this brief essay, written many years before the "Death of God" became a catchword and a label, Père Teilhard de Chardin expressed his conviction that something has gone wrong between Man and God as represented to Man in the present age. At the same time, Père Teilhard was confident that a view of history which excluded a divine consummation in the end could not sustain the heart of man.

Just as in Père Teilhard's view the religious community had in practice deformed itself by effective separation from some fundamental human values of earthly life, so the marxist movement, in fastening on to these human values, deformed itself in another way by divorcing them from the sacredness of the human person. On the one hand a religious faith failing to acknowledge mankind's development toward a mature and reflective relation to the natural world; on the other hand a secular humanism so indifferent to the person that it demands its own forms of thought control: such seemed to Père Teilhard the shape of the conflict at the heart of modern unrest.

Unwilling to entertain seriously a view of the human soul as "so badly devised that it contradicts within itself its own profoundest aspirations," Père Teilhard was convinced and attempted to convince others that the Christ of the Gospels, once seen as the focus and Omega of our evolving universe, provides the movement of the world with a direction which ultimately accords with man's deepest concerns both terrestrial and religious. He attempted to show that secular humanism and religious faith both find in Christ—through the idea of evolution—a focus which matches their rhythm and prolongs their extent beyond the vicissitudes of chance, tragedy, and destruction.

In this way he sought to explain the relation between religious consciousness and the sphere of human action upon the world.

The heart of the problem

Introduction

Among the most disquieting aspects of the modern world is its general and growing state of dissatisfaction in religious matters. Except in a humanitarian form, which we shall discuss later, there is no present sign anywhere of Faith that is expanding: there are only, here and there, creeds that at the best are holding their own, where they are not positively retrogressing. This is not because the world is growing colder: never has it generated more psychic warmth! Nor is it because Christianity has lost anything of its power to attract: on the contrary, everything I am about to say goes to prove its extraordinary power of adaptability and mastery. But the fact remains that for some obscure reason something has gone wrong between Man and God *as in these days He is represented to Man.* Man would seem to have no clear picture of the God he longs to worship. Hence (despite certain positive indications of re-birth which are, however, still largely obscured) the impression one gains from everything taking place around us is of an irresistible growth of atheism—or more exactly, a mounting and irresistible de-Christianisation.

For the use of those better placed than I, whose direct or indirect task it is to lead the Church, I wish to show candidly *where*, in my view, the root of the trouble lies, and *how*, by means of a simple readjustment at this particular, clearly localised point, we may hope to procure a rapid and complete rebound in the religious and Christian evolution of Mankind.

I say "candidly." It would be presumptuous on my part to deliver a lecture, and criticism would be out of place. What I have to offer is simply the testimony of my own life, a testimony which I have the less right to suppress since I am one of the few beings who can offer it. For more than fifty years it has been my lot (and my good fortune) to live in close and intimate professional contact, in Europe, Asia and America, with what was and still is most humanly valuable, significant and influential—"seminal" one might say—among the people of many countries. It is natural that, by reason of the exceptional contacts which have enabled me, a Jesuit (reared, that is to say, in the bosom of the Church) to penetrate and move freely in active spheres of thought and free research, I should have been very forcibly struck by things scarcely apparent to those who have lived only in one or other of the two opposed worlds, so that I feel compelled to cry them aloud.

It is this cry, and this alone, which I wish to make heard here—the cry of one who thinks he sees.

1. A major event in human consciousness: the emergence of the 'ultra-human'

Any effort to understand what is now taking place in the human conscience must of necessity proceed from the fundamental change of view which since the sixteenth century has been steadily exploding and rendering fluid what had seemed to be the ultimate stability—our concept of the world itself. To our clearer vision the universe is no longer a State but a Process. The cosmos has become a Cosmogenesis. And it may be said without exaggeration that, directly or indirectly, all the intellectual crises through which civilisation has passed in the last four centuries arise out of the successive stages whereby a static *Weltanschauung* has been and is being transformed, in our minds and hearts, into a *Weltanschauung* of movement.

In the early stage, that of Galileo, it may have seemed that the stars alone were affected. But the Darwinian stage showed that the cosmic process extends from sidereal space to life on earth; with the result that, in the present phase, Man finds himself overtaken and borne on the whirlwind which his own science has discovered and, as it were, unloosed. From the time of the Renaissance, in other words, the cosmos has looked increasingly like a cosmogenesis; and now we find that Man in his turn is identified with an anthropogenesis. This is a major event which must lead, as we shall see, to the profound modification of the whole structure not only of our Thought but of our Beliefs.

Many biologists, and not the least eminent among them (all being convinced that Man, like everything else, emerged by evolutionary means, i.e. was *born* in Nature) undoubtedly still believe that the human species, having attained the level of *Homo sapiens*, has reached an upper organic limit beyond which it cannot develop, so that anthropogenesis is only of retrospective interest. But I am convinced that, in opposition to this wholly illogical and arbitrary idea of arrested hominisation, a new concept is arising, out of the growing accumulation of analogies and facts, which must eventually replace it. This is that, under the combined influence of two irresistible forces of planetary dimensions (the geographical curve of the Earth, by which we are physically compressed, and the psychic curve of Thought, which draws us closer together), the power of reflection of the human mass, which means its degree of *humanisation*, far from having come to a stop, is entering a critical period of intensification and renewed growth.

What we see taking place in the world today is not merely the multiplication of *men* but the continued shaping of *Man*. Man, that is to say, is not yet zoologically mature. Psychologically he has not spoken his last word. In one form or another something ultra-human is being born which, through the direct or indirect effect of socialisation, cannot fail to make its appearance in the near future: a future that is not simply the unfolding of Time, but which is being constructed in advance of us . . . Here is a vision which Man, we may be sure, having first glimpsed it in our day, will never lose sight of.

This being postulated, do those in high places realise the revolutionary power of so novel a concept (it would be better to use the word "doctrine") in its effect on religious Faith? For the spiritually minded, whether in the East or the West, one point has hitherto not been in doubt: that Man could only attain to a fuller life by rising "vertically" above the material zones of the world. Now we see the possibility of an entirely different line of progress. The Higher Life, the Union, the long dreamed-of consummation that has hitherto been sought *Above,* in the direction of some kind of transcendency: should we not rather look for it *Ahead,* in the prolongation of the inherent forces of evolution?

Above or ahead—or both? . . .

This is the question that must be forced upon every human conscience by our increasing awareness of the tide of anthropogenesis on which we are borne. It is, I am convinced, the vital question, and the fact that we have thus far left it unconfronted is the root cause of all our religious troubles; whereas an answer to it, which is perfectly possible, would mark a decisive advance on the part of Mankind towards God. That is the heart of the problem.

2. At the source of the modern religious crisis: a conflict of faith between upward and forward

Arising out of what I have said, the diagram at the end of this chapter represents the state of tension which has come to exist more or less consciously in every human heart as a result of the seeming conflict between the modern forward impulse (ox), induced in us all by the newly-born force of trans-hominisation, and the traditional upward impulse of religious worship (OY). To render the problem more concrete it is stated in its most final and recognisable terms, the co-ordinate OY simply representing the Christian impulse and OX the Communist or Marxist impulse [1] as these are commonly manifest in the present-day world. The question is, how does the situation look, here and now, as between these opposed forces?

We are bound to answer that it looks like one of conflict that may even be irreconcilable. The line OY, faith in God, soars upward, indifferent to any thought of an ultra-evolution of the human species, while the line OX, faith in the World, formally denies (at least in words) the existence of any transcendent God. Could there be a greater gulf, or one more impossible to bridge?

Such is the appearance: but let me say quickly that it cannot be true, not finally true, unless we accept the absurd position that the human soul is so badly devised that it contradicts within itself its own profoundest aspirations. Let us look more closely at OX and OY and see how these two vectors or currents appear and are at present behaving in their opposed state. Is it not apparent that both suffer from it acutely, and therefore that there must be some way of overcoming their mutual isolation?

Where OX is concerned the social experiment now in progress abundantly demonstrates how impossible it is for a *purely immanent* current of hominization to live wholly, in a closed circuit, upon itself. With no outlet ahead offering a way of escape from total death, no supreme centre of personalisation to radiate love among the human cells, it is a frozen world that in the end must disintegrate entirely in a Universe without heart or ultimate purpose. However powerful its impetus in the early stages of the course of biological evolution into which it has thrust itself, the Marxist anthropogenesis, because it rules out the existence of an irreversible Centre at its consummation, can neither justify nor sustain its momentum to the end.

Worldly faith, in short, is not enough in itself to move the earth forward: but can we be sure that the Christian faith in its ancient interpretation is still sufficient of itself to carry the world upward?

By definition and principle it is the specific function of the Church to Christianise all that is human in Man. But what is likely to happen (indeed, is happening already) if at the very moment when an added component begins to arise in the *anima naturaliter christiana,* and one so compelling as the awareness of a terrestrial "ultra-humanity," ecclesiastical authority ignores, disdains and even condemns this new aspiration without seeking to understand it? This authority, which is no more nor less than Christianity, will lose, to the extent that it fails to embrace as it should *everything that is human on earth,* the keen edge of its vitality and its full power to attract. Being for the time *incompletely human* it will no longer fully satisfy even its own disciples. It will be less able to win over the unconverted or to resist its adversaries. We wonder why there is so much unease in the hearts of members of Christian orders and of priests, why so few deep conversions are effected in China despite the flood of missionaries, why the Christian Church, with all its superiority of benevolence and devotion, yet makes so little appeal to the working masses. My answer is simply this, that it is because at present our magnificent Christian charity lacks what it needs to make it decisively effective, the sensitizing ingredient of *Human* faith and hope without which, in reason and in fact, no religion can henceforth appear to Man other than colorless, cold and inassimilable.

OY and OX, the Upward and the Forward: two religious forces, let me repeat, now met together in the heart of every man; forces which, as we have seen, weaken and wither away in separation . . . Therefore, as it remains for me to show, forces which await one thing alone—not that we should choose between them, but that we should find the means of combining them.

3. The rebound of the christian faith: upward by way of forward

It is generally agreed that the drama of the present religious conflict lies in the apparent irreconcilability of two opposed kinds of faith—Christian faith, which disdains the primacy of the ultra-human and the Earth, and "natural" faith, which is founded upon it. But is it certain that these two forces, neither of which, as we have seen, can achieve its full development without the other, are really so mutually exclusive (the one so antiprogressive and the other so wholly atheist) as we assume? Is this so if we look to the very heart of the matter? Only a little reflection and psychological insight is required to see that it is not.

On the one hand, neo-human faith in the World, to the extent that it is truly a Faith (that is to say, entailing sacrifice and the final abandonment of self for something greater) necessarily implies an element of worship, the acceptance of something "divine."[2] Every conversation I have ever had with communist intellectuals has left me with a decided impression that Marxist atheism is not absolute, but that it simply rejects an "extrinsical" God, a *deux ex machina* whose existence can only undermine the dignity of the Universe and weaken the springs of human endeavor—a "pseudo-God," in short, whom no one in these days any longer wants, least of all the Christians.

And on the other hand Christian faith (I stress the word Christian, as opposed to those "oriental" faiths for which spiritual ascension often expressly signifies the negation or condemnation of the phenomenal world), by the very fact that it is rooted in the idea of Incarnation, has

always based a large part of its tenets on the tangible values of the World
and of Matter. A too humble and subordinate part, it may seem to us now
(but was not this inevitable in the days when Man, not having become
aware of the genesis of the Universe in progress, could not apprehend the
spiritual possibilities still buried in the entrails of the Earth?) yet still a
part so intimately linked with the essence of Christian dogma that, like a
living bud, it needed only a sign, a ray of light, to cause it to break into
flower. To clarify our ideas let us consider a single case, one which sums
up everything. We continue from force of habit to think of the Parousia,
whereby the Kingdom of God is to be consummated on Earth, as an event
of a purely catastrophic nature—that is to say, liable to come about at any
moment in history, irrespective of any definite state of Mankind. But why
should we not assume, in accordance with the latest scientific view of
Mankind in a state of anthropogenesis,[3] that the parousiac spark can, of
physical and organic necessity, only be kindled between Heaven and a
Mankind which has biologically reached a certain critical evolutionary
point of collective maturity?

For my own part I can see no reason at all, theological or traditional,
why this "revised" approach should give rise to any serious difficulty. And
it seems to me certain, on the other hand, that by the very fact of making
this simple readjustment in our "eschatological" vision we shall have
performed an operation having incalculable consequences. For if truly,
in order that the Kingdom of God may come (in order that the Pleroma
may close in upon its fullness), it is necessary, as an essential physical
condition,[4] that the human Earth should already have attained the natural
completion of its evolutionary growth, then it must mean that the ultra-
human perfection which neo-humanism envisages for Evolution will
coincide in concrete terms with the crowning of the Incarnation awaited
by all Christians. The two vectors, or components as they are better called,
veer and draw together until they give a possible resultant. The super-
naturalizing Christian Upward is incorporated (not immersed) in the
human Forward! And at the same time Faith in God, in the very degree
in which it assimilates and sublimates within its own spirit the spirit of
Faith in the World, regains all its power to attract and convert!

I said at the beginning of this paper that the human world of today
has not grown cold, but that it is ardently searching for a God propor-
tionate to the newly discovered immensities of a Universe whose aspect
exceeds the present compass of our power of worship. And it is because
the total Unity of which we dream still seems to beckon in two different
directions, towards the zenith and towards the horizon, that we see the
dramatic growth of a whole race of "spiritual expatriates"—human beings
torn between a Marxism whose depersonalizing effect revolts them and a
Christianity so lukewarm in human terms that it sickens them.

But let there be revealed to us the possibility of believing *at the same*

time and wholly in God *and* the World, the one through the other;[5] let
this belief burst forth, as it is ineluctably in process of doing under the
pressure of these seemingly opposed forces, and then, we may be sure of
it, a great flame will illumine all things: for a Faith will have been born (or
re-born) containing and embracing all others—and, inevitably, it is the
strongest Faith which sooner or later must possess the Earth.

*Diagram illustrating the conflict between the two kinds of Faith in the heart of modern
man*

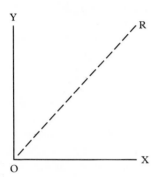

OY: Christian Faith, aspiring Upward, in a personal transcendency, towards the
    Highest.
OX: Human Faith, driving Forward to the ultra-human.
OR: Christian Faith, 'rectified' or 'made explicit', reconciling the two: salvation
    (outlet) at once Upward and Forward in a Christ who is both Saviour and
    Mover, not only of individual men but of anthropogenesis as a whole.

Let it be noted that by its construction OR is not a half-measure, a compromise
between Heaven and Earth, but a resultant combining and fortifying, each through
the other, two forms of detachment—that is to say, of 'sacrifice to that which is
greater than self'.

NOTES

[1] An unfavourable simplification where OX is concerned, inasmuch as Marxism and Communism (the latter a thoroughly bad, ill-chosen word, it may be said in passing) are clearly no
more than an embryonic form, even a caricature, of a neo-humanism that is still scarcely born.
[2] As in the case of biological evolutionary theory which also bore a materialist and atheist
aspect when it appeared a century ago, but of which the spiritual content is now becoming
apparent.
[3] And, it may be added, in perfect analogy with the mystery of the first Christmas which (as
everyone agrees) could only have happened between Heaven and an Earth which was *prepared*,
socially, politically and psychologically, to receive Jesus.
[4] But not, of course, sufficient in itself!
[5] In a Christ no longer seen only as the Saviour of individual souls, but (precisely because He
is the Redeemer in the fullest sense) as the ultimate Mover of anthropogenesis.

Evolutionary humanism and the faith

Raymond J. Nogar

CONTEXTUALIZING COMMENTS

The Christian faith lays before man ultimate answers concerning his personal existence and collective destiny. One of the great weaknesses of Christianity, historically speaking, has been the effort of many of its protagonists to regard these ultimate resolutions as practical norms, infallible criteria for assessing and resolving problems of personal relationships, social change, cultural conflicts, etc.[1] Thus the evolutionary inquiry into human origins was ridiculed and condemned in the light of a "common theological teaching" concerning the condition of the unique couple, Adam and Eve, in their unique garden home, Paradise; the "war of our members," flesh pitted against spirit, found instant theological justification in a conception of "original sin" as an ontological defect in human nature transmitted entitatively, i.e., physically,[2] from parent to

Reprinted from *Is God Dead?*, *Concilium*, Volume 16, edited by Johannes Metz. Copyright ©
1966 by Paulist Fathers, Inc. and Stichting Concilium. By permission of Paulist Press, 304
West 58th St., New York, New York, 10019.
[1] To say that the great ideas of the Christian faith do not provide infallible criteria for the resolution of the issues which rise from history to confront man, however, is very far from an endorsement of the position of Sydney Hook, which is but the other side of the extreme we are here referring to: "*If* authentic Christianity comprises a set of general, eternal, and immutable truths beyond history, and time, they cannot serve as a guide to *specific* problems of history and time." (Sydney Hook, *Reason, Social Myths, and Democracy* [New York: Humanities, 1950], p. 87).
[2] Cf. John N. Deely, *The Tradition Via Heidegger* (The Hague: Martinus Nijhoff, 1971), Appendix I, "The Thought of Being and Theology," pp. 184–188, for a suggestion of the underlying philosophical issues.

progeny—yet not until the assumption that man *must* be at war with himself was challenged, rather than passively interpreted, was the door to modern conceptions of healing, especially of psychotherapy, opened. And so it has gone. The timeless, ultimate aspect of Revelation became often a blinder to the timely, immediate challenges of the human world.

In a pluralistic, relatively free-thinking social order, someone was bound to notice that man does not need God if His function is to sanction the status quo, or conjure away the very real possibilities and responsibilities of man for understanding and transcending many of the destructive conflicts within and oppressive outward obstacles to the unfolding of human potentialities. Behind all the equivocations and logical contradictions of the "Christian atheists" and "God is Dead Theologians" lies this very pressing message.

In fact, it has been in this dimension of thought for the human future that the impact of evolution on Christian thought has been most direct and in general salutary. It has forced us to realize as never before the irreducibility of God's ways to the ways of man.

Evolutionary Humanism and the Faith

THEORY OF EVOLUTION AND "EVOLUTIONISM"

Christian faith and theology have nothing to fear from biological and anthropological evolution. Scientific evolution is a highly documented, well-supported fact.[1] It is rather the ideological projection of some forms of *evolutionism*, in the name of science, which directly confronts the Christian faith. Because these ideologies are formulated and promoted, not by professional philosophers, but by novelists, dramatists and scientists, theologians are slow to take them seriously. One of the most completely developed forms of atheistic evolutionism today, promoted in the name of science and directly opposing the Christian revelation, is *evolutionary humanism*. The prophet of this new religion is Julian Huxley.

EVOLUTIONARY HUMANISM (JULIAN HUXLEY): A "NEW" RELIGION

Many scientists have espoused some form of evolutionary humanism.[2] Its influence is felt in almost every area of human endeavor where the scientific method has become a source of guidance to human destiny. But no one has done more than Julian Huxley to systematize and project evolutionary humanism into a cosmic view, an ideology, an ethic and a religion. More than Charles Darwin, he has been one of the chief contributors to, and formulators of, contemporary evolutionary theory.[3] More than Herbert Spencer, he has organized evolutionary humanism into a philosophy of life.[4] His grandfather, Thomas H. Huxley, created an ideology of agnosticism and became its protagonist; Julian Huxley has created a religion of atheism and is its champion in the world of science.[5] More than that, Huxley has spread his ideas politically and culturally through UNESCO in the United Nations, offering a solution to world peace which opposes Marxian dialectics on the one hand and every form of traditional religion on the other. This extreme doctrine of evolutionary humanism claims that scientific knowledge has rendered the Christian faith obsolete and valueless to man's future—a claim which will be examined in this article.

Evolutionary science has underscored the problem of the future of man, and serious conflicts throughout the world threaten the survival of *homo sapiens*. Man is a unique animal who creates his own destiny. If he does not discover his position and role in the universe and the correct means to fulfill that role, he will become extinct. Contrary to the materialism of the Western world and of Marxism, which would extricate religion from the hearts of men, Huxley sees religion as the necessary noetic organ of evolving man which serves the function of revealing man's role in the universe, his relations to the rest of the cosmos, and, in particular, his attitude toward the powers or forces operating in it.[6]

Prior to the 20th century, according to Huxley, man groped in darkness for an interpretation of his position in the universe. In his ignorance, he tried to explain in three ways the powers of destiny toward which he had a sacred awe: by the "magic hypothesis," by the "spirit hypothesis," and by the "God hypothesis." The Catholic Church, for example, uses all three hypotheses in her sacramental system, her veneration of the saints and her belief in a transcendent God and a supernatural revelation. Evolutionary science of the 20th century, Huxley continues, has demolished all three hypotheses and, therefore, all the religions of man constructed upon them, including Christianity.[7]

FOUNDATIONS OF EVOLUTIONARY HUMANISM

Evolutionary science has discovered a new cosmology, a new biology and a new anthropology in which the timely (space-time contingency) has replaced the timeless and the divine. Primitive and classical man sought meaning in the timeless, in the "myth of the eternal return,"[8] in the "other world" of transcendent being of spirits and gods, and in revelations with divine sanction. However, Huxley argues that contemporary evolutionary science has shown a material cosmos dynamically unfolding with self-sufficiency and natural order. Evolution of matter and of man, therefore, reveals their own immanent powers and meaning. Human destiny need not be sought elsewhere; indeed, there is no elsewhere.

For Huxley, the evolutionary science of Freud and Jung has dealt a death blow to the dualism of matter and spirit, the natural and the supernatural. Evolutionary progression and process unify the fragmentation of existence made by erroneous philosophies of dualism in the past. Huxley freely grants that the cosmos and man are still mysterious and that much about existence is unknown to modern science. However, he denies, in principle, anything basically occult, i.e., transcendent to scientific scrutiny. The meaning of man is to be found exclusively within the space-time unfolding of evolution. Religion is the necessary interpretive human organ of man's destiny guiding him toward future survival. Therefore, man must discover a new religion without revelation. That religion, proclaims Julian Huxley, is evolutionary humanism.

Huxley's humanism shares the man-centered orientation of every other form of humanism. What is distinctive is that allegedly it is entirely constructed upon the facts and inferences of evolutionary science. In broadest outlines, the following are the guidelines provided by science for the future destiny of man.[9] The universe consists of a single world-stuff of matter/energy in a single process of self-transformation and progressive self-creation of novelty. After two billion years of organic evolution, man has emerged as the only product still capable of advance. Man is evolution made conscious of itself, and it is through this new psychosocial process that man has the capacity to create his own future and, in large part, control the future evolution of the planet. Man's unique evolutionary endowment is a composite of freedom, creativity and personality.

From this capacity springs all spiritual value as well as the sacred duty to enhance to the fullest the inherent possibilities of freedom, creativity and personality. However, man can choose either to fulfill or to impede these possibilities, individually or socially. This is the essence of virtue or moral evil. Man's solidarity with nature gives rise to enjoyment and the duty to conserve and beautify his surroundings. Freedom and creativity demand greater awareness, extension and organization of knowledge by science and technology. Art is the personal exercise of human creativity and control. From these general guidelines of natural evolutionary orientation, Huxley and his fellow evolutionary humanists have constructed an entire ethics, aesthetics and religious program.[10] His system has the ring of novelty, for it is completely 20th-century humanism; it has the crisp ring of authenticity, for it is allegedly based upon the latest word in scientific biology, anthropology and psychology. Furthermore, it is totally opposed to the crude materialism of Marxism, yet openly atheistic and opposed to Christianity and every form of traditional religion.

DISCUSSION

There is much of value to be learned from this form of scientific atheism, because, as is the case with so much of contemporary atheism, it proceeds in part from a valid rejection of the weaknesses of traditional theism and theology.[11] Evolutionary humanism is concerned with the meaning of man in time and space, with his future survival. It is therefore preoccupied with the timely, not the timeless—with this world, not with eternity. Christian theology has largely neglected the timely, the historical space-time dimension of reality. The failure to present Christ's redemptive action to man "in this world" is a scandal which even Vatican Council II has attempted to rectify.[12]

Huxley's evolutionism also points out the folly of unenlightened opposition to scientific progress, an opposition displayed in the reluctance of many to accept the fact of evolution and its important bearing upon man's future. The argument for the existence of God from ignorance of natural processes is also exposed by Huxley. To assert that God is necessary in order to explain by miracle what may well have a natural, though yet unexplained, series of causes is always vulnerable to Huxley's valid criticism. For many, the synthesis of life in the laboratory is unthinkable because they erroneously believe the event possible only to the creative act of God. A close reading of Huxley's criticism of Christianity reveals that he is often reacting not to the authentic faith but to excesses consequent upon the cultural accretions which inevitably accompany Christian belief in every age. It must be admitted that a superstitious veneration of the saints, an exaggerated fear of the mysterious and a preoccupation with petitions to the Father without a fitting and courageous confidence in personal creativity and responsibility are caricatures of the true faith. Belief in the redemption of Christ is never a substitute for human ineptitude and cowardice.

Let us grant that Huxley's demand for a new religion stems in part from a reaction to the shortcomings of traditional religion. The Christian religion has to be purified constantly and placed fully into the developmental world of man if it is to be relevant to his future destiny on this planet. Nevertheless, Huxley's evolutionary humanism falls by the weight of its inherent illusions. As is the case with many forms of historicism and existentialism, preoccupation with the timely often obliterates the timeless and openness to God. Huxley misses the necessity for transcendent being, i.e., existence not fully contained by the materials of the universe, because he mistakes the apparent dynamic order of evolutionary unfolding for existential self-sufficiency. He is so enamored with the marvelous order of evolution in the universe—one which he has spent the greater part of his life describing and organizing into a coherent scientific theory—that he misses the absolute requirement of its *contingency*, its fundamental *creatureliness*. That requirement is God.

Huxley sees the universe as a self-contained, self-transforming, self-creative process, but what he fails to see is that the *existence* of the reality he describes is a communicated one. Consequently, the evolving universe, whatever be the dynamic order of its materials and agents, is basically a gift, a subject which lies completely outside the domain of science to analyze. That openness to a creator which the gift of existence commands is closed to Huxley. Hence, he is shut off from the possibility of true belief in the divine, and his own analysis of the nature of Christian belief in a revelation from God is entirely myopic and unreliable.[13]

However, the greatest illusion of evolutionary humanism is the expectation that the pattern of man's evolution will reveal man's destiny and provide a blueprint of survival in terms of that destiny. Man has emerged from some family of hominids and has advanced by evolutionary stages to the present *homo sapiens* by becoming free, creative and personal. How are those capacities to serve as a specific orientation for the future? All that can be projected from the evolutionary past is a preoccupation with ways by which man can become more free, more creative and more personal. But to what end? What is it that man should choose with greater freedom, creativity and personality, and in terms of what destiny? Evolutionary humanism is mute. To say that man should prepare himself to be ever more free, more creative and more personal without discovering a meaning and a destiny is like continuously whetting the appetite of a starving man and never providing him with food.

When the mystery of moral evil is faced squarely, the superficiality of evolutionary humanism is apparent not only to the Christian thinker but also to men like Jean-Paul Sartre and Albert Camus who hold other atheistic positions. For Huxley, evil is simply man's choice not to fulfill his maximum evolutionary capacity. Like the primitive "noble savage", the Greek "virtuous man" and the mid-Victorian "gentleman", Huxley's evolutionary man has no meaning beyond aspiring to an arbitrary, sophisticated, aristocratic ideal, an ideal which the deep absurdities of existence reveal to be illusory. From within his misery, frustration, failure and degradation, how can man know of the spiritual solidarity between man and God which has been irremediably fractured by personal and corporate choice of egoism? How can he know of the consequent reverberations of that alienation from God in the personal, creative and free communication among men in their struggle for happiness?

Without revelation from God, man cannot discover the moral and religious meaning of his destiny or raise a finger to extricate himself from the moral pathology into which he has fallen. Belief in the revelation of God and the redemption of Christ, accessible to the man who is open to the timeless and the divine, is the only release available to the despair of evolutionary humanism. No plan for the space-time future existence of man is realistic until it solves this fundamental moral and religious issue of absurdity. Huxley's instinct that the future of man must be thought out and *acted out in the timely* is an authentic one, but the *meaning* of that future can only be found in the timeless, in the gift of the divine.

SALVATION HISTORY

The challenge of evolutionary humanism to Christianity raises this important and difficult problem: How can the search for the meaning of man's destiny avoid the illusion of regarding man exclusively in the light of history and the unrealism inherent in contemplating man solely in the light of eternity? The presentation of the Christian revelation through the traditional forms of theology, with their reliance upon Greek specula-tion on the timeless and the divine, is seriously suspect today. These theological forms, so useful in harmonizing the revelation with accepted strains of classical wisdom, are quite inadequate to interpret 20th-century man's preoccupation with his history, his space-time existence. A vehicle is needed which places eternal value and meaning squarely in the historical unfolding of man's future. Perhaps the form which needs exploring today, in the light of the evolutionary problem, like those of existentialism and historicism, is *salvation history*.[14]

It is man's time and space which has been redeemed by Christ, the first-born of all creation, through whom, for whom and by whom all things hold together. Because the Son of God has entered man's time and space and has chosen to act out the drama of his incarnation-redemption-resurrection in that time and space, human history has been endowed with intrinsic value. It is true that the cosmic future is not identical with Christian eschatology—just as our planet is not the cosmic but the theo-logical center of existence, made so by the incarnation of Christ. However, man's unfolding from day to day, from moment to moment, can no longer be viewed as incidental or irrelevant. It is his free, creative and personal choice to enter into the drama of Christ's life as it unfolds in man's own day that places God in the timely and brings timeless meaning to the future of man. Thus, both the excesses of the "myth of the eternal return" in which time means nothing and the excesses of the "myth of evolutionism" in which time means everything are avoided. Through salvation history, time and the timeless are harmonized in the divine.

NOTES

[1] R. Nogar, *The Wisdom of Evolution* (New York: Doubleday, 1963).

[2] C. Judson Herrick, *The Evolution of Human Nature* (Austin: University of Texas Press, 1956); G. Simpson, *This View of Life* (New York: Harcourt, Brace and World, 1964); C. Waddington, *The Ethical Animal* (New York: Atheneum, 1961).

[3] His *Evolution: The Modern Synthesis* (New York: Harper and Brothers, 1942) is a classic of modern evolutionary theory.

[4] See the application of evolutionary humanism to every major department of human endeavor in *The Humanist Frame* (New York: Harper and Brothers, 1961), edited by J. Huxley.

[5] On Thanksgiving Day, November 26, 1959, Huxley delivered the Darwin Centennial Convocation Address in the Rockefeller Memorial Chapel to the international assembly of scientists, in which he proclaimed, in the name of scientific evolution, the death of old religions (and belief in the supernatural) as well as the birth of the new religion—evolutionary humanism: "The Evolutionary Vision," in *Evolution after Darwin*, III (Chicago: University of Chicago Press, 1960), pp. 249–261.

[6] J. Huxley, *Religion without Revelation* (New York: Mentor Books, 1957), p. 182.

[7] *Ibid.*, pp. 183–184.

[8] From the excellent book by M. Eliade, *The Myth of the Eternal Return* (New York: Pantheon Books, 1954).

[9] The following summary of evolutionary humanism is taken from J. Huxley, *Religion without Revelation*, pp. 190–212, but similar development of his view can be found in his *Essays of a Humanist* (New York: Harper and Row, 1964), *Man Stands Alone* (Harper and Brothers, 1941), and *Man in the Modern World* (New York: Mentor Books, 1962).

[10] Cf. *supra*, footnote 4.

[11] Cf. J. LaCroix, "The Meaning and Value of Atheism Today," in *Cross Currents*, V (Summer, 1955), pp. 203–219; also cf. his *The Meaning of Modern Atheism* (Dublin: Gill and Son, 1965); J. Murray, *The Problem of God* (New Haven: Yale University Press, 1964); M. Buber, *Eclipse of God* (New York: Harper and Row, 1952).

[12] To this end Teilhard de Chardin bent his apostolic energies, especially in his *The Phenomenon of Man* (New York: Harper and Brothers, 1959) and *The Divine Milieu* (New York: Harper and Brothers, 1960).

[13] The necessity of man's realization of his basic creatureliness for the openness of spirit required for the gift of divine belief is lucidly treated by J. Pieper in *Belief and Faith* (New York: Pantheon Books, 1963).

[14] The application of the salvation-history theme to the evolutionism, historicism and existentialism of contemporary thought is made possible by the growing literature on the theology of history. Cf. especially H. Urs von Balthasar, *A Theology of History* (New York: Sheed and Ward, 1963); J. Mouroux, *The Mystery of Time* (New York: Desclee Co., 1964); J. Daniélou, *Christ and Us* (New York: Sheed and Ward, 1961); *idem*, *The Lord of History* (Chicago: Henry Regnery, 1964); J. Connolly, *Human History and the Word of God* (New York: Macmillan, 1965).

6

Toward an evolutionary world view

Rationale of this section

There are few serious students of mankind evolving who would not endorse the late Pope John XXIII's contention that we are entering upon a new order of human relationships.[1] We have reached a stage in the life of nations where it becomes clear that only through the establishment of a world community justly ordered under law, with no super-wealthy individuals and no penuriously poor ones, can we hope to channel the energies of man's spirit toward a continued unfolding of human potentialities. In this sense, "the bloody wars that have followed one on the other in our times, the spiritual ruins caused by many ideologies, and the fruits of so many bitter experiences have not been without useful teachings."[2] They have at least made increasingly unavoidable the acknowledgement—reluctant and grudging as yet in many quarters—that "the destiny of the human community has become all of a piece, where once the various groups of men had a kind of private history of their own."[3] Yet how is a world-community to be realized?

[1] "In the present order of things, Divine Providence is leading us to a new order of human relations which, by men's own efforts and even beyond their very expectations, are directed toward the fulfillment of God's superior and inscrutable designs."—From "Pope John's Opening Speech to the Council," 11 October 1962, as published in *The Documents of Vatican II*, Walter M. Abbott, General Editor (New York: The America Press, 1966), pp. 712–713.
[2] From Pope John's address, "Humanae Salutis," dated 25 December 1961, convoking the Second Vatican Council for sometime in 1962, as published in *The Documents of Vatican II*, p. 704.
[3] From the "Pastoral Constitution on the Church in the Modern World," in *The Documents of Vatican II*, pp. 203–204.

Probably the single most important clue has come from the tragic spectacle of *the spiritual ruins caused by many ideologies.* This spectacle, terrifying in its extent, has at least made it possible for us to begin to realize that the organization and interrelations of the human group as such are always defined and dominated by the culture of the group. "The fact is that culture has been developing as an unconscious, blind, bloody, brutal, tropismatic process so far," observes Leslie White,[4] and no one can gainsay his observation. Evolution in its properly human dimension has not yet reached the point where intelligence, self-consciousness, and just dealing are very conspicuous. Men on the whole are still dominated by partial ideologies refractory to the exigencies of the ongoing emergence of the humanness of man. Yet the successful realization of an order of human relations securing the welfare of each at the expense of none pre-supposes an acknowledgement of the developmental openness of our humanity based upon our specific biological unity.[5] Can such a realization of what it means to be human be secured by anything less than an integration of the fragmented cultural perspectives of national life into a truly supra-national world-view? A world-society requires a specifically, not parochially, human frame of reference.

There is every reason to doubt the human adequacy of any cultural perspective achieved so far, and even more reason to doubt the adequacy of any world-view the root metaphor of which is based on a fixed delineation of "class struggles" or of social roles and norms of personality development. "Thus, the human race has passed from a rather static concept of reality to a more dynamic, evolutionary one. In consequence, there has arisen a new series of problems, a series as important as can be, calling for new efforts of analysis and synthesis."[6]

---

[4] Leslie White, *The Science of Culture* (New York: Grove Press, 1949), pp. 356–357.
[5] See the reading selection on "The Consequences for Action" by M. J. Adler, esp. sec. (2) and fn. 8, in Section III of this volume, The Moral Issues.
[6] "Pastoral Constitution on the Church in the Modern World," in *The Documents of Vatican II,* p. 204.

The evolutionary vision

Julian S. Huxley

CONTEXTUALIZING COMMENTS

Jung in particular has pointed out that society consists of many historical layers. There are men today whose basic psychological life belongs to the nineteenth, eighteenth, and seventeenth centuries; and in fact only a minority who live rooted in the cultural perspectives of present developments. One of those few is Julian Huxley. Sir Julian Huxley is one of the remarkable men and true citizens of the twentieth century. In many ways he is unsurpassed as a biological scientist and an author of scope, insight, and responsibility. His deep concern over human suffering and responsibility is reflected in his writings and public activities—for example, he was the first Director-General of UNESCO. He has attempted as few men have to face squarely the philosophical and theological implications of scientific evolution. Though one may often wonder at the extent to which basic assumptions are left unexamined in his writings, one must respect and take in all seriousness the tireless efforts he has made to fashion the elements of evolutionary science into an instrument of conscious purpose, into a world-view accepting rather than turning away from the full weight of man's responsibility for man.

As the leading articulator of a thorough-going evolutionary naturalism, Huxley speaks for many of the keenest students of human evolution

Reprinted from *Issues in Evolution*, Volume III of *Evolution After Darwin*, the University of Chicago Centennial, edited by Sol Tax and Charles Callendar, by permission of the University of Chicago Press. Copyright © 1960 by the University of Chicago.

as he raises the question whether the mind can be forced with logical necessity to recognize the existence of a spiritual order of being, transcendent to but ontologically involved in the existence of the universe, seeing that, within the evolutionary scheme, the survival of man depends upon courageous, free, personal, creative effort, and that upon this remarkably simple interpretation of human existence an entire morality, a complete "spirituality" (if the term may be so used) can be constructed.

The evolutionary vision

Future historians will perhaps take this Centennial week as epitomizing an important critical period in the history of this earth of ours—the period when the process of evolution, in the person of inquiring man, began to be truly conscious of itself. This is, so far as I am aware, the first time that authorities on the evolutionary aspects of the three great branches of scientific study—the inorganic sciences, the life-sciences, and the human sciences—have been brought together for mutual criticism and joint discussion. We participants who are assembled here, some of us from the remotest parts of the globe, by the magnificently intelligent enterprise of the University of Chicago, include representatives of astronomy, physics, and chemistry; of zoology, botany, and paleontology; of physiology, ecology, and ethology; of psychology, anthropology, and sociology. We have all been asked to contribute an account of our knowledge and understanding of evolution in our special fields to the Centennial's common pool, to submit our contributions to the criticism and comments of our fellow participants in quite other fields, to engage in public discussion of key points in evolutionary theory, and to have our contributions and discussions published to the world at large.

This is one of the first public occasions on which it has been frankly faced that all aspects of reality are subject to evolution, from atoms and stars to fish and flowers, from fish and flowers to human societies and values—indeed, that all reality is a single process of evolution. And ours is the first period in which we have acquired sufficient knowledge to begin to see the outline of this vast process as a whole.

Our evolutionary vision now includes the discovery that biological advance exists, and that it takes place in a series of steps or grades, each grade occupied by a successful group of animals or plants, each group sprung from a pre-existing one and characterized by a new and improved pattern of organization.

Improved organization gives biological advantage. Accordingly, the new type becomes a successful or dominant group. It spreads and multiplies and differentiates into a multiplicity of branches. This new biological success is usually achieved at the biological expense of the older dominant group from which it sprang or whose place it has usurped. Thus the rise of the placental mammals was correlated with the decline of the terrestrial reptiles, and the birds replaced the pterosaurs as dominant in the air.

Occasionally, however, when the breakthrough to a new type of organization is also a breakthrough into a wholly new environment, the new type may not come into competition with the old, and both may continue to coexist in full flourishment. Thus the evolution of land vertebrates in no way interfered with the continued success of the teleost bony fish.

The successive patterns of successful organization are stable patterns: they exemplify continuity and tend to persist over long periods. Reptiles have remained reptiles for three hundred million years: tortoises, snakes, lizards, and crocodiles are all still recognizably reptilian, all variations on one organizational theme.

It is difficult for life to transcend this stability and achieve a new successful organization. That is why breakthroughs to new dominant types are so rare—and also so important. The reptilian type radiated out into well over a dozen important groups or orders; but all of them remained within the reptilian framework except two, which broke through to the new and wonderfully successful patterns of bird and mammal.

In the early stages, a new group, however successful it will eventually become, is few and feeble and shows no signs of the success that it may eventually achieve. Its breakthrough is not an instantaneous matter but has to be implemented by a series of improvements which eventually become welded into the new stabilized organization.

With mammals, there was first hair, then milk, then partial and later on full-temperature regulation, then brief and finally prolonged internal development, with evolution of a placenta. Mammals of a small and insignificant sort had existed and evolved for over a hundred million years before they achieved a full breakthrough to their explosive dominance in the Cenozoic.

Something very similar occurred during our own breakthrough from mammalian to psychosocial organization. Our prehuman ape ancestors were never particularly successful or abundant. There was not just one "missing link" between them and us. For their transformation into man a series of steps was needed. Descent from the trees; erect posture; some enlargement of brain; more carnivorous habits; the use and then the making of tools; further enlargement of brain; the discovery of fire; true speech and language; elaboration of tools and rituals. These steps took the better part of half a million years: it was not until less than a hundred thousand years ago that man could begin to deserve the title of dominant type and not until less than ten thousand years ago that he became fully dominant.

After man's emergence as truly man, the same sort of thing continued to happen, but with an important difference. Man's evolution is not biological but psychosocial; it operates by the mechanism of cultural tradition, which involves the cumulative self-reproduction and self-variation of mental activities and their products. Accordingly, major steps in the human phase of evolution are achieved by breakthroughs to new dominant patterns of mental organization, of knowledge, ideas, and beliefs—ideological instead of physiological or biological organization.

There is a succession of successful idea-systems instead of a succession of successful bodily organizations. Each new, successful idea-system spreads and dominates some important sector of the world, until it is superseded by a rival system or itself gives birth to its successor by a breakthrough to a new organization-system of thought and belief. We need only think of the magic pattern of tribal thought; the god-centered medieval pattern, organized round the concept of divine authority and revelation; and the rise in the last three centuries of the science-centered pattern, organized round the concept of human progress, but progress somehow under the control of supernatural Authority. In 1859, Darwin opened the passage leading to a new psychosocial level, with a new pattern of ideological organization—an evolution-centered organization of thought and belief.

Through the telescope of our scientific imagination, we can discern the existence of this new and improved ideological organization; but its details are not clear, and we can also see that the necessary steps upward toward it are many and hard to take.

Let me change the metaphor. To those who did not deliberately shut their eyes or who were not allowed to look, it was at once clear that the fact and concept of evolution was bound to act as the central germ or living template of a new dominant thought organization. And in the century since the *Origin of Species*, there have been many attempts to understand the implications of evolution in many fields, from the affairs of the stellar universe to the affairs of men, and to integrate the facts of evolution and our knowledge of its processes into the over-all organization of our general thought.

All dominant thought organizations are concerned with the ultimate, as well as with the immediate, problems of existence or, I should rather say, with the most ultimate problems that the thought of the time is capable of formulating or even envisaging. They are all concerned with giving some interpretation of man, of the world which he is to live in, and of his place and role in that world—in other words, some comprehensible picture of human destiny and significance.

The broad outlines of the new evolutionary picture of ultimates are beginning to be visible. Man's destiny is to be the sole agent for the future evolution of this planet. He is the highest dominant type to be produced by over two and a half billion years of the slow biological improvement effected by the blind opportunistic workings of natural selection; if he does not destroy himself, he has at least an equal stretch of evolutionary time before him to exercise his agency.

During the later part of biological evolution, mind—our word for the mental activities and properties of organisms—emerged with greater clarity and intensity and came to play a more important role in the individual lives of animals. Eventually it broke through, to become the basis for further evolution, though the character of evolution now became cultural instead of genetic or biological. It was to this breakthrough, brought about by the automatic mechanism of natural selection and not by any conscious effort on his own part, that man owed his dominant evolutionary position.

Man is therefore of immense significance. He has been ousted from his self-imagined centrality in the universe to an infinitesimal location in a peripheral position in one of a million of galaxies. Nor, it would appear, is he likely to be unique as a sentient being. On the other hand, the evolution of mind or sentiency is an extremely rare event in the vast meaninglessness of the insentient universe, and man's particular brand of sentiency may well be unique. But in any case he is highly significant. He is a reminder of the existence, here and there, in the quantitative vastness of cosmic matter and its energy equivalents, of a trend toward mind, with its accompaniment of quality and richness of existence—and, what is more, a proof of the importance of mind and quality in the all-embracing evolutionary process.

It is only through possessing a mind that he has become the dominant portion of this planet and the agent responsible for its future evolution; and it will be only by the right use of that mind that he will be able to exercise that responsibility rightly. He could all too readily be a failure in the job; he will succeed only if he faces it consciously and if he uses all his mental resources—of knowledge and reason, of imagination, sensitivity, and moral effort.

And he must face it unaided by outside help. In the evolutionary pattern of thought there is no longer either need or room for the supernatural. The earth was not created; it evolved. So did all the animals and plants that inhabit it, including our human selves, mind and soul as well as brain and body. So did religion. Religions are organs of psychosocial man concerned with human destiny and with experiences of sacredness and transcendence. In their evolution, some (but by no means all) have given birth to the concept of gods as supernatural beings endowed with mental and spiritual properties and capable of intervening in the affairs of nature, including man. Such supernaturally centered religions are early organizations of human thought in its interaction with the puzzling, complex world with which it has to contend—the outer world of nature and the inner world of man's own nature. In this, they resemble other early organizations of human thought confronted with nature, like the doctrine of the Four Elements, Earth, Air, Fire and Water, or the Eastern concept of rebirth and reincarnation. Like these, they are destined to disappear in competition with other, truer, and more embracing thought organizations which are handling the same range of raw or processed experience—in this case, with the new religions which are surely destined to emerge on this world's scene.

Evolutionary man can no longer take refuge from his loneliness in the arms of a divinized father-figure whom he has himself created, nor escape from the responsibility of making decisions by sheltering under the umbrella of Divine Authority, nor absolve himself from the hard task of meeting his present problems and planning his future by relying on the will of an omniscient, but unfortunately inscrutable, Providence.

On the other hand, his loneliness is only apparent. He is not alone as a type. Thanks to the astronomers, he now knows that he is one among the many organisms that bear witness to the trend toward sentience, mind, and richness of being, operating so widely but so sparsely in the cosmos. More important, thanks to Darwin, he now knows that he is not an isolated phenomenon, cut off from the rest of nature by his uniqueness. Not only is he made of the same matter and operated by the same energy as all the rest of the cosmos, but, for all his distinctiveness, he is linked by genetic continuity with all the other living inhabitants of his planet. Animals, plants, and micro-organisms, they are all his cousins or remoter kin, all parts of one single evolving flow of metabolizing protoplasm.

Nor is he individually alone in his thinking. He exists and has his being in the intangible sea of thought which Teilhard de Chardin has christened the "noösphere," in the same sort of way that fish exist and have their being in the material sea of water which the geographers include in the term "hydrosphere." Floating in the noösphere there are, for his taking, the daring speculations and aspiring ideals of man long dead, the organized knowledge of science, the hoary wisdom of the ancients, the creative imaginings of all the world's poets and artists. And in his own nature there is, waiting to be called upon, an array of potential helpers—all the possibilities of wonder and knowledge, of delight and reverence, of creative belief and moral purpose, of passionate effort and embracing love.

Turning the eye of an evolutionary biologist upon this situation, I would compare the present stage of evolving man to the geological moment, some three hundred million years ago, when our amphibian ancestors were just establishing themselves out of the world's water. They had created a bridgehead into a wholly new environment—no longer buoyed up by water, they had to learn how to support their own weight; debarred from swimming with their muscular tail, they had to learn to crawl with clumsy limbs. The newly discovered realm of air gave them direct access to the oxygen they needed to breathe, but it also threatened their moist bodies with desiccation. And though they managed to make do on land during their adult lives, they found themselves still compulsorily fishy during the early part of their lives.

On the other hand, they had emerged into completely new freedoms. As fish, they had been confined below a bounding surface; now the air above them expanded out into the infinity of space. Now they were free of the banquet of small creatures prepared by the previous hundred million years of life's terrestrial evolution. The earth's land surface provided a greater variety of opportunity than did its waters and, above all, a much greater range of challenge to evolving life. Could the early Stegocephalians have been gifted with imagination, they might have seen before them the possibility of walking, running, perhaps even flying over the earth; the probability of their descendants escaping from bondage to winter cold by regulating their temperature and escaping from bondage to the waters by constructing private ponds for their early development; the inevitability of an upsurge of their dim minds to new levels of clarity and performance. But meanwhile they would see themselves tied to an ambiguous existence, neither one thing nor the other, on the narrow moist margin between water and air. They could have seen the promised land afar off, though but dimly through their bleary, newtish eyes. But they would also have seen that, to reach it, they would have to achieve many difficult and arduous transformations of their being and way of life.

So with ourselves. We have only recently emerged from the biological to the psychosocial area of evolution, from the earthy biosphere into the freedom of the noösphere. Do not let us forget how recently: we have been truly men for perhaps a tenth of a million years—one tick of evolution's clock; even as protomen, we have existed for under one million years— less than a two-thousandth fraction of evolutionary time. No longer supported and steered by a framework of instincts, we try to use our conscious thought and purposes as organs of psychosocial locomotion and direction through the tangles of our existence—but so far with only moderate success and with the production of much evil and horror, as well as of some beauty and glory of achievement. We too have colonized only an ambiguous margin between an old bounded environment and the new territories of freedom. Our feet still drag in the biological mud, even when we lift our heads into the conscious air. But, unlike those remote ancestors of ours, we can truly see something of the promised land beyond. We can do so with the aid of our new instrument of vision—our rational, knowledge-based imagination. Like the earliest pre-Galilean telescopes, it is still a very primitive instrument and gives a feeble and often distorted view. But, like the early telescopes, it is capable of immense improvement and could reveal many secrets of our noöspheric home and destiny.

Meanwhile, no mental telescope is required to see the immediate evolutionary landscape and the frightening problems which inhabit it. All that is needed—but that is plenty!—is for us to cease being intellectual and moral ostriches and take our heads out of the sand of wilful blindness. If we do so, we shall soon see that the alarming problems are two-faced and are also stimulating challenges.

What are those alarming monsters in our evolutionary path? I would list them as follows. The threat of superscientific war, nuclear, chemical, and biological; the threat of overpopulation; the rise and appeal of Communist ideology, especially in the underprivileged sectors of the world's people; the failure to bring China, with nearly a quarter of the world's population, into the world organization of the United Nations; the erosion of the world's cultural variety; our general preoccupation with means rather than ends, with technology and quantity rather than creativity and quality; and the revolution of expectation caused by the widening gap between the haves and the have-nots, between the rich and the poor nations.

Today is Thanksgiving Day. But millions of people now living have little cause to give thanks for anything. When I was in India this spring, a Hindu man was arrested for the murder of his small son. He explained that his life was so miserable that he had killed the boy as a sacrifice to the goddess Kali, in the hope that she would help him in return. That is an extreme case. But let us remember that two-thirds of the world's people are underprivileged—underfed, underhealthy, undereducated—and that millions of them live in squalor and suffering. They have little to be thankful for, save hope that they will be helped to escape from this misery. If we in the West do not give them aid, they will look to other systems for help—or even turn from hope to destructive despair.

We attempt to deal with these problems piecemeal, often half-heartedly; sometimes, as with population, we refuse to recognize it officially as a world problem (just as we refuse to recognize Communist China as a world power). In reality, they are not separate monsters, to be dealt with by a series of separate ventures, however heroic or saintly. They are all symptoms of a new evolutionary situation; and this can be successfully met only in the light and with the aid of a new organization of thought and belief, a new dominant pattern of ideas.

It is hard to break through the firm framework of an accepted belief system and build a new acceptable successor, but it is necessary. It is necessary to organize our *ad hoc* ideas and scattered values into a unitive pattern, transcending conflicts and divisions in its unitary web. Only by such a reconciliation of opposites and disparates can our belief-system release us from inner conflicts; only so can we gain that peaceful assurance that will help unlock our energies for development in strenuous practical action.

Somehow or other, we must make our new pattern of thinking evolution-centered. It can give us assurance by reminding us of our long evolutionary rise; how this was also, strangely and wonderfully, the rise of the mind; and how that rise culminated in the eruption of mind as the dominant factor in evolution and led to our own spectacular, but precarious, evolutionary success. It can give us hope by pointing to the eons of evolutionary time that lie ahead of our species if it does not destroy itself or damage its own chances; by recalling how the increase in man's understanding and the improved organization of his knowledge have in fact enabled him to make a whole series of advances, such as control of infectious disease or efficiency of telecommunication, and to transcend a whole set of apparently unbridgeable oppositions, like the conflict between Islam and Christendom or that between the seven Kingdoms of the Heptarchy; and by reminding us of the vast stores of human possibility —of intelligence, imagination, co-operative good will—which still remain untapped.

Our new organization of thought—belief-system, framework of values, ideology, call it what you will—must grow and be developed in the light of our new evolutionary vision. So, in the first place, it must, of course, itself be evolutionary. That is to say, it must help us to think in terms of an overriding process of change, development, and possible improvement; to have our eyes on the future rather than on the past; to find support in the growing body of our knowledge, not in fixed dogma or ancient authority.

Equally, of course, the evolutionary outlook must be scientific, not in the sense that it rejects or neglects other human activities, but in believing in the value of the scientific method for eliciting knowledge from ignorance and truth from error and in basing itself on the firm ground of scientifically established knowledge. Unlike most theologies, it accepts the inevitability and, indeed, the desirability of change, and it advances by welcoming new discovery even when this conflicts with old ways of thinking.

The only way in which the present split between religion and science could be mended would be through the acceptance by science of the fact and value of religion as an organ of evolving man and the acceptance by religion that religions do and must evolve.

Next, the evolutionary outlook must be global. Man is strong and successful insofar as he operates in interthinking groups, which are able to pool their knowledge and beliefs. To have any success in fulfilling his destiny as the controller or agent of future evolution on earth, he must become one single interthinking group, with one general framework of ideas; otherwise his mental energies will be dissipated in ideological conflict. Science gives us a foretaste of what could be. It is already global, with scientists of every nation contributing to its advance; and, because it is global, it is advancing fast. In every field we must aim to transcend nationalism, and the first step toward this is to think globally—how could this or that task be achieved by international co-operation rather than by separate action?

But our thinking must also be concerned with the individual. The well-developed, well-patterned individual human being is, in a strictly scientific sense, the highest phenomenon of which we have any knowledge, and the variety of individual personalities is the world's highest richness.

The individual need not feel just a meaningless cog in the social machine or merely the helpless prey and sport of vast impersonal forces. He can do something to develop his own personality, to discover his own talents and possibilities, to interact personally and fruitfully with other individuals. If so, in his own person, he is effecting an important realization of evolutionary possibility: he is contributing his own personal quality to the fulfilment of human destiny. He has assurance of his own significance in the greater and more enduring whole of which he is part.

I spoke of quality. This must be the dominant concept of our new belief-system—quality and richness as against quantity and uniformity.

Though our new idea-pattern must be unitary, it need not and should not impose a drab or boring cultural uniformity. A well-organized system, whether of thought, expression, social life, or anything else, has both unity and richness. Cultural variety, both in the world as a whole and within its separate countries, is the spice of life; yet it is being threatened and indeed eroded away by mass production, mass communications, mass conformity, and all the other forces making for uniformization—an ugly word for an ugly thing! We have to work hard to preserve and foster it.

One sphere where individual variety could and should be encouraged is education. In many school systems, under the pretext of so-called democratic equality, variety of gifts and capacity is now actually being discouraged. The duller children become frustrated by being rushed too fast, the brighter become frustrated by being held back and bored.

Our new idea-system must jettison the democratic myth of equality. Human beings are *not* born equal in gifts or potentialities, and human progress stems largely from the very fact of their inequality. "Free but unequal" should be our motto, and diversity of excellence, not conforming normalcy or mere adjustment, should be the aim of education.

Population is people in the mass; and it is in regard to population that the most drastic reversal or reorientation of our thinking has become necessary. The unprecedented population explosion of the last half-century has strikingly exemplified the Marxist principle of the passage of quantity into quality. Mere increase in quantity of people is increasingly affecting the quality of their lives, and affecting it almost wholly for the worse.

Population increase is already destroying or eroding many of the world's resources, both those for material subsistence and those—equally essential but often neglected—for human enjoyment and fulfillment. Early in man's history the injunction to increase and multiply was right. Today it is wrong, and to obey it will be disastrous. The Western world, the United States in particular, has to achieve the difficult task of reversing the direction of its thought about population. It has to begin thinking that we should aim—not at increase but at decrease—certainly and quickly a decrease in the *rate* of population growth and, in the long run equally certainly, a decrease in the absolute number of people in the world, including our own countries.

The spectacle of explosive population increase is prompting us to ask the simple but basic question *What are people for?* And we see that the answer has something to do with their quality as human beings and the quality of their lives and their achievements.

We must make the same reversal of ideas about our economic system. At the moment (and again I take the United States as most representative) our Western economic system (which is steadily invading new regions) is based on expanding production for profit, and production for profit is based on expanding consumption. As one writer has put it, the American economy depends on persuading more people to believe they want to consume more products.

But, like population explosion, this consumption explosion cannot continue much longer; it is an inherently self-defeating process. Sooner, rather than later, we must get away from a system based on artificially increasing the number of human wants and set about constructing one aimed at the qualitative satisfaction of real human needs, spiritual and mental as well as material and physiological. This means abandoning the pernicious habit of evaluating every human project solely in terms of its utility—by which the evaluators mean solely its material utility and especially its utility in making a profit for somebody.

Once we truly believe (and true belief, however necessary, is rarely easy)—once we truly believe that man's destiny is to make possible greater fulfilment for more human beings and fuller achievement by human societies, utility in the customary sense becomes subordinate. Quantity of material production is, of course, necessary as the basis for the satisfaction of elementary human needs—but only up to a certain degree. More than a certain number of calories or cocktails or TV sets or washing machines per person is not merely unnecessary but bad. Quantity of material production is a means to a further end, not an end in itself.

The important ends of man's life include the creation and enjoyment of beauty, both natural and man-made; increased comprehension and a more assured sense of significance; the preservation of all sources of pure wonder and delight, like fine scenery, wild animals in freedom, or unspoiled nature; the attainment of inner peace and harmony; the feeling of active participation in embracing and enduring projects, including the cosmic project of evolution. It is through such things that individuals attain greater fulfilment.

As for nations and societies, they are remembered not for their wealth or comforts or technologies but for their great buildings and works of art, their achievements in science or law or political philosophy, their success in liberating human thought from the shackles of fear and ignorance.

Although it is to his mind that man owes both his present dominant position in evolution and any advances he may have made during his tenure of that position, he is still strangely ignorant and even superstitious about it. The exploration of the mind has barely begun. It must be one of the main tasks of the coming era, just as was the exploration of the world's surface a few centuries ago. Psychological exploration will doubtless reveal as many surprises as did geographical exploration and will make available to our descendants all kinds of new possibilities of fuller and richer living.

Finally, the evolutionary vision is enabling us to discern, however incompletely, the lineaments of the new religion that we can be sure will arise to serve the needs of the coming era. Just as stomachs are bodily organs concerned with digestion and involving the biochemical activity of special juices, so are religions psychosocial organs of man concerned with the problems of destiny and involving the emotion of sacredness and the sense of right and wrong.

Religion of some sort is certainly a normal function of psychosocial existence. It seems to be necessary to man. But it is not necessarily a good thing. It was not a good thing when the Hindu I read about this spring killed his son as a religious sacrifice. It is not a good thing that religious pressure has made it illegal to teach evolution in Tennessee because it conflicts with fundamentalist beliefs. It is not a good thing that in Connecticut and Massachusetts women should be subject to grievous suffering because Roman Catholic pressure refuses to allow even doctors to give information on birth control even to non-Catholics. It was not a good thing for Christians to persecute and even burn heretics; it is not a good thing when communism, in its dogmatic-religious aspect, persecutes and even executes deviationists.

The emergent religion of the near future could be a good thing. It will believe in knowledge. It should be able to take advantage of the vast amount of new knowledge produced by the knowledge explosion of the last few centuries to construct what we may call its "theology"—the framework of facts and ideas which provide it with intellectual support; it should be able, with our increased knowledge of mind, to define our sense of right and wrong more clearly so as to provide a better moral support; it should be able to focus the feeling of sacredness onto fitter objects, instead of worshipping supernatural rulers, so as to provide truer spiritual support, to sanctify the higher manifestations of human nature in art and love, in intellectual comprehension and aspiring adoration, and to emphasize the fuller realization of life's possibilities as a sacred trust.

Thus the evolutionary vision, first opened up to us by Charles Darwin a century back, illuminates our existence in a simple, but almost overwhelming, way. It exemplifies the truth that truth is great and will prevail, and the greater truth that truth will set us free. Evolutionary truth frees us from subservient fear of the unknown and supernatural and exhorts us to face this new freedom with courage tempered with wisdom and hope tempered with knowledge. It shows us our destiny and our duty. It shows us mind enthroned above matter, quantity subordinate to quality. It gives our anxious minds support by revealing the incredible possibilities that have already been realized in evolution's past and, by pointing to the hidden treasure of fresh possibilities that could be realized in its long future, it gives man a potent incentive for fulfilling his evolutionary role in the universe.

Necessity and freedom

Theodosius Dobzhansky

CONTEXTUALIZING COMMENTS

Like Huxley, Dobzhansky is a leading student of the life sciences who has made fundamental contributions within his scientific specialty while remaining a concerned citizen of the human city. Thus Dobzhansky was among those researchers who finally established and made public the conclusion—debatable until very recent years—that there is absolutely no ground biologically speaking for singling out on the basis of ethnic origin any group of human beings as inherently superior. On the contrary, mankind was and is a single mendelian population, i.e., a single specific life-form. But, again like Huxley, Dobzhansky has not contributed solely to the empirical or "material" basis for an evolutionary humanism, but has focused his attention as well on the larger philosophical issues involved in the establishment of an evolutionary world-view. But whereas Dobzhansky no less than Huxley has entered into the world-view based on a time-centered, space-centered root-metaphor, Dobzhansky cannot admit that preclusive openness to the timely is any more adequate in the end than was the tendency of a pre-evolutionary world-view which encouraged exclusive preoccupation with the timeless questions of the meaning of it all. In this way, Dobzhansky has exercised a salutary constraint on certain enthusiastic moderns whose reaction to the scientistic culture has been no less tropismatic than the response of children to parental attitudes.

Necessity and freedom

EVOLUTIONARY HUMANISM

It is unavoidable that the foregoing discussion will be regarded by some people as failing to come to grips with the really fundamental problem of man's uniqueness. We have occupied ourselves with matters as base and arid as genes and genotypes. But, do genes have any bearing at all on the higher values which make man's estate truly unique? Is there any biological basis, explanation, or justification for the ideas of right and wrong which only man possesses? No doubt the question is legitimate, and perhaps the greatest challenge which a biologist encounters has been stated as follows by Julian Huxley: "Medieval theology urged man to think of human life in the light of eternity—*sub specie aeternitatis*. I am attempting to re-think it *sub specie evolutionis*." To which the present writer would like to add that evolution, too, will have to be thought about in the light of eternity, eternity in the light of evolution, and human life in the lights of both.

Some attempts in this direction have been made repeatedly since the time of Herbert Spencer. Darwin's theory came as a crowning achievement to the spectacular development of natural science in the nineteenth century. Following on the heels of the great discoveries in physics and chemistry, Darwin appeared, to some of his contemporaries, to have successfully done away with the mystery of life. (Darwin himself was under no such illusion.) After that no problem seemed too difficult for mechanistic science. Spencer's *Principles of Ethics* stands as a monument to this surfeit of optimism. The decades that followed ushered in some difficulties and also some needed sophistication. The current of evolutionary naturalism, begun with Darwin and Spencer, became, however, an integral part of the scientific movement. The most active modern exponent of evolutionary naturalism or of evolutionary humanism, as he prefers to call it, is Julian Huxley. To Huxley, Waddington, Chauncey Leake, and several others, we owe modern versions of this viewpoint. Huxley has even asserted that his evolutionary humanism "is capable of becoming the germ of a new religion, not necessarily supplanting existing religions but supplementing them." This is probably claiming far too much. The decision may well be left to the future.

The evolutionary approach has become an integral part of the intellectual equipment of modern man because it is indeed capable of yielding some ideas of broad philosophical interest and significance. About a century ago Darwin proposed a theory of the evolutionary development of the biological species. Since then the idea of evolution has become applied much more widely than in the field of biology. On the one hand, the Cosmos itself is now being thought about as a product of an evolutionary development. On the other, it is not only man's bodily frame but the whole Man that must be regarded as a result of an evolutionary process. Evolution is the method whereby Creation is accomplished.

EVOLUTIONARY ETHICS

It has been said that "Man was formed for society." The seclusion of an anchorite is assuredly neither the usual nor the normal human environment. The environment of a member of our species is set by the society of which he is a member. He is influenced by his relations to other persons in the same society; in fact, interpersonal relations constitute the most important aspect of human environment. These relations often determine the ability of an individual to survive and to leave progeny. The interpersonal relations which prevail in a society or in a group within a society affect its chances for survival and perpetuation. These simple considerations have led many thinkers, from Darwin to our day, to suppose that man's behavior as a member of society is, like the structure and the physiology of his body, molded in the evolutionary process and controlled by natural selection. The basic assumption to this view has been stated most concisely by Leake as follows: "The probability of survival of a relationship between individual humans or groups of humans increases with the extent to which that relationship is mutually satisfying."

The greatest, although by no means the only, difficulty for this utilitarian explanation of ethics lies in understanding the nature and origin of the moral sense which every man, to some degree, has. Admiration of the good and disapprobation of the bad is ineradicable from the human mind. Regardless of whether a person does or does not live up to the standards of ethics accepted in his social environment, he usually feels that these standards have some validity which he cannot gainsay. Yet the kinds of behavior which many ethical codes hold most praiseworthy do not always promote the survival and welfare of the person who adopts them. Indeed, the selfish often prosper more than the generous, the cunning more than the truthful, and the cowards more than the brave. The rule that whatever promotes the survival and welfare of the individual is good, and whatever puts them in jeopardy is bad, is no safe guide in ethics. Yet such a rule would be expected to hold if ethics were a product of what nineteenth-century evolutionists called the survival of the fittest in the struggle for existence. There is a glaring conflict between the "gladiatorial theory of existence," seemingly implied in the evolutionary theory, and the Golden Rule, which, to most people, still stands as the most trenchant statement of the guiding principle of ethics. In 1893, T. H. Huxley admitted with admirable courage and sincerity that he was thwarted by the contradiction.

Ways towards a resolution of this apparent conflict were not opened up until it was realized that the "gladiatorial theory" is not only not a necessary part of the theory of natural selection, let alone a part of the theory of evolution, but is, in fact, invalid on purely biological grounds. Biological fitness is by no means always promoted by the ability to win in combat. It is much more likely to be furthered by the inclination to avoid combat, and in any case, it is measured in terms of reproductive success rather than in terms of the numbers of enemies destroyed. Moreover, not only individuals, but also groups of individuals, such as tribes and races, are units of natural selection. It is, then, at least conceivable that evolutionary processes may promote the formation of codes of ethics which, under some conditions, may operate against the interests of a few individuals but which favor the group to which these individuals belong. Biological analogies are readily available. Among ants, termites, and other social insects, an individual often sacrifices his life for the sake of the colony. In social insects most individuals do not reproduce, and natural selection favors the forms of behavior which increase the chances of survival of the colony, and particularly of its sexual members.

This, certainly, does not dispose of all the difficulties in understanding ethics on an evolutionary basis. In man, codes of morality and ethics vary from group to group. In a given group they may undergo changes with time, as history abundantly shows. Are we to conclude that, in man, natural selection favors the ethical codes which benefit the group at the expense of the individual? Such a view would leave unresolved the ethical paradox of conflicting interests between the individual and the society to which he belongs. We saw that the behavior of an ant is largely determined by its biological heredity, and that the ant has no choice but to act as it does. On the other hand, man is free to choose between various courses of action, and feels himself responsible for the choices he makes. He can choose to work for or against the society. Should he always sacrifice himself to the interest of his group, and does the group always have the right to expect its members to do so? This is, of course, one of the greatest problems facing mankind. All the great literatures and philosophies have struggled to resolve this conflict, and most of them have found that the only solution is to accept a divine sanction as the foundation of ethics. The crumbling of this foundation in our day leaves a terrible void in the human soul.

Julian Huxley has made a valiant attempt to find a substitute for divine sanction as the foundation of ethics. He proposes an evolutionary sanction instead. According to him, "The function of social ethics is, in biological terminology, phylogenetic, helping society to persist, to reproduce itself, and in some cases to change and to advance." The basis of "scientific" ethics is, then, the degree of agreement between the consequences of a given action and the evolutionary trend of the development of the human species. The rule becomes: "Anything which permits or promotes open development is right; anything which restricts or frustrates development is wrong. It is a morality of evolutionary direction," and: "In this light the highest and most sacred duty of man is seen as the proper utilization of the untapped resources of human beings."

IS EVOLUTION ALWAYS RIGHT?

No theory of evolutionary ethics can be acceptable unless it gives a satisfactory explanation of just why the promotion of evolutionary development must be regarded as the *summum bonum*. Indeed, any morality of evolutionary direction rests of necessity on our ability to determine what this direction has been and what it is at present. There is no single trend in evolution, but rather many different trends in different organisms. To be sure, some trends predominate in the evolution of some groups. Thus, the development of the cerebral functions, of intelligence, seems to have been the dominant trend in human evolution. Continuation of intellectual development becomes, then, a moral duty of mankind. But this is surely not the whole story. Human evolution has also involved many other trends, some of which were more constant than others, some which reversed themselves, and some which are of a degenerative character. Thus, there was for some time in the history of mankind a trend towards the differentiation of the human species into races, and towards biological divergence of these races. More recently, this trend was reversed because of the increasing mobility of human populations and of intermarriage between them, and human races began to converge. Which of these two trends should, then, be the basis of our ethics? And how about the allegedly great fertility of the intellectually less well-endowed persons and groups within human populations? If such a trend actually exists, surely it is more reasonable to think of combating it rather than assisting it.

Suppose, however, that future studies of human biology and evolution tell us exactly what the direction of evolution in general, and of human evolution in particular, has been. Just why should we take for granted that this direction, which we have not chosen, is good? The very fact that man knows that he has evolved and is evolving means that he is able to contemplate speeding up his evolution, slowing it down, stopping it altogether, or changing its direction. And his increasing knowledge and understanding of evolution may enable him to translate his thoughts into reality. Despite any exhortations to the contrary, man will not permanently deny himself the right to question the wisdom of anything, including the wisdom of his evolutionary direction. He may rebel against this direction, even though it may be shown to be a beneficial one. Just such an "unreasonable" rebellion was envisaged by Dostoevsky in his *Letters from the Underworld*. Man is likely to prefer to be free rather than to be reasonable. As Simpson put it: "There is no ethics but human ethics, and a search that ignores the necessity that ethics be human, relative to man, is bound to fail."

DETERMINISM AND EDUCABILITY

As we have shown, there is ample evidence that the behavior of an individual animal towards other individuals of the same or of different species is in part genetically conditioned. Conspecific individuals have to cooperate with each other, at least at certain times, such as during periods of reproduction. Cooperation is especially prominent in social organisms, because social cohesion demands a certain minimum of cooperation among members of a society. Since social organization may increase the biological fitness of the species, natural selection may be expected to promote cooperative behavior.

The problem is to what extent will the behavior be inborn, resulting from biological heredity, and to what extent will it be conditioned in every generation by learning from other members of the society. Different solutions of this problem have been achieved by different organisms. Social insects offer many examples of behavior which is to a large extent inborn, and only to a limited extent learned. Conversely, human behavior is in the main genetically unfixed; it shows a remarkably high degree of phenotypic plasticity. It is acquired in the process of socialization, of training received from other individuals. Its base is set by the genes, but the direction and extent of its development are, for the most part, culturally, rather than biologically, determined.

The uncertainties of the nature-nurture problem are due precisely to the intricate blending of the inborn or genetic, and the acquired or cultural stimuli of behavior. The tender, protective, self-denying, passionate devotion felt by parents for their children may well be genetically conditioned. So may be the fact that the devotion felt by children, at least by grown-up ones, for their parents is usually less ardent than that felt by the parents towards the children. This is in no way contradicted by the fact that expressions of parental and filial devotion vary greatly from culture to culture, and that within a culture some parents may, under some circumstances, display a deficiency of parental feelings.

The manifestation of inborn drives in man depends on the cultural setting, just as the manifestation of any genetic trait may, to a greater or lesser extent, depend on the environment in which development takes place. The anthropologist Coon maintains that human nature was shaped by natural selection chiefly during the long, formative stage of the history of our species when men obtained their sustenance by hunting wild animals. "Our biological make-up is the same as theirs [Paleolithic hunters] and our biological needs were determined by natural selection over hundreds of thousands of years." This may well be right, but the progeny of these hunters developed agriculture, settled on land, built cities, invented industries and innumerable machines and gadgets, and finally rose to a point when they hold their own future evolution in their own brains and hands.

Attempts to discover a biological basis of ethics suffer from mechanistic oversimplification. Human acts and aspirations may be morally right or morally wrong, regardless of whether they assist the evolutionary process to proceed in the direction in which it has been going, or whether they assist it in any direction at all. But the matter is more subtle than that. Dostoevsky makes his Ivan Karamazov spurn the promise of the universe's evolution towards perfection and eternal harmony if this evolution must be promoted by the torture of just one innocent child. Ethics are a part of the cultural heritage of mankind, and consequently belong to the new human evolution, rather than to the old biological evolution. Moral rightness and wrongness have meaning only in connection with persons who are free agents, and who are consequently able to choose between different ideas and between possible courses of action. Ethics presuppose freedom.

ETHICS AND FREEDOM

Ethics, as such, have no genetic basis and are not the product of biological evolution. Natural selection has not propagated genes for ethics, or genes for inventing Euclidean geometry, propounding evolutionary theories, composing musical symphonies, painting landscapes, making a million dollars on Wall Street, loving the soil, or becoming a military leader. Such genes simply do not exist. Genes do not transmit and do not determine specific components of our cultural heredity.

The ability to study geometry, let alone higher mathematics, had no selective value in the ancestors of our species, and it is not certain that such abilities are biologically advantageous even at present. However, the ability of abstract thinking, or of perceiving causal relationships between events, did confer tremendous adaptive advantages on the human species. This basic ability has enabled man gradually to achieve a mastery over his environments, to develop ways and means to alleviate hunger, poverty, and disease. And the same basic ability has made possible the birth and development of science, art, philosophy, and religion. If natural selection has not developed genes for philosophy, it has favored genetic endowments which enable their carriers to become, among other things, philosophers. Science, art, philosophy, and religion are products of the basic intellectual powers of the human species. But their development, let alone the development of particular forms of science, art, philosophy, and religion which actually came into existence, was not contained in the human genotype or predestined by it.

This does not mean, of course, that biological evolution has reached its end and has been replaced by cultural evolution. Man of the Old Stone Age already had the genetic equipment needed to develop a culture; this fact has led some anthropologists to conclude that no further genetic progress has taken place in human evolution since the dawn of mankind. Surely, such a conclusion is not necessitated by the facts, and is most improbable on biological grounds. Our remote ancestors were biologically successful precisely because they came to possess the genetic wherewithal to acquire and to transmit rudiments of culture. Because of this adaptive success, the genetic equipment which makes culture possible is being maintained and reinforced by continuous selection going on in most human populations most of the time. We have discussed briefly the fears that, in modern times, natural selection no longer favors high intelligence in some human societies. Whether these fears are justified or not, the very possibility of their existence shows that the genetic basis of culture is not something immutable or incapable of improvement or deterioration.

The relative autonomy of cultural evolution from biological evolution lies in a different plane. Biological evolution has produced the genetic basis which made the new, specifically human, phase of the evolutionary process possible. But this new evolution, which involves culture, occurs according to its own laws, which are not deducible from, although also not contrary to, biological laws. The ability of man to choose freely between ideas and acts is one of the fundamental characteristics of human evolution. Perhaps freedom is even the most important of all the specifically human attributes. Human freedom is wider than "necessity comprehended," which is the only kind of freedom recognized by Marxists. Man has freedom to defy necessity, at least in his imagination. Ethics emanate from freedom and are unthinkable without freedom.

Ethics are, consequently, a human responsibility. We cannot rely on genes or on natural selection to guarantee that man will always choose the right direction of his evolution. This has been beautifully stated by Simpson in the following paragraph:

"Man has risen, not fallen. He can choose to develop his capacities as the highest animal and to try to rise still farther, or he can choose otherwise. The choice is his responsibility, and his alone. There is no automatism that will carry him upward without choice or effort and there is no trend solely in the right direction. Evolution has no purpose; man must supply this for himself. The means to gaining right ends involve both organic evolution and human evolution, but human choice as to what *are* the right ends must be based on human evolution. It is futile to search for an absolute ethical criterion retroactively in what occurred before ethics themselves evolved. The best human ethical standard must be relative and particular to man and is to be sought rather in the new evolution, peculiar to man, than in the old, universal to all organisms. The old evolution was and is essentially amoral. The new evolution involves knowledge, including the knowledge of good and evil."

Aquinas, Sartre, and the Lemmings

Raymond J. Nogar

CONTEXTUALIZING COMMENTS

It may seem the first time around that this speculative preoccupation about the meaning of cosmic epigenesis has little to do with man's practical future. Similarly, a grand attempt to sketch a place and role for man in nature, and to fashion quasi-deliberately an evolutionary world-view, may seem pretty far removed from the hard social and political traumas which threaten the here and now. But a short meditation upon the fact that man is the only animal which does not have its future mapped out in its genes and physical surroundings makes it evident that awareness of his free, creative, and personal capacities is absolutely necessary for long-term survival. Meaning of existence makes all the difference in man's attitude toward the future, especially if technology and economic patterns are ever to be integrally subordinated to the authentic welfare of the human group.

For man could easily—has he not already gone far in this very direction?—elaborate conditions for his physical, mental, and moral constitution which would ultimately lead to his complete disability to adapt. "Everything is subordinated to the increase in industrial production and to armaments . . . while men concentrate on the serious business of killing each other and eating."[1]

Reprinted from *Chicago Studies*, 4 (Summer 1965), pp. 161–170, by permission of the Editor.
[1] Pierre Teilhard de Chardin, *The Phenomenon of Man*, p. 279.

The problem of the future of man has disclosed itself as simply too large and interarticulated with the nonhuman, subhuman, nonrational and irrational facets of reality to be answered with anything like adequacy by those who would foreclose the issue by discounting *any* avenue of inquiry as irrelevant to the world of human concern. If an evolutionary world-view is to establish itself as more than just one more divisive ideology, then it must provide a cultural setting at the ideational level wherein concern for understanding will be the dominating component.

Let us stop to think for a moment of the proportion of human energy devoted, here and now, to the pursuit of truth. Or, in still more concrete terms, let us glance at the percentage of a nation's revenue allotted in its budget for the investigation of clearly-defined problems whose solution would be of vital consequence for the world. If we did we should be staggered. . . . Surely our great-grandsons will not be wrong if they think of us as barbarians?

The truth is that, as children of a transition period, we are neither fully conscious of, nor in full control of, the new powers that have been unleashed. . . . But the moment will come. . . .

We can envisage a world whose constantly increasing 'leisure' and heightened interest would find their vital issue in fathoming everything, trying everything, extending everything; a world in which, not only for the restricted band of paid research-workers, but also for the man in the street, the day's ideal would be the wresting of another secret or another force from corpuscles, stars, or organized matter; a world in which, as happens already, one gives one's life to be and to know, rather than to possess. That, on an estimate of the forces engaged, is what is being relentlessly prepared around us.[2]

For such a social order, respect must be learned for every authentic insight into man and his world, be it speculative or practical, poetic or scientific, traditional or contemporary. In such a social order, summarily, ways must be found to prevent that hardening of insights we know so well as pre-judgment—prejudice, that loss of the curiosity, interest, and wonder which animates most human beings in the early years of life, but which is somehow left behind by most of us as we move into adolescence and adulthood, content henceforward and for the most part to subordinate the goods of the spirit to physical appearance and properties, and to replace moral judgment with absolute demands for "in-group" solidarity, for acceptance and promotion, that is, of the tribal mentality. It is just here, in terms of the need for a new order of human relationships founded on insight and understanding, attentive to the hard facticity of things on one side and to the equally revealing aspirations of the human spirit on the other, and founded—most importantly—beyond ideological dogmatism (be it racist, nationalist, economic and dialectical, or whatever), the sophisticated tribal spirit, just here that the indispensable

[2] *Ibid.* pp. 279–280.

role and highest function of our great university centers most clearly delineates itself.

But if this function is to be fulfilled adequately, then our ultramodern instrumentalist conception of education as subordinated to sophisticated "problem solving," technical training, and technological research must be tempered anew by the ancient conception of study as before all else a search for truth. There have been consequences enough to our abandonment of this view, as Harry Rubin, Professor of Molecular Biology at the University of California at Berkeley, has recently pointed out: "There exists ample knowledge to provide a decent life for everyone in the world but we refuse to apply this knowledge. The shortage is not of facts; it is of human and political insight. The great and heartening examples for our society and for the world are coming, not from the scientists nor from others of highly polished I.Q.'s but from . . . the Negroes crying out for justice."[3] What we lack is just what technology cannot provide—a total spiritual environment for man. Upon its establishment the future of man depends: for since the natural law is nothing other than the application of the law of specific adaption to the human species, it is the inclination of the *whole of man* that must be fulfilled if man is to adapt in the specific sense.

Aquinas, Sartre, and the Lemmings

> Every authentic insight into reality whether traditional or contemporary must be brought to bear upon the issue: to fashion the future of man.

Jean-Paul: "It is a fearsome thing to watch the golden lemmings gather for their dash to the Sea."

Thomas: "Yes, once the signal is given they must obey . . . They can do nothing but follow the plan of migration so deeply ingrained in their life-stuff . . ."

In 1914 the blue passenger pigeon became extinct. The wolverine is disappearing from the Michigan northlands. The giant shy condor, once a familiar shadow upon California's Sierra Nevadas, will soon die out.

---

[3] Letter from Harry Rubin, Professor of Molecular Biology at the University of California, Berkeley, in response to the article on "The Biological Future of Man," by Dwight J. Ingle, Professor and Chairman of the Department of Physiology at the University of Chicago, which article appeared in the spring, 1966, issue of *Chicago Today*, III (1966), pp. 36–42.

Such is the story of plant and animal species: they emerge, flourish for a time, pass away. What is the future of the human species? Man arose upon this planet about a million years ago. He now flourishes. Will he survive? Why should he be an exception to the rule of nature?

Man is an exceptional species, unique in this. All other animals adapt, in a very limited way, to an environment which they did not fashion. *Homo sapiens* is a free and creative animal which survives by adapting to an environmental niche which he designs for himself. Barring catastrophe like the bursting of our sun, and imbecility like atomic warfare, are we designing a future to which we can hope to adapt? Or are we on the way to the Sea? The answer is as frightening as it is simple. If man discovers who he is and what is his destiny in time, he can blueprint his future. If not, he has no future.

For the first time in human history, it is clear to all that intellectual and spiritual solidarity is absolutely necessary if free and creative man is to survive. Every individual has a grave responsibility to contribute to our survival; yet only a total world-wide solidarity of human understanding and love can meet the momentous issue which suddenly looms up before the entire species. Perhaps the most hopeful instrument of the consolidation of mankind is the university. By definition, the university is the storehouse of knowledge and understanding, conserving and projecting the finest insights of man. The university is the only available intellectual organization which can assume total detachment from local culture, a prerequisite to its invaluable function of critical reflection upon the present age and future survival.

In the university, the college shares this urgent responsibility by bringing forth its special unique contribution to the destiny of man. The college of law, of medicine, of fine arts conserve, create, and project the best of their professional traditions to the university. Nor is it necessary that the college be located on a university campus, though this is often desirable. The private college, whether secular or religious, whatever else may be its role in a free society, exists for the future of man by its contribution to the university of those living traditions which mankind cannot afford to lose. The existence of our species depends upon the living traditions which project man's best insights into himself, his universe, and his God.

REQUIREMENTS FOR THE FUTURE

Man's future is in the hands of two agencies: his biological evolution and his psychosocial evolution. Genetic and natural environmental forces continue to influence the development of man; yet there has been little biological change in our species in the last 30,000 years. Psychosocial control now dominates the direction of human evolution. Population problems, inherited pathology, and other biological concerns certainly pertain to man's future, but the life-death issue cannot be reduced to an actuarial table. Preoccupation with the significance of death is the mere threshold of the thinking of Sartre and Marcel; even the naturalistic humanism of Muller and Huxley presupposes a basic world-view. Man's personal potential is biologically based, but how he freely and creatively designs his future depends upon his understanding and attitudes toward reality, in a word, upon his psychosocial or spiritual ecology.

The prehistory and the history of man reveal that human survival depends upon his ability to face reality on three levels. The first level is the *world of the timely*, the cosmological, work-a-day world of sensation, the historically unfolding human existence of wonder, of work, and of interpersonal relations. The second level is that same universe of reality, now viewed in its entirety as a simultaneous whole. This *timeless world* of the human spirit responds to man's quest for ultimate meaning, for an answer to the question: "Why not just nothing?" as Heidegger puts it. The impingement of the world as timeless, though more difficult to isolate in human adaptation, is basic to all motivation, all decision, all commitment, all affirmation in the order of the timely. Man cannot long survive without underpinning himself with the "meaning of it all." Finally, the most illusive yet the most important level of human spiritual ecology (and man has ever judged it so) is man's *openness to divine communication* of the deeper meanings of the mysterious, unfolding reality about him.

It is the condition of a free and creative human spirit to move swiftly and probingly in and out of these three distinguishable though not separate worlds. It hovers where it wills, yet tries to avoid fragmentation. Man seeks to blend harmoniously all his insights into reality, but he is suspicious of the tiniest illusion which might be imposed by unrealistic preoccupation with the timely, the timeless, or the divine. Because of the danger of illusion, men have attempted, in the name of human freedom, to eliminate one or other of these three levels of experience. To no avail. They are irreducible elements of human adaptability.

Man will be creative and free in projecting his future only if he can, at one and the same time, realistically immerse himself in the history of his age and culture, discover trans-historical meaning and affirm divine communication and adore. In the interest of man's survival then, the university must conserve, promote, and project the specialized insights of man into these three areas. From these experiences with reality, man formulates his world-view. To that world-view of reality, he must adapt all that he is—his art, his technology, his life. Man cannot afford to lose a single authentic intuition of reality, and it is the solemn responsibility of every college to contribute its unique tradition to the university. Today, it is no longer merely a question of sophistication; it is a matter of human survival.

THE EXPLORATION INTO AUTHENTICITY

Most human insights are fragmentary. Not every tradition is wholly authentic. The role of intellectual criticism is perhaps the most important and the most difficult the university must assume. The ultimate test must be man's grasp of reality itself, and the illusions of any world-view can be smashed only by showing that "reality just isn't like that." Yet human survival feeds upon the insights which positively reveal what the world "is really like." Consequently, the university has a very valuable built-in intellectual attrition which tends to conserve the authentic insights and prunes away the illusions: the dialogue between traditional and contemporary thought. Interdepartmental attrition, like the recent clashes between the scientific and humanistic cultures, and intradepartmental attrition, like the philosophical reactions of phenomenology to all traditional metaphysics, work toward that purification so necessary for realistic thought and action. With the critical tool of intellectual attrition, the university has an invaluable instrument for preparing man's future with the best available design of the future.

In the American Catholic college's contribution to man's future, this intellectual attrition between tradition and contemporary thought is dramatically illustrated by the current dialogue between existentialism and a major element of the Catholic tradition, neo-Thomism. In Europe, this confrontation has been underway for over a half-century in the work of Heidegger, Merleau-Ponty, Scheler, and others, but recently many American Catholic colleges have experimentally revamped their philosophy courses to fit the existentialist frame of reference. Relative strength and weakness are keenly felt by this edge-to-edge contrast of view, and both the promise of greater authenticity and the dangers of loss of insight have been clarified.

Aside from polemics, which generate more heat than light, we are beginning to see the emerging value of this attrition to the role of the university and the survival of man. Keeping in mind that the aim of university criticism is to assess the realism of man's world-view, the Catholic college has brought two veins of thought into dialogue, the insights of Thomas Aquinas and those of Jean-Paul Sartre. Although neither Catholic tradition nor contemporary thought are represented *in toto* by these two thinkers, from this confrontation examples of strength and weakness can be dramatically portrayed. Out of similar contrasts of representatives of tradition and contemporary thought should emerge the purified insights of both, those urgently needed bases for a blueprint of human survival.

THOMAS AQUINAS: STRENGTH AND WEAKNESS

The true philosopher is not preoccupied with the opinions of men, the needs of pedagogy, nor the polemics of the day. He is only interested in how the truth of things stands. From this detached vantage point, Thomas Aquinas penetrated the reality of the universe with remarkable, enduring insights. The first had to do with the *timely*, the cosmological. He acknowledged that the world of reality had its own light, that the human spirit could lay hold of the interior of things, not perfectly but securely enough to repudiate all forms of scepticism and agnosticism. He asserted that what a man thinks about creatures and the *timely* conditions not only his estimate of the *timeless* but even what he thinks about the Creator and his revelation.

Aquinas' second major insight was that the human spirit could transcend all the partial aspects of the universe of reality and grasp something of the *timeless*, the totality of existing things. Man can ask the question "Why not just nothing?" as Heidegger later formulated it; and he can understand something of the "meaning of it all." The human spirit, he asserted, was at home in the work-a-day world of sensible experience, but from this world it could simultaneously penetrate through and beyond the fragments of this world to pin-points of timelessness, the meaning of existence itself. This latter intuition threw open the human spirit to an even more transcendent light upon reality. Aquinas saw that man was open to affirmation and homage towards the very source of reality and could, with full freedom and creativity, hear divine speech and respond with belief and love. For him, this homage and adoration would, in its turn, open up meaning of reality in a way far exceeding the mere human efforts of the mind.

But his strength was his weakness. The spirit of man ever hovers where its insights are surest. Thomas Aquinas, preoccupied with the divine revelation and deeply orientated in the theocentric thought of the Greeks, was surest in his contemplation of the timeless and the divine. The timely of the cosmos, the orientation of creation in time and space, did not enter substantially into his scientific thought. He insisted that man's understanding of the timeless and of God had to be drawn from authentic analogies to creation. Yet without a science of reality which laid open the cosmos *sub ratione evolutionis* (a fact totally unsuspected in 1250), space-time contingency remained incidental to this thought and his tradition. He too quickly (for our tastes) transcended creation with his world-view. He was precipitously caught up by the transhistorical, made ready by theocentric traditions, and *sub ratione aeternitatis*, abandoned the imperfections and the grandeur of a universe of space-time, the very being of which is unfolding contingency.

This failure of space-time orientation proved serious. By Aquinas' own insistence, it is folly to think that man's metaphysical and theological views are not conditioned by his cosmic world-view. The tradition of Aquinas failed to respond to the space-time orientation demanded by scientific progress and as a consequence the sciences of every area of reality suffered: its cosmology, its psychology, its moral views, and even some of its theological expressions of revelation itself. By modern standards, the thought of Aquinas labors under the effects of the "myth of the eternal return" imposed by earlier philosophical cultures in which the epigenic unfolding space-time history of reality plays no significant role.[1] His basic insights that the human spirit can and must confront the cosmic realities as the test of all understanding, can arise thence to realistic understanding of supratemporal meaning of existence, can and must remain open to divine communication of ultimate meaning of human existence, stand firm and must not be lost. But his realism must be brought into the focus of epigenic contingency of space and time. Upon this nerve contemporary philosophy has placed its finger.

JEAN-PAUL SARTRE: STRENGTH AND WEAKNESS

Like Aquinas, Sartre interrogates reality on three levels: the timely, the timeless, and the divine. But the whole thrust of his thought, like that of Hegel and Marx, of Heidegger and Dilthey, of Scheler, Merleau-Ponty, and Marcel, is orientated to the space-time contingency of reality. The "worldhood of the world," the "facticity of things," the "immersion of being in history," all signify that new look at reality with "rinsed eyes" which crushes the illusions of traditional world-views ensnared by the "myth of the eternal return."

Contemporary thought begins where traditional thought was weakest. It places created material being in time and space so securely that assertions about reality can no longer prescind from space-time and remain authentic. Within this century whole areas of science, philosophy, theology, the arts and humanities, even Scripture study itself, have been recast into the critical mold of historicity. The epigenic evolution of created being has refashioned our fundamental thoughts about the universe, about man and his destiny. Historical orientation has altered the analogies for Divine Being and divine communication. With Sartre, unfolding creation has become the pruning knife, the smasher of illusions, the remover of cultural accretions of traditional systems. Where Aquinas was weak, Sartre and the other contemporary philosophers are strong.

But Sartre knows that total meaning of reality is not achieved merely by immersing oneself in the timely. It would be a mistake of the grossest kind to think that the ultimate considerations of contemporary space-time philosophies are history-bound. For Hegel, the destiny of a people preserved a transhistorical significance, a more perfect manifestation of the Universal Spirit. For Marx, the end of history was "salvation" which exorcises its terror. Dilthey tried to find a way to transcend mere historicity. Meinecke thought he found a way through the "examination of conscience." Nietzsche said: ". . . always one thing which makes for happiness: the capacity to feel unhistorically." The chief issue for Marcel is the "metaproblematic." In his autobiography *The Words*, Sartre slashes illusions right and left with his sword of the timely, only because he is searching for the timeless "meaning of it all."

Sartre's strength, indeed the brilliant intuition of contemporary thought, is that the timeless in reality is guaranteed only through the iconoclasm of interrogating creatures in their space-time contingency. Sartre discovered a means of separating what belongs to "the way things are" from mere cultural determinants. But his tool proved to be a double-edged sword. As in the case of Aquinas, Sartre's strength is his weakness. In order to get at existing reality without illusion, he cut away all tradition and closed himself off from divine communication by dialectical denial. In his consequent panic to take the terror out of history by restraining the movement of the spirit to the timeless, and by deliberately closing his ears to the possibilities of divine speech, he despairs. Absurdity becomes the theme of the timely, and in the name of freedom he finds no escape from the morass of space and time. He succeeded where Aquinas had failed; what Aquinas had gained, he lost.

BLUEPRINT FOR THE FUTURE

It is urgent that the timeliness of Sartre and the timelessness of Aquinas, the strength of tradition and that of contemporary thought, be harmonized. For man's future rests upon the fullness and authenticity of his world-view. We cannot afford to allow a single valid insight into man, his universe, and his God to be lost. Such a project of intellectual and spiritual solidarity can be accomplished only by the university roundtable. Its critical role is to correct the myopia of world-views, smash illusions, and prune away dead wood without losing a single authentic insight into the future of man.

The invaluable role of the college is to contribute the basic insights of its unique tradition to the university round-table which must ultimately propose a blueprint for the future of the species. The Catholic college has a unique living tradition which must not be lost. Yet to be of value to man, it must undergo the attrition, the pruning and renewal upon which the contemporary philosophies of space-time insist. Conversely, contemporary thought urgently needs the balance and enrichment which traditional insights into the timeless and the divine can give.

Extinction of man, failure to creatively and freely adapt, can come from following blindly in the path of either traditional or contemporary intuitions, from following uncritically either Aquinas or Sartre. Failure to orient man's world-view in space and time, to see how epigenic unfolding enters into all that man is and does will result in illusion. Total immersion in the timely without discovering the supernatural meaning of existence and without spiritual openness to the divine will just as surely threaten man's ability to survive. It is not simply a question of updating or providing bibliographies of Sartre. Polemics which barter a word from Sartre for a word from Aquinas are futile, and they make spiritual solidarity based upon authentic insight doubly difficult.

Yet there is great expectation. Man must look to the future of *Homo sapiens*. It is no longer simply a question of the Greek Academy or Lyceum, of a scholastic synthesis, of cultivating manners, of technical prowess or even a liberal education. It is a question of survival by creative, free, spiritual solidarity. The solution is to call forth every authentic insight into harmonious bearing upon the issue: to fashion the future of man. We cannot afford to lose a single insight into reality whether traditional or contemporary. Neither Aquinas nor Sartre suffices. With the best of both, we stand a chance of survival.

Thomas: "Too late to change; the die is cast."

Jean-Paul: "All the same, it is grim to watch the lovely lemmings scrambling with such glee, unaware that they leap into the yawning jaws of oblivion."

NOTES

[1] St. Thomas' scientific thought was constructed upon the Greek idea of archetypes and a cyclical history. He believed that individuals developed, came to be and passed away, but not species. Species differed like numbers, and time and space could not efface them. Thus time and space were incidental and could not enter into science. History could be abstracted from because time regenerated the species and cosmic natures remained perennially the same. Camels are camels are camels . . . To this day the typological idea that natural "essences" are "eternal, necessary and immutable" forms the basis of Thomistic cosmology in the manuals. Cyclical history, the eternal return, has been completely overthrown by evolution studies, but in every area of the Thomistic tradition, whether it be about the cosmos, man, or God, the thought labors under the myopic, static, deterministic fixity of "the eternal return."

The word *epigenic* is drawn from usage in the philosophy of biology. *Epigenesis* and *preformation* represent two different attitudes to the problem of biological development in the individual organism. Preformation is the view that in the fertilized ovum there is present a fully actualized "little man"; it is just a matter of growing bigger. The epigenic view is that there is a real unfolding of being with novelty at every turn. It is dynamic and open to development depending upon the contingent circumstances of space and time. Applied to the development not only of individuals but of species, "epigenic unfolding space-time history" is opposed to a view of the cosmos in which no new specific novelties arise, where history (time and space) is incidental, and where static, deterministic, finalized order admits of no basic development.

How human is man?

Loren Eiseley

Over a hundred years ago a Scandinavian philosopher, Sören Kierkegaard, made a profound observation about the future. Kierkegaard's remark is of such great, though hidden, importance to our subject that I shall begin by quoting his words. "He who fights the future," remarked the philosopher, "has a dangerous enemy. The future is not, it borrows its strength from the man himself, and when it has tricked him out of this, then it appears outside of him as the enemy he must meet."

We in the western world have rushed eagerly to embrace the future—and in so doing we have provided that future with a strength it has derived from us and our endeavors. Now, stunned, puzzled and dismayed, we try to withdraw from the embrace, not of a necessary tomorrow, but of that future which we have invited and of which, at last, we have grown perceptibly afraid. In a sudden horror we discover that the years now rushing upon us have drained our moral resources and have taken shape out of our own impotence. At this moment, if we possess even a modicum of reflective insight, we will give heed to Kierkegaard's concluding wisdom: "Through the eternal," he enjoins us, "we can conquer the future."

The advice is cryptic; the hour late. Moreover, what have we to do with the eternal? Our age, we know, is littered with the wrecks of war, of outworn philosophies, of broken faiths. We profess little but the new and study only change.

Three hundred years have passed since Galileo, with the telescope, opened the enormous vista of the night. In those three centuries the phenomenal world, previously explored with the unaided senses, has undergone tremendous alteration in our minds. A misty light so remote as to be scarcely sensed by the unaided eye has become a galaxy. Under the microscope the previously unseen has become a cosmos of both beautiful and repugnant life, while the tissues of the body have been resolved into a cellular hierarchy whose constituents mysteriously produce the human personality.

Similarly, the time dimension, by the use of other sensory extensions and the close calculations made possible by our improved knowledge of the elements, has been plumbed as never before, and even its dead, forgotten life has been made to yield remarkable secrets. The great stage, in other words, the world stage where the Elizabethans saw us strutting and mouthing our parts, has the skeletons of dead actors under the floor boards, and the dusty scenery of forgotten dramas lies abandoned in the wings. The idea necessarily comes home to us then with a sudden chill: What if we are not playing on the center stage? What if the Great Spectacle has no terminus and no meaning? What if there is no audience beyond the footlights, and the play, in spite of bold villains and posturing heroes, is a shabby repeat performance in an echoing vacuity? Man is a perceptive animal. He hates above all else to appear ridiculous. His explorations of reality in the course of just three hundred years have so enlarged his vision and reduced his ego that his tongue sometimes fumbles for the proper lines to speak, and he plays his part uncertainly, with one dubious eye cast upon the dark beyond the stage lights. He is beginning to feel alone and to hear nothing but echoes reverberating back.

It will do no harm then, if in this moment of hesitation we survey the history of our dilemma. Man's efforts to understand his predicament can be compassed in the simple mechanics of the theatre. We have examined the time allowed the play, the nature of the stage, and what appears to be the nature of the plot. All else is purely incidental to this drama, and it may well be that we can see our history in no other terms, being mentally structured to look within as well as without, and to be influenced within by what we consider the nature of the "without" to be. It is for this reason that the "without," and our modes of apprehending it, assume so pressing an importance. Nor is it fully possible to understand the human drama, the drama of the great stage, without a historical knowledge of how the characters have interpreted their parts in the play, and in doing so perhaps affected the nature of the plot itself.

This, in brief, epitomizes the role of the human mind in history. It has looked through many spectacles in the last several centuries, and each time the world has appeared real, and the plot has been played accordingly. Strange colorings have been given to reality and the colors have come mostly from within. As science extends itself, the colors, and through them the nature of reality, continue to change. The "within" and "without" are in some strange fashion intermingled. Perhaps, in a sense, the great play is actually a great magic, and we, the players, are a part of the illusion, making and transforming the plot as we go.

If the play has its magical aspect, however, there is an increasing malignancy about it. A great Russian novelist ventures to remark mildly that the human heart, rather than the state, is the final abode of goodness. He is immediately denounced by his colleagues as a heretic. In the West, psychological studies are made of human "rigidity," and although there is a dispassionate scientific air about them the suggestion lingers that the "normal" man should conform; that the deviant is pathological. The television networks seek the lowest denominator which will entrance their mass audience. There is a muted intimation that we can do without the kind of intellectual individualists who used to declaim along the edges of the American wilderness and who have left the world some highly explosive literature in the shape of *Walden* and *Moby Dick*. It is obvious that the whole of western ethic, whether Russian or American, is undergoing change, and that the change is increasingly toward conformity in exterior observance and, at the same time, toward confusion and uncertainty in deep personal relations. In our examination of this phenomenon there will emerge for us the meaning of Kierkegaard's faith in the eternal as the only way of achieving victory against the corrosive power of the human future.

II

If we examine the living universe around us which was before man and may be after him, we find two ways in which that universe, with its inhabitants, differs from the world of man: first, it is essentially a stable universe; second, its inhabitants are intensely concentrated upon their environment. They respond to it totally, and without it, or rather when they relax from it, they merely sleep. They reflect their environment but they do not alter it. In Browning's words, "It has them, not they it."

Life, as manifested through its instincts, demands a security guarantee from nature that is largely forthcoming. All the release mechanisms, the instinctive shorthand methods by which nature provides for organisms too simple to comprehend their environment, are based upon this guarantee. The inorganic world could, and does, exist in a kind of chaos, but before life can peep forth, even as a flower, or a stick insect, or a beetle, it has to have some kind of unofficial assurance of nature's stability, just as we have read that stability of forces in the ripples impressed in stone, or the rain marks on a long-vanished beach, or the unchanging laws of light in the eye of a four-hundred-million-year-old trilobite.

The nineteenth century was amazed when it discovered these things, but wasps and migratory birds were not. They had an old contract, an old promise, never broken till man began to interfere with things, that nature, in degree, is steadfast and continuous. Her laws do not deviate, nor the seasons come and go too violently. There is change, but throughout the past life alters with the slow pace of geological epochs. Calcium, iron, phosphorus, could exist in the jumbled world of the inorganic without the certainties that life demands. Taken up into a living system, however, *being* that system, they must, in a sense, have knowledge of the future. Tomorrow's rain may be important, or tomorrow's wind or sun. Life, in contrast to the inorganic, is historic in a new way. It reflects the past, but must also expect something of the future. It has nature's promise—a guarantee that has not been broken in three billion years—that the universe has this queer rationality and "expectedness" about it. "Whatever interrupts the even flow and luxurious monotony of organic life," wrote Santayana, "is odious to the primeval animal."

This is a true observation, because on the more simple levels of life, monotony is a necessity for survival. The life in pond and thicket is not equipped for the storms that shake the human world. Its small domain is frequently confined to a splinter of sunlight, or the hole under a root. What life does under such circumstances, how it meets the precarious future (for even here the future can be precarious), is written into its substance by the obscure mechanisms of nature. The snail recoils into his house, the dissembling caterpillar who does not know he dissembles, thrusts stiffly, like a budding twig, from his branch. The enemy is known, the contingency prepared for. But still the dreaming comes from below, from somewhere in the molecular substance. It is as if nature in a thousand forms played games against herself, but the games were each one known, the rules ancient and observed.

It is with the coming of man that a vast hole seems to open in nature, a vast black whirlpool spinning faster and faster, consuming flesh, stones, soil, minerals, sucking down the lightning, wrenching power from the atom, until the ancient sounds of nature are drowned in the cacophony of something which is no longer nature, something instead which is loose and knocking at the world's heart, something demonic and no longer planned —escaped, it may be—spewed out of nature, contending in a final giant's game against its master.

Yet the coming of man was quiet enough. Even after he arrived, even after his strange retarded youth had given him the brain which opened up to him the dimensions of time and space, he walked softly. If, as was true, he had sloughed instinct away for a new interior world of consciousness,

he did something which at the same time revealed his continued need for
the stability which had preserved his ancestors. Scarcely had he stepped
across the border of the old instinctive world when he began to create the
world of custom. He was using reason, his new attribute, to remake, in
another fashion, a substitute for the lost instinctive world of nature. He
was, in fact, creating another nature, a new source of stability for his conflicting erratic reason. Custom became fixed: order, the new order imposed
by cultural discipline, became the "nature" of human society. Custom
directed the vagaries of the will. Among the fixed institutional bonds of
society man found once more the security of the animal. He moved in a
patient renewed orbit with the seasons. His life was directed, the gods had
ordained it so. In some parts of the world this long twilight, half in and
half out of nature, has persisted into the present. Viewed over a wide
domain of history this cultural edifice, though somewhat less stable than
the natural world, has yet appeared a fair substitute—a structure, like
nature, reasonably secure. But the security in the end was to prove an
illusion. It was in the West that the whirlpool began to spin. Ironically, it
began in the search for the earthly Paradise.

The medieval world was limited in time. It was a stage upon which the
great drama of the human Fall and Redemption was being played out.
Since the position in time of the medieval culture fell late in this drama,
man's gaze was not centered scientifically upon the events of an earth
destined soon to vanish. The ranks of society, even objects themselves,
were Platonic reflections from eternity. They were as unalterable as the
divine Empyrean world behind them. Life was directed and fixed from
above. So far as the Christian world of the West was concerned, man was
locked in an unchanging social structure well nigh as firm as nature. The
earth was the center of divine attention. The ingenuity of intellectual men
was turned almost exclusively upon theological problems.

As the medieval culture began to wane toward its close, men turned
their curiosity upon the world around them. The era of the great voyages,
of the breaking through barriers, had begun. Indeed, there is evidence that
among the motivations of those same voyagers, dreams of the recovery of
the earthly Paradise were legion. The legendary Garden of Eden was
thought to be still in existence. There were stories that in this or that far
land, behind cloud banks or over mountains, the abandoned Garden still
survived. There were speculations that through one of those four great
rivers which were supposed to flow from the Garden, the way back might
still be found. Perhaps the angel with the sword might still be waiting at
the weed-grown gateway, warning men away; nevertheless, the idea of that
haven lingered wistfully in the minds of captains in whom the beliefs of the
Middle Ages had not quite perished.

There was, however, another, a more symbolic road into the Garden.
It was first glimpsed and the way to its discovery charted by Francis

Bacon. With that act, though he did not intend it to be so, the philosopher had opened the doorway of the modern world. The paradise he sought, the dreams he dreamed, are now intermingled with the count-down on the latest model of the ICBM, or the radioactive cloud drifting downwind from a megaton explosion. Three centuries earlier, however, science had been Lord Bacon's road to the earthly Paradise. "Surely," he wrote in the *Novum Organum*, "it would be disgraceful if, while the regions of the material globe, that is, of the earth, of the sea, and of the stars—have been in our times laid widely open and revealed, the intellectual globe should remain shut up within the narrow limits of the old discoveries."

Instead, Bacon chafed for another world than that of the restless voyagers. "I am now therefore to speak touching Hope," he rallied his audience, who believed, many of them, in a declining and decaying world. Much, if not all, that man lost in his ejection from the earthly Paradise might, Bacon thought, be regained by application, so long as the human intellect remained unimpaired. "Trial should be made," he contends in one famous passage, "whether the commerce between the mind of men and the nature of things . . . might by any means be restored to its perfect and original condition, or if that may not be, yet reduced to a better condition than that in which it now is." To the task of raising up the new science he devoted himself as the bell ringer who "called the wits together."

Bacon was not blind to the dangers in his new philosophy. "Through the premature hurry of the understanding," he cautioned, "great dangers may be apprehended . . . against which we ought even now to prepare." Out of the same fountain, he saw clearly, could pour the instruments of beneficence or death.

Bacon's warning went unheeded. The struggle between those forces he envisaged continues into the modern world. We have now reached the point where we must look deep into the whirlpool of the modern age. Whirlpool or flight, as Max Picard has called it, it is all one. The stability of nature on the planet—that old and simple promise to the living, which is written in every sedimentary rock—is threatened by nature's own product, man.

Not long ago a young man—I hope not a forerunner of the coming race on the planet—remarked to me with the colossal insensitivity of the new asphalt animal, "Why can't we just eventually kill off everything and live here by ourselves with more room? We'll be able to synthesize food pretty soon." It was his solution to the problem of overpopulation.

I had no response to make, for I saw suddenly that this man was in the world of the flight. For him there was no eternal, nature did not exist save as something to be crushed, and that second order of stability, the cultural world, was, for him, also ceasing to exist. If he meant what he said, pity had vanished, life was not sacred, and custom was a purely useless impediment from the past. There floated into my mind the penetrating statement of a modern critic and novelist, Wright Morris. "It is not fear of the bomb that paralyzes us," he writes, "not fear that man has no future. Rather, it is the *nature* of the future, not its extinction, that produces such foreboding in the artist. It is a numbing apprehension that such future as man has may dispense with art, with man as we now know him, and such as art has made him. The survival of men who are strangers to the nature of this conception is a more appalling thought than the extinction of the species."

There before me stood the new race in embryo. It was I who fled. There was no means of communication sufficient to call across the roaring cataract that lay between us, and down which this youth was already figuratively passing toward some doom I did not wish to see. Man's second rock of certitude, his cultural world, that had gotten him out of bed in the morning for many thousand years, that had taught him manners, how to love, and to see beauty, and how, when the time came, to die—this cultural world was now dissolving even as it grew. The roar of jet aircraft, the ugly ostentation of badly designed automobiles, the clatter of the supermarkets could not lend stability nor reality to the world we face.

Before us is Bacon's road to Paradise after three hundred years. In the medieval world, man had felt God both as exterior lord above the stars, and as immanent in the human heart. He was both outside and within, the true hound of Heaven. All this alters as we enter modern times. Bacon's world to explore opens to infinity, but it is the world of the outside. Man's whole attention is shifted outward. Even if he looks within, it is largely with the eye of science, to examine what can be learned of the personality, or what excuses for its behavior can be found in the darker, ill-lit caverns of the brain.

The western scientific achievement, great though it is, has not concerned itself enough with the creation of better human beings, nor with self-discipline. It has concentrated instead upon things, and assumed that the good life would follow. Therefore it hungers for infinity. Outward in that infinity lies the Garden the sixteenth-century voyagers did not find. We no longer call it the Garden. We are sophisticated men. We call it, vaguely, "progress," because the word in itself implies the endless movement of pursuit. We have abandoned the past without realizing that without the past the pursued future has no meaning, that it leads, as Morris has anticipated, to the world of artless, dehumanized man.

III

Some time ago there was encountered, in the litter of a vacant lot in a small American town, a fallen sign. This sign was intended to commemorate the names of local heroes who had fallen in the Second World War. But that war was over, and another had come in Korea. Probably the population of that entire town had turned over in the meantime. Tom and Joe and Isaac were events of the past, and the past of the modern world is short. The names of yesterday's heroes lay with yesterday's torn newspaper. They had served their purpose and were now forgotten.

This incident may serve to reveal the nature of what has happened, or seems to be happening, to our culture, to that world which science was to beautify and embellish. I do not say that science is responsible except in the sense that men are responsible, but men increasingly are the victims of what they themselves have created. To the student of human culture, the rise of science and its dominating role in our society presents a unique phenomenon.

Nothing like it occurs in antiquity, for in antiquity nature represented the divine. It was an object of worship. It contained mysteries. It was the mother. Today the phrase has disappeared. It is nature we shape, nature, without the softening application of the word mother, which under our control and guidance hurls the missile on its path. There has been no age in history like this one, and men are increasingly brushed aside who speak of the possibility of another road into the future.

Some time ago, in a magazine of considerable circulation, I spoke about the role of love in human society, and about pressing human problems which I felt, rightly or wrongly, would not be solved by the penetration of space. The response amazed me, in some instances, by its virulence. I was denounced for interfering with the colonization of other planets, and for corruption of the young. Most pathetically of all, several people wrote me letters in which they tried to prove, largely for their own satisfaction, that love did not exist, that parents abused and murdered their children and children their parents. They concentrated upon sparse incidents of pathological violence, and averted their eyes from the normal.

It was all too plain that these individuals were seeking rationalizations behind which they might hide from their own responsibilities. They were in the whirlpool, that much was evident. But so are we all. In 1914 the London *Times* editorialized confidently that no civilized nation would bomb open cities from the air. Today there is not a civilized nation on the face of the globe that does not take this aspect of warfare for granted. Technology demands it. In Kierkegaard's deadly future man strives, or rather ceases to strive, against himself.

But crime, moral deficiencies, inadequate ethical standards, we are prone to accept as part of the life of man. Why, in this respect, should we be regarded as unique? True, we have had Buchenwald and the Arctic slave camps, but the Romans had their circuses. It is just here, however, that the uniqueness enters in. After the passage of three hundred years from Bacon and his followers—three hundred years on the road to the earthly Paradise—there is a rising poison in the air. It crosses frontiers and follows the winds across the planet. It is man-made; no treaty of the powers has yet halted it.

Yet it is only a symbol, a token of that vast maelstrom which has caught up states and stone-age peoples equally with the modern world. It is the technological revolution, and it has brought three things to man which it has been impossible for him to do to himself previously.

First, it has brought a social environment altering so rapidly with technological change that personal adjustments to it are frequently not viable. The individual either becomes anxious and confused or, what is worse, develops a superficial philosophy intended to carry him over the surface of life with the least possible expenditure of himself. Never before in history has it been literally possible to have been born in one age and to die in another. Many of us are now living in an age quite different from the one into which we were born. The experience is not confined to a ride in a buggy, followed in later years by a ride in a Cadillac. Of far greater significance are the social patterns and ethical adjustments which have followed fast upon the alterations in living habits introduced by machines.

Second, much of man's attention is directed exteriorly upon the machines which now occupy most of his waking hours. He has less time alone than any man before him. In dictator-controlled countries he is harangued and stirred by propaganda projected upon him by machines to which he is forced to listen. In America he sits quiescent before the flickering screen in the living room while horsemen gallop across an American wilderness long vanished in the past. In the presence of so compelling an instrument, there is little opportunity in the evenings to explore his own thoughts or to participate in family living in the way that the man from the early part of the century remembers. For too many men, the exterior world with its mass-produced daydreams has become the conqueror. Where are the eager listeners who used to throng the lecture halls; where are the workingmen's intellectual clubs? This world has vanished into the whirlpool.

Third, this outward projection of attention, along with the rise of a science whose powers and creations seem awe-inspiringly remote, as if above both man and nature, has come dangerously close to bringing into existence a type of man who is not human. He no longer thinks in the old terms; he has ceased to have a conscience. He is an instrument of power. Because his mind is directed outward upon this power torn from nature, he does not realize that the moment such power is brought into the human domain it partakes of human freedom. It is no longer safely *within* nature; it has become violent, sharing in human ambivalence and moral uncertainty.

At the same time that this has occurred, the scientific worker has frequently denied personal responsibility for the way his discoveries are used. The scientist points to the evils of the statesmen's use of power. The statesmen shrug and remind the scientist that they are encumbered with monstrous forces that science has unleashed upon a totally unprepared public. But there are few men on either side of the Iron Curtain able to believe themselves in any sense personally responsible for this situation. Individual conscience lies too close to home, and is archaic. It is better, we subconsciously tell ourselves, to speak of inevitable forces beyond human control. When we reason thus, however, we lend powers to the whirlpool; we bring nearer the future which Kierkegaard saw, not as the *necessary* future, but one just as inevitable as man has made it.

IV

We have now glimpsed, however briefly and inadequately, the fact that modern man is being swept along in a stream of things, giving rise to other things, at such a pace that no substantial ethic, no inward stability, has been achieved. Such stability as survives, such human courtesies as remain, are the remnants of an older Christian order. Daily they are attenuated. In the name of mass man, in the name of unionism, for example, we have seen violence done and rudeness justified. I will not argue the justice or injustice of particular strikes. I can only remark that the violence to which I refer has been the stupid, meaningless violence of the rootless, nonhuman members of the age that is close to us. It is the asphalt man who defiantly votes the convicted labor boss back into office and who says: "He gets me a bigger pay check. What do I care what he does?" This is a growing aspect of modern society that runs from teen-age gangs to the corporation boards of amusement industries that deliberately plan the further debauchment of public taste.

It is, unfortunately, the "ethic" of groups, not of society. It cannot replace personal ethic or a sense of personal responsibility for society at large. It is, in reality, group selfishness, not ethics. In the words of Max Picard, "Spirit has been divided, fragmented; here is a spirit belonging to this and to that sociological group, each group having its own peculiar little spirit, exactly what one needs in the Flight, where, in order to flee more easily, one breaks the whole up into parts; and as always happens when one separates the part from the whole . . . one magnifies the tiny part, making it ridiculously important, so that no one may notice that the tiny part is not the whole."

All over the world this fragmentation is taking place. Small nationalisms, as in Cyprus or Algiers, murder in the name of freedom. In America, child gangs battle in the streets. The group ethic as distinct from personal ethic is faceless and obscure. It is whatever its leaders choose it to mean; it destroys the innocent and justifies the act in terms of the future. In

Russia this has been done on a colossal scale. The future is no more than the running of the whirlpool. It is not divinely ordained. It has been wrought by man in ignorance and folly. That folly has two faces; one is our secularized conception of progress; the other is the mass loss of personal ethic as distinguished from group ethic.

It would be idle to deny that progress has its root in the Christian ethic, or that history, viewed as progression toward a goal, as unique rather than cyclic, is also a product of Christian thinking. There is a sense in which one can say that man entered into history through Christianity, for as Berdyaev somewhere observes, it is this religion, par excellence, which took God out of nature and elevated man above nature. The struggle for the realization of the human soul, the attempt to lift it beyond its base origins, became, in the earlier Christian centuries, the major preoccupation of the Church.

When science developed, in the hands of Bacon and his followers, the struggle for progress ceased to be an interior struggle directed toward the good life in the soul of the individual. Instead, the enormous success of the experimental method focused attention upon the power which man could exert over nature. Now he found, through Bacon's road back to the Garden, that he could share once more in that fruit of the legendary tree. With the rise of industrial science, "progress" became the watch-word of the age, but it was a secularized progress. It was the increasing whirlpool of goods, cannon, bodies and yielded-up souls that an outward concentration upon the mastery of material nature was sure to bring.

Let us admit at once that the interpretation of secular progress is two-sided. If this were not so, men would more easily recognize their dilemma. Science has brought remedies for physical pain and disease; it has opened out the far fields of the universe. Gross superstition and petty dogmatism have withered under its glance. It has supplied us with fruits unseen in nature, and given an opportunity, has told us dramatically of the paradise that might be ours if we could struggle free of ancient prejudices that still beset us. No man can afford to ignore this aspect of science, no man can evade those haunting visions.

It is the roar of the whirlpool, nonetheless, that breaks now most constantly upon our listening ears, increasingly instructing us upon the most important aspect of progress—that which in secularizing the concept we have forgotten. Its sound marks the dangerous near-dissolution of man's second nature, custom. Ideas, heresies, run like wildfire and death over the crackling static of the air. They no longer pick their way slowly through the experience of generations. Tax burdens multiply and reach upward year by year as man pays for his engines of death and underwrites ever more wearily the cost of the "progress" to which this road has led him. There is no retreat. The great green forest that once surrounded us Americans and behind which we could seek refuge has been consumed.

And thus, though more symbolically, has it been everywhere for man. We have re-entered nature, not like a Greek shepherd on a hillside hearing joyfully the returning pipes of Pan, but rather as an evil and precocious animal who slinks home in the night with a few stolen powers. The serenity of the gods is not disturbed. They know well on whose head the final lightning will fall.

Progress secularized, progress which pursues only the next invention, progress which pulls thought out of the mind and replaces it with idle slogans, is not progress at all. It is a beckoning mirage in a desert over which stagger the generations of men. Because man, each individual man among us, possesses his own soul and by that light must live or perish, there is no way by which Utopias—or the lost Garden itself—can be bought out of the future and presented to man. Neither can he go forward to such a destiny. Since in the world of time every man lives but one life, it is in himself that he must search for the secret of the Garden. With the fading of religious emphasis and the growth of the torrent, modern man is confused. The tumult without has obscured those voices that still cry desperately to man from somewhere within his consciousness.

V

One hundred years ago last autumn, Charles Darwin published the *Origin of Species*. Epic of science though it is, it was a great blow to man. Earlier, man had seen his world displaced from the center of space; he had seen the Empyrean heaven vanish to be replaced by a void filled only with the wandering dust of worlds; he had seen earthly time lengthen until man's duration within it was only a small whisper on the sidereal clock. Finally, now, he was to be taught that his trail ran backward until in some lost era it faded into the night-world of the beast. Because it is easier to look backward than to look forward, since the past is written in the rocks, this observation, too, was added to the whirlpool.

"I am an animal," man considered. It was a fair judgment, an outside judgment. Man went into the torrent along with the steel of the first iron-clads and a new slogan, "the survival of the fittest." There would be one more human retreat when, in the twentieth century, human values themselves would fall under scrutiny and be judged relative, shifting and uncertain. It is the way of the torrent—everything touched by it begins to circle without direction. All is relative, there is nothing fixed, and of guilt there can, of course, remain but little. Moral responsibility has difficulty in existing consistently beside the new scientific determinism.

I remarked at the beginning of this discussion that the "within," man's subjective nature, and the things that come to him from without often bear a striking relationship. Man cannot be long studied as an object without his cleverly altering his inner defenses. He thus becomes a very difficult creature with which to deal. Let me illustrate this concretely.

Not long ago, hoping to find relief from the duties of my office, I sought refuge with my books on a campus bench. In a little while there sidled up to me a red-faced derelict whose parboiled features spoke eloquently of his particular weakness. I was resolved to resist all blandishments for a handout.

"Mac," he said, "I'm out of a job. I need help."

I remained stonily indifferent.

"Sir," he repeated.

"Uh huh," I said, unwillingly looking up from my book.

"I'm an alcoholic."

"Oh," I said. There didn't seem to be anything else to say. You can't berate a man for what he's already confessed to you.

"I'm an alcoholic," he repeated. "You know what that means? I'm a sick man. Not giving me alcohol is ill-treating a sick man. I'm a sick man. I'm an alcoholic. I have to have a drink. I'm telling you honest. It's a disease. I'm an alcoholic. I can't help myself."

"Okay," I said, "you're an alcoholic." Grudgingly I contributed a quarter to his disease and his increasing degradation. But the words stayed in my head. "I can't help myself." Let us face it. In one disastrous sense, he was probably right. At least at this point. But where had the point been reached, and when had he developed this clever neo-modern, post-Freudian panhandling lingo? From what judicious purloining of psychiatric or social-work literature and lectures had come these useful phrases?

And he had chosen his subject well. At a glance he had seen from my book and spectacles that I was susceptible to this approach. I was immersed in the modern dilemma. I could have listened, gazing into his mottled face, without an emotion if he had spoken of home and mother. But he was an alcoholic. He knew it and he guessed that I might be a scientist. He had to be helped from the outside. It was not a moral problem. He was ill.

I settled uncomfortably into my book once more, but the phrase stayed with me, "I can't help myself." The clever reversal. The outside judgment turned back and put to dubious, unethical use by the man inside.

"I can't help myself." It is the final exteriorization of man's moral predicament, of his loss of authority over himself. It is the phrase that, above all others, tortures the social scientist. In it is truth, but in it also is a dreadful, contrived folly. It is society, a genuinely sick society, saying to its social scientists, as it says to its engineers and doctors: "Help me. I'm rotten with hate and ignorance that I won't give up, but you are the doctor; fix me." This, says society, is *our* duty. We are social scientists. Individuals, poor blighted specimens, cannot assume such responsibilities. "True, true," we mutter as we read the case histories. "Life is dreadful, and yet—"

Man on the inside is quick to accept scientific judgments and make use of them. He is conditioned to do this. This new judgment is an easy one; it deadens man's concern for himself. It makes the way into the whirlpool easier. In spite of our boasted vigor we wait for the next age to be brought to us by Madison Avenue and General Motors. We do not prepare to go there by means of the good inner life. We wait, and in the meantime it slowly becomes easier to mistake longer cars or brighter lights for progress. And yet—

Forty thousand years ago in the bleak uplands of southwestern Asia, a man, a Neanderthal man, once labeled by the Darwinian proponents of struggle as a ferocious ancestral beast—a man whose face might cause you some slight uneasiness if he sat beside you—a man of this sort existed with a fearful body handicap in that ice-age world. He had lost an arm. But still he lived and was cared for. Somebody, some group of human things, in a hard, violent and stony world, loved this maimed creature enough to cherish him.

And looking so, across the centuries and the millennia, toward the animal men of the past, one can see a faint light, like a patch of sunlight moving over the dark shadows on a forest floor. It shifts and widens, it winks out, it comes again, but it persists. It is the human spirit, the human soul, however transient, however faulty men may claim it to be. In its coming man had no part. It merely came, that curious light, and man, the animal, sought to be something that no animal had been before. Cruel he might be, vengeful he might be, but there had entered into his nature a curious wistful gentleness and courage. It seemed to have little to do with survival, for such men died over and over. They did not value life compared to what they saw in themselves—that strange inner light which has come from no man knows where, and which was not made by us. It has followed us all the way from the age of ice, from the dark borders of the ancient forest into which our footprints vanish. It is in this that Kierkegaard glimpsed the eternal, the way of the heart, the way of love which is not of today, but is of the whole journey and may lead us at last to the end. Through this, he thought, the future may be conquered. Certainly it is true. For man may grow until he towers to the skies, but without this light he is nothing, and his place is nothing. Even as we try to deny the light, we know that it has made us, and what we are without it remains meaningless.

We have come a long road up from the darkness, and it well may be— so brief, even so, is the human story—that viewed in the light of history, we are still uncouth barbarians. We are potential love animals, wrenching and floundering in our larval envelopes, trying to fling off the bestial past. Like children or savages, we have delighted ourselves with technics. We have thought they alone might free us. As I remarked before, once launched on this road, there is no retreat. The whirlpool can be conquered, but only by placing it in proper perspective. As it grows, we must learn to cultivate that which must never be permitted to enter the maelstrom— ourselves. We must never accept utility as the sole reason for education. If all knowledge is of the outside, if none is turned inward, if self-awareness fades into the blind acquiescence of the mass man, then the personal responsibility by which democracy lives will fade also.

Schoolrooms are not and should not be the place where man learns only scientific techniques. They are the place where selfhood, what has been called "the supreme instrument of knowledge" is created. Only such deep inner knowledge truly expands horizons and makes use of technology; not for power, but for human happiness. As the capacity for self-awareness is intensified, so will return that sense of personal responsibility which has been well-nigh lost in the eager yearning for aggrandizement of the asphalt man. The group may abstractly desire an ethic, theologians may preach an ethic, but no group ethic ever could, or should replace the personal ethic of individual, responsible men. Yet it is just this which the Marxist countries are seeking to destroy; and we, in a vague, good-natured indifference, are furthering its destruction by our concentration upon material enjoyment and our expressed contempt for the man who thinks, to our mind, unnecessarily.

Let it be admitted that the world's problems are many and wearing, and that the whirlpool runs fast. If we are to build a stable cultural structure above that which threatens to engulf us by changing our lives more rapidly than we can adjust our habits, it will only be by flinging over the torrent a structure as taut and flexible as a spider's web, a human society deeply self-conscious and undeceived by the waters that race beneath it, a society more literate, more appreciative of human worth than any society that has previously existed. That is the sole prescription, not for survival—which is meaningless—but for a society worthy to survive. It should be, in the end, a society more interested in the cultivation of noble minds than in change.

There is a story about one of our great atomic physicists—a story for whose authenticity I cannot vouch, and therefore I will not mention his name. I hope, however, with all my heart that it is true. If it is not, then it ought to be, for it illustrates well what I mean by a growing self-awareness, a sense of responsibility about the universe.

This man, one of the chief architects of the atomic bomb, so the story runs, was out wandering in the woods one day with a friend when he came upon a small tortoise. Overcome with pleasurable excitement, he took up the tortoise and started home, thinking to surprise his children with it. After a few steps he paused and surveyed the tortoise doubtfully.

"What's the matter?" asked his friend.

Without responding, the great scientist slowly retraced his steps as precisely as possible, and gently set the turtle down upon the exact spot from which he had taken him up.

Then he turned solemnly to his friend. "It just struck me," he said, "that perhaps, for one man, I have tampered enough with the universe." He turned, and left the turtle to wander on its way.

The man who made that remark was one of the best of the modern men, and what he had devised had gone down into the whirlpool. "I have tampered enough," he said. It was not a denial of science. It was a final recognition that science is not enough for man. It is not the road back to the waiting Garden, for that road lies through the heart of man. Only when man has recognized this fact will science become what it was for Bacon, something to speak of as "touching upon Hope." Only then will man be truly human.

Conclusion

Comments on the passing of the Picture People.[1]

"There is something wrong with our world-view," Loren Eiseley tells us. "It is still Ptolemaic, though the sun is no longer believed to revolve around the earth.

"We teach the past, we see farther backward into time than any race before us, but we stop at the present, or, at best, we project idealized versions of ourselves. All that long way behind us we see, perhaps inevitably, through human eyes alone. We see ourselves as the culmination and the end, and if we do indeed consider our passing, we think that sunlight will go with us and the earth be dark. We are the end. For us continents rose and fell, for us the waters and the air were mastered, for us the great living web has pulsated and grown more intricate.

"To deny this, a man once told me, is to deny God. This puzzled me."[2] And with some reason, for Darwin saw clearly that the succession of life on this planet has not been a formal pattern imposed from without, or moving exclusively in one direction.[3] "It is quite apparent, therefore, that there is an aspect of Darwin's discoveries which has never penetrated to the mind of the general public. It is the fact that once undirected variation and natural selection are introduced as the mechanism controlling the development of plants and animals, the evolution of every world in space becomes a series of unique historical events."[4]

And why has this aspect of the world into which Darwin led us never yet been widely acknowledged? What is the bitter truth, as Nogar puts it, that is so hard to accept? Simply this: That a deeply penetrating look at the unfolding cosmos and the history of man reveals a world at odds with

---

[1] "I realized," wrote Nogar (*The Lord of the Absurd* [St. Louis: B. Herder, 1966], pp. 16–17), "that there were two modes of human existence, two kinds of people: *picture people* and *drama people*, depending upon how they looked upon life. The first looked for a picture of things in order, a cosmic and personal world-view; the second looked for a drama unfolding. The first were restless until they felt that they had grasped something timeless, some eternal verity, and had placed themselves securely into changeless order. The second were impatient and distrustful until they found themselves in the space-time story of reality unfolding, a drama in which they had a part. They wanted to be involved; dynamically identified with all reality. But the picture people wanted just the opposite: detachment, escape from the effort and effect of the sorrows and joys of biographical and temporal existence."

[2] Loren Eiseley, *The Immense Journey* (London: Victor Gollancz, 1958), p. 57.

[3] *Ibid.*, p. 160.

[4] *Ibid.*, pp. 157–158.

itself and a mankind in a desperate situation lacking every possible remedy.[5]

If we are to have any hope of collectively facing mankind's destiny intelligently, of shouldering man's responsibility for man without being crushed by the burden, we shall have to learn to act in terms of evidence—"to engage in non-wishful thinking," as Adler puts it—far more constantly and broadly than have any previous societies.

To this end, it is high time that we begin to take seriously what scientific experience has long since testified, but what has only just begun to be acknowledged in philosophy, and as yet not at all in theology, namely, that the evolution of life, and the emergence of man, is a natural process in which chance, failure, waste, disorder and death may ultimately prevail. No use to rely on a supposed cosmic order (and which cosmic order is the truly "natural" one—the muskrat's or the man's[6]), or to "let things take care of themselves" under the divine governance of the everyday and the divine providence of the long haul.

For if in the world's evolution "there is no scenario prepared in advance," so that "we must purge our thought of this spurious idea of a play written in advance, in a time anterior to time—a play in which time unfolds, and the characters of time read, the parts," in other words, "if everything is adventure and improvisation . . . the plan of which is fixed . . . simultaneously with the entire unfolding of time—then it is indeed necessary that in the essential contingency of the spectacle thus immutably ordained and directed by God, the characters of the drama [be they inorganic, biological, or human,] have also their share of initiative."[7]

---

[5] Raymond Nogar, *The Lord of the Absurd*, p. 143. "The reality of chance, unpredictability, disorder, waste, frustration and absurdity interwoven throughout the space-time unfolding of the universe and the human biography," wrote Nogar—and this was truly his final testament ("I began to see why my argument from dynamic order to an intelligent God" in *The Wisdom of Evolution*, pp. 387–397, "seemed so puerile to Dr. Simpson—*and why it really had no force for me either*")—"remains an important insight into the world of evolutionary humanism," that of Teilhard de Chardin and Robert Francoeur no less than that of Julian Huxley and G. Ledyard Stebbins. "For that world of scientific evolutionism is far too orderly; its expectation is overconfident. Its cosmos is too harmonious; its view is pasted together with far too many illusory extrapolations." "The failure of evolutionary humanism . . . is its superficial rationalism. Its contemplation is not sufficiently realistic; it doesn't go deep enough. It relies too heavily upon a neat picture of the cosmos, upon a superficial, almost romantic idea of human existence." (*Ibid.*, pp. 79–80: cf. Ch. VIII, "The Strange World of Father Teilhard," esp. pp. 117–126.)

[6] Pondering this question, and with a view to its realistic implications, Eiseley was moved to state a most profound issue: "In so many worlds, I thought, how natural is 'natural'—and is there anything we can call a natural world at all?" (*The Firmament of Time*, p. 157).

[7] Jacques Maritain, *God and the Permission of Evil*, trans. by Joseph Evans (Milwaukee: Bruce, 1966), p. 79: passim. Some remarks of Bergson are particularly arresting in this connection: "Evolution is not only a movement forward; in many cases we observe a marking-time, and still more often a deviation or turning back. It must be so, as we shall show further on, and the same causes that divide the evolution movement often cause life to be diverted from itself, hypnotized by the form it has just brought forth. Thence results an increasing disorder. No doubt there is progress, if progress means a continual advance in the general direction determined by a first impulsion; but this progress is accomplished only on the two or three great

It is not necessarily therefore a question of "denying God;" it is just that it may very well be that creation is truly a gift, so prodigal that "there is no logical reason for a snowflake any more than there is for evolution. It is an apparition from that mysterious shadow world beyond nature, that final world which contains—if anything contains—the explanation of men and catfish and green leaves."[8]

On the one hand, from the standpoint of dependency in being—that is, of metaphysics—evolution presents a spectacle the greatness and universality of which make the activating omnipresence of God more tellingly sensed by our minds than was generally possible for minds contemplating the cycles of the unchanging heavens. For it is the waste, mess, terror, and frustration built into a radically evolutionary and pluralist universe which, far more than order and design or any purposeful pattern with its appearance of self-sufficiency, brings us face to face with the mystery of contingent existence; and precisely "that being upon whom all contingency finally rests we call God."[9]

On the other hand, from the standpoint of science itself and that of natural philosophy, the available evolutionary evidences (and they are abundant) do not foster a particularly optimistic view of nature.

Every progress in evolution is dearly paid for: miscarried attempts, merciless struggle everywhere. The more detailed our knowledge of nature becomes, the more we see, together with the element of generosity and progression which radiates from being,

---

lines of evolution on which forms ever more and more complex, ever more and more high, appear; between these lines run a crowd of minor paths in which, on the contrary, deviations, arrests, and set-backs, are multiplied. The philosopher, who begins by laying down as a principle that each detail is connected with some general plan of the whole, goes from one disappointment to another as soon as he comes to examine the facts; and, as he had put everything in the same rank, he finds that, as the result of not allowing for accident, he must regard everything as accidental. For accident, then, an allowance must first be made, and a very liberal allowance. We must recognize that all is not coherent in nature. By so doing, we shall be led to ascertain the centers around which the incoherence crystallizes. This crystallization itself will clarify the rest; the main directions will appear, in which life is moving whilst developing the original impulse. True, we shall not witness the detailed accomplishment of a plan. Nature is more and better than a plan in course of realization. A plan is a term assigned to a labor: it closes the future whose form it indicates. Before the evolution of life, on the contrary, the portals of the future remain wide open. It is a creation that goes on forever in virtue of an initial movement. This movement constitutes the unity of the organized world—a prolific unity, of an infinite richness, superior to any that the intellect could dream of, for the intellect is only one of the aspects or products." (Bergson, *Creative Evolution*, pp. 115–116 of Modern Library ed.).
[8] Eiseley, *The Immense Journey*, p. 27.
[9] Cf. Nogar, *The Lord of the Absurd*, pp. 78–80. "Just pull the thread of contingency, the delicate temporal-spacial condition of the whole of it, and the imaginary cosmos [imaginary "in the ancient sense of a universe conceived as an orderly and harmonious system," since for science today, as Whitrow makes so clear, "the universe as a whole is but an hypothesis"] will unravel into nothingness. The universe, no matter how indefinite its endurance, is constructed of stuff that is ever in need of existential support, and no amount of order and dynamism can cover for long that vulnerable spot. Creatureliness can be hopeful, expectant of promise, only so long as the Creator remains in sight." (*Ibid.*, p. 80).

the law of degradation, the powers of destruction and death, the implacable voracity which are also inherent in the world of matter. And when it comes to man, surrounded and invaded as he is by a host of warping forces, psychology and anthropology are but an account of the fact that while being essentially superior to all of them, he is the most unfortunate of animals. So it is that when its vision of the world is enlightened by science, the intellect . . . realizes still better that nature, however good in its own order, does not suffice, and that if the deepest hopes of mankind are not destined to turn to mockery, it is because a God-given energy better than nature is at work in us.[10]

This, of course, was the essential contention of Teilhard de Chardin: "Either nature is closed to our demands for futurity, in which case thought, the fruit of millions of years of effort, is stifled, still-born in a self-abortive and absurd universe; or else an opening exists—that of the super-soul above our souls."[11]

But it should be clear to the reader that in concluding thus we are neither slipping off into a pseudo-philosophical mysticism nor advocating a view of human destiny as predetermined. We are simply taking up in the evolutionary context the traditional view that the human species has not just the rationality that other species lack, but a freedom not possessed by other species—the freedom to pursue a course of life to a self-appointed end and to pursue it through a free choice among means for reaching that end.[12]

What is certain is that if man has not a core of freedom, there is no reason for hope in the simple discovery that human nature is open to change, since there is nothing that would guarantee that evolutionary changes are always directed towards the good: some in fact are regressive, and the final one for most evolutionary lines is the biological-existential disaster of extinction.

Without a core of freedom, the behavior of the human species cannot be autonomously altered in its broad lines and basic tendencies; we would at bottom be dependent on the forces of evolution working to arrange everything for our benefit. Since, as Dobzhansky so well says, "there are no such forces, and in neither the biological nor in cultural evolution is betterment guaranteed,"[13] the passive wait for a world made human for man would be longer than the future of the species.

On the other hand, it is also certain that a realistic view of nature, just because it cannot be particularly optimistic, need not be completely

---

[10] Jacques Maritain, essay on "God and Science" in *On The Use of Philosophy* (New York: Atheneum, 1965), pp. 70–71.

[11] Teilhard de Chardin, *The Phenomenon of Man*, pp. 231–232.

[12] See Section III of Part II *supra*, The Moral Issues, esp. the reading by M. J. Adler on "The Consequences For Action."

[13] Theodosius Dobzhansky, "Man in the Light of Evolution," paper presented in Ottawa, Canada, on October 2, 1968 (cited from p. 8 of mimeographed text).

pessimistic. For however desperate our present condition, however beset we are by warping forces, if man has a core of freedom, his future, *pace* Marx and Darlington, cannot be mapped out in his genetic and economic systems. As Platt has well said, "over the long run, the intellectual or leader will be successful only if he leads, only if he is not a dictator but a counsellor, not the sole designer of the system but the seer and explainer of the consequences of doing things one way rather than another." [14] If man has a core of freedom, his future is outside the determined systems of culture as well as those of biology:

It has to do with the peculiar logical and causal status of the "self-knowledge" which characterizes any decision-system that belongs to our universe of interpersonal communication. Predict to a ball that it will fall, and it makes no difference to its falling; but predict to a man that he will fall, and he may take extra precautions to stay upright, or he may fall more absurdly to show it is true. Self-predictions of our conscious choices are not real predictions in the noninterfering sense like predicting rain—which is the only sense in which we can speak of determinism. By the same token, predictions of the behavior of a decision-system are not real predictions if they are communicated to the system. With full interpersonal communication, a decision-system ceases to be an "object" and becomes a "co-subject," with all the subjective and goal-directed freedom of choice that that implies. . . . This means that our whole society is in principle unpredictable as a deterministic object, except perhaps statistically, because interpersonal communication is precisely the basis on which a society is constructed. Society is capable of changing to a new course at any time because of some objectively unforeseeable act of insight or decision . . . —some act of leadership or violence or invention—that carries the whole society along with it.[15]

Our world today is a world of divided peoples and conflicting social systems. It extends the pattern of pre-atomic and even stone-age times, wherein each group defined humanity in terms of itself. The great revelation of the idea of evolution in this regard is that we can in principle transcend these destructive conflicts if we will but establish the means for building the sort of socio-cultural unity that our biological unity of itself

---

[14] John R. Platt, "The New Biology and the Shaping of the Future," in *The Great Ideas Today 1968*, ed. by R. M. Hutchins and M. J. Adler (Chicago: Britannica, 1968), p. 166.

[15] *Ibid.*, pp. 153–154. "This conclusion goes considerably beyond the concept of cybernetics, as developed by Norbert Wiener; that is, the concept of feedback which can guide goal-directed behavior in animate or inanimate systems. . . . It is the basic principle of teleological behavior, or action with 'purpose', and therefore it is one of the important ways in which biology goes far beyond physics and chemistry. Goal-directed feedback is our way of understanding the biological phenomenon of 'homeostasis', or stabilization responses, which Walter Canon emphasized, as well as the more important phenomenon of internally directed growth. . . . Such ideas make Aristotle's discussion of purpose in natural systems seem much more scientific and less objectionable than they seemed to nineteenth-century determinists." (*Ibid.*, p. 154. See the reading by C. H. Waddington, "The Shape of Biological Thought," in Section IV *supra*, The Metaphysical Issues, and esp. the Contextualizing Comments on this reading.)

calls for. The present age faces the choice: life in the old, increasingly dangerous patterns involving pollution of nature and subjugation of peoples, or international organization for survival and advance. "The decision of whether we live or die forever will be determined by the efforts of men in the next few years." [16]

In the evolutionary perspective, the imperative suddenness with which this time of decision has come and will pass is well illustrated by a comparison, first made by Julian Huxley and repeated by Platt, between the history of life and the height of St. Paul's Cathedral.

If we took the height of the cathedral, over 300 feet, as representing life's 2–3 billion year history, then each 10 million year period would be represented by 1 foot. A two-inch block on the roof of the cathedral would represent man's two million years, while a postage stamp atop that block would by its thickness represent the twenty thousand years since man's discovery of agriculture. The four hundred years of mathematical interpretation of nature structuring the rise of modern science and industry is represented by the thickness of the ink on the stamp; while the thirty to fifty years from the introduction of the atomic and space age to the time when we must achieve a stable world organization of peoples or perish through conflict would be represented by the film of moisture on the ink!

If the need for informed decision is pressing, and if it is true that the specifically human interaction system depends on interpersonal communication for pursuit of the human good, then it follows that the task of modern thought and of evolutionary thinkers is pre-eminently the task of dialogue, dialogue with the thinkers of the past and the researchers of the present. Only in terms of cumulative wisdom can we hope to recognize the requirements of an environment for man and discern the means for making the world human.

In a famous passage of his *Nicomachean Ethics* Aristotle pointed out that before a man reaches a hopeless condition of health or character, it is ordinarily possible for him to pursue a different course:

To the unjust and to the self-indulgent man it was open at the beginning not to become men of this kind, and so they are unjust and self-indulgent voluntarily; but now that they have become so it is not possible for them not to be so. But not only are the vices of the soul voluntary, but sometimes those of the body also . . .; while no one blames those who are ugly by nature, we blame those who are so owing to want of exercise and care. So it is too with respect to weakness and infirmity. [17]

An evolutionary humanism must take up this insight and develop it beyond its application to individuals by pointing to its implications for mankind as a whole. Since mankind as a species is now known to be caught

[16] *Ibid.*, p. 130.
[17] Bk. III, ch 5, 1114a 20–25.

up in an ongoing process of change—man being precisely "evolution become conscious of itself"—it follows that even as the human individual must through his personal choices and on his own responsibility weave the moral fabric of his character, or in the end lose the ability to make himself; so also the human species must accept responsibility for determining the quality of the environment in which its development takes place, or lose the ability to dominate the conditions of its survival.

Just as a slum environment can undermine the autonomous growth of the individual personality, so we may eventually allow such planetary conditions to accumulate as will effectively undermine the potentialities of our entire species. As individual men have always had to face the crisis of maturation wherein growth and change were inevitable but would be for the better only if mediated by knowledgeable personal choices, so today does mankind as a species face a crisis of maturation wherein continued change is inevitable, but will be for the better only through choices based on a degree of insight into our place in nature and a degree of willingness to undertake responsible collective actions, higher than anything past or present world leadership would lead us to expect as forthcoming.

Whatever shape our world may take in the next generation or in the next ten generations, for post-Darwinian man there will be no escape from responsibility. With man evolution has passed from a drift to a conscious destiny. We now know that it is we who are responsible for shaping the future. We have passed from drift to choice; and even if our choice shall be to continue drifting, it remains our choice.

Let it be admitted that the world's problems are many and wearing, and that the whirlpool runs fast. If we are to build a stable cultural structure above that which threatens to engulf us by changing our lives more rapidly than we can adjust our habits, it will only be by flinging over the torrent a structure as taut and flexible as a spider's web, a human society deeply self-conscious and undeceived by the waters that race beneath it, a society more literate, more appreciative of human worth than any society that has previously existed. That is the sole prescription, not for survival —which is meaningless—but for a society worthy to survive. It should be, in the end, a society . . . interested in the cultivation of noble minds. . . .[18]

It is this increasingly clear realization of being left in the hands of their own counsel, caught up the while in an inexorable process of change which can be guided for the better but is not widely being so, that is reflected in the fact that everywhere in America and Europe, yes, and in Africa and Asia too, the spirit of man is speaking strange new words— words about the illusions of former philosophies, the uselessness of

[18] Loren Eiseley, "How Human Is Man?", ch. V of *The Firmament of Time* (New York: Atheneum, 1960), p. 147.

cultural baggage, the irrelevance of old world-pictures and set ways of the value systems of yesterday. Post-Darwinian men are being forced by the very pressure of daily events to pass from naive confidence to doubt and denial.

Agnosticism and atheism are not just name-tags for a handful of renegades, they are essential conditions of the spirit of our times (and is not the Christian revelation of the God man kills?[19]). Men are speaking, and speaking men are existing men. They are speaking of a new search for man, one that is not rooted in the moral commitments of this or that culture group and imposed on the rest of men, one that is not irretrievably set in esoteric myths, but rather of one that is rooted in the free movement of the human spirit. It must be a search which begins and ends here.

Whatever reality there is must be the reality which responds to the free, creative, and personal spirit of man. His relation to reality, to himself, to other men, to God, must be authentically human and spiritual in this sense. The static, impersonal pictures and world-views of former times must go. The sun has set, at least temporarily, on the day of the picture people. Without accusation of betrayal or bitter repudiation of the past,

---

[19] "How can I be so sure that God is dead? How do I know that it is not just a superficial passing of the *name* of God? Just the death of a superstitious understanding foisted upon us by tribal traditions? No, it is the death of the Christian God, One who is now, in our age, absolutely, totally and irrevocably dead! By what right have I to say *how* he was killed, to describe the method in such detail? I'll tell you how I can do all these things: *because I was there! Because I killed Him!*

"I am a man, and man has ever been about killing his God. He has to destroy illusion in order to survive, to live with himself, to hope in anything. He cannot bear to live a lie, to see himself in the mirror of pure fiction. He must ask himself the question daily: is it meaningful to live at all? Why should I bother to face my tomorrow? That is why he is ever destroying something of himself. Self-destruction is, at base, a desire to root out illusion. So a man destroys his God.

"It is true that I killed God, in the real sense of cutting Him down. But it has been in the name of pruning away illusions. Let us not be deceived. Our existence just doesn't find God relevant. There seems to be no place for Him in our organization today. He had to go. Man will not play junior partner in a firm he thinks he can run better himself. That is why I killed God.

"However . . . I have said that man is forever killing his God. But this could not be *if his God were not forever rising again*, like the phoenix out of the ashes of the last cremation. The Jews killed the tribal gods before them, only to fall down before the face of the God who rose again, this time to lead them personally out of the desert into the promised land. Yet they immediately got busy to plot His death. There were idolaters, unfaithful Judases all through the history of that Journey. The chosen people killed the messengers of God, the prophets. And out of that mayhem arose a new God—or was it a new God, really?

"That God claimed to be the God of always, this time come into the very history of man, as Man Himself. He exposed the false god of law and revealed the true God of love. He became the Christian God and His name was Christ. You say that the Christian God is dead. I say that this is true, we have killed Him. Yet we have just killed Him again. Has he not revealed Himself as *the God man kills?* Yes, we killed Him then and we have killed Him again today. And I have no doubt but what the future holds a killing more horrible than ever before. . . . But I have no doubt, either, that this will only mean another Easter. . . .

"That is why I laugh." (From "Laughter and the Death of God," part of R. J. Nogar's unfinished *Journal of a Clown.*)

without promise of permanence in the future, and with a deep sense of the risk involved in evolution, we are witnessing a new dawn, the day of the drama people, and of the planetary dialogue.

For many reasons, some of the most basic of which (thanks to man's animal heritage) come to light only in the dimension of depth psychology, the task of dialogue is an arduous task. The neat dismissal of differences by dilettantes and superficial "ecumenists" no less shirks this task than does the unremitting "pedagogical" labors of those who pander an ideology —be it religious or social.

It has been the purpose of the co-fashioners of this volume to take up the modern task, and to do so over *the* issue of contemporary reflection: the "meaning" for man of a world in evolution. What are the authentic implications of the freshly acquired realization of the brevity, relativity, and dependency of human existence upon a peculiar set of segmental laws of time and space—which laws are themselves only special cases of more general time-space laws which we will apprehend, it is certain, no more than fitfully?

This has been to ask what are the key issues raised by evolutionary thought crucial for all areas of humanistic thought—a large order indeed. We have tried to fill it, at least virtually, by arranging readings in a way which seems to respect the natural structure and development of the full evolutionary problematic. The point most worthy of note is that the time for theologians and philosophers to question whether evolution is a fact now demonstrated has gone. What is thrust upon all thinking people is that we have entered upon a new world-view, based on root metaphors which are time-centered and space-centered: to miss this is to miss the communication of an era which has redefined all philosophy, all theology, all Christianity, all concern for man or *humanism,* in terms of the secular (*in saeculo*), mundane (*in mundo*), "profane" or everyday setting of the human condition.

Thus we have entered the age of the round-table discussion, wherein men must seek to acknowledge every authentic insight into "the way things are," and recognize that no one and no "school" can determine in advance from what quarter such insight will spring. Such an attitude is not easy to maintain, and it is certain that no one ever achieves it integrally. Yet it is speaking with others which allows the egotism of the spirit to be cracked open and penetrated by the understanding and concern of another. This dialogue is the only hope of correcting our myopic visions, our illusions, and the essential untidiness of our thought. Speaking *with* others and not always *to* others, breaking out of a perpetual monologue, is the safety-valve against madness and dogmaticism, the pathology of playing endlessly the same broken record. Being alone, thinking alone, loving alone and speaking alone is very dangerous, and a man has to have an overriding excuse to be thus habitually occupied. *Genesis* revealed the mind of God on the point: "It is not good for man to be alone."

That is why we need the collage of the round-table, because the meaning of existence makes all the difference in man's design for the future; and no point of view has a monopoly on existential meaning.

III
Bibliography

Part I:

"Historical perspective": The Impact of Evolution on Scientific Method

A. Books

Adler, Mortimer J. *The Conditions of Philosophy*. New York: Atheneum, 1965.
————. *The Difference of Man and the Difference It Makes*. New York: Holt, Rinehart and Winston, 1967.
————. *St. Thomas and the Gentiles*. Milwaukee: Marquette, 1938.
————. *What Man Has Made of Man*. New York: Ungar, 1937.
Aquinas, Thomas. *In libros Aristotelis de caelo et mundo expositio*, ed. by R. M. Spiazzi. Rome: Marietti, 1952.
————. *In octos libros physicorum Aristotelis Expositio*, ed. by P. M. Maggiolo. Rome: Marietti, 1954.
————. *In Aristotelis duodecim libros metaphysicorum expositio*, ed. by M. -R. Cathala and R. Spiazzi. Rome: Marietti, 1950. English trans. by J. P. Rowan, *Commentary on the Metaphysics of Aristotle*. Chicago: Regnery, 1961, in 2 Vols.
————. *Summa contra gentiles* (Textus Leoninus). Rome: Herder, 1934. English trans. by A. C. Pegis, J. F. Anderson, V. J. Bourke and C. J. O'Neil, *On the Truth of the Catholic Faith*. New York: Image, 1955, in 5 Vols.
————. *Summa Theologica*. (Textus Leoninus). Rome: Marietti, 1952. English trans., *Summa theologica*, New York: Benziger, 1947, in 3 Vols.
Ardley, Gavin. *Aquinas and Kant*. New York: Longmans, 1950.
Aristotle. *De anima (Treatise on the Soul)*, trans. by J. A. Smith, in *The Basic Works of Aristotle*, ed. by R. McKeon. New York: Random House, 1941, pp. 533–603.
————. *De caelo (Treatise on the Heavens)*, trans. by J. L. Stock, in *The Basic Works of Aristotle*, ed. by Richard McKeon. New York: Random House, 1941, pp. 395–466.

――. *De generatione et corruptione* (*Treatise on Coming to Be and Passing Away*), trans. Harold H. Joachim, in *The Basic Works of Aristotle*, ed. by R. McKeon. New York: Random House, 1941, pp. 467–531.

――. *Metaphysics*, trans. by W. D. Ross, in *The Basic Works of Aristotle*, ed. by R. McKeon. New York: Random House, 1941, pp. 681–926.

――. *Physics*, trans. by R. P. Hardie and R. K. Gaye, in *The Basic Works of Aristotle*, ed. by R. McKeon, in *The Basic Works of Aristotle*. New York: Random House, 1941, pp. 213–394.

Ashley, Benedict M. *Are Thomists Selling Science Short?* River Forest, Ill.: Albertus Magnus Lyceum, n.d.

――. *The Arts of Learning and Communication*. Chicago: Priory Press, 1961.

Barnett, Lincoln. *The Universe and Dr. Einstein*. New York: Signet, 1957.

Bergson, Henri. *Creative Evolution*, authorized trans. by Arthur Mitchell. New York: Modern Library, 1941.

――. *The Creative Mind*, trans. by Mabelle L. Andison. New York: Philosophical Library, 1946.

――. *Time and Free Will*, trans. by F. L. Pogson. New York: Harper, 1960.

Bidney, David. *Theoretical Anthropology*. New York: Columbia, 1953.

Burtt, E. A. *The Metaphysical Foundations of Modern Physics*. New York: Anchor Books, 1954.

Carlo, William E. *Philosophy, Science, and Knowledge*. Milwaukee: Bruce, 1967.

Cassirer, Ernst. *Language and Myth*, trans. by Susanne K. Langer. New York: Dover, 1946.

Cohen, I. Bernard. *The Birth of a New Physics*. New York: Anchor Books, 1960.

Deely, John N. *The Tradition Via Heidegger*. The Hague: Martinus Nijhoff, 1971.

De Koninck, Charles. *Natural Science as Philosophy*. Quebec: Laval University, 1959.

Dobzhansky, Theodosius. *Evolution, Genetics, and Man*. New York: Science Editions, 1963.

――. *Genetics and the Origin of Species*. 3rd ed., rev.; New York: Columbia, 1951.

――. *Mankind Evolving*. New Haven: Yale, 1962.

Dray, W. *Laws and Explanations*. Oxford University Press, 1957.

Duhem, Pierre. *Le Système du Monde*. New ed.; Paris, 1954, Vol. I.

Einstein, A., and Infeld, L. *The Evolution of Physics*. New York: Simon and Schuster, 1950.

Eiseley, Loren. *Darwin's Century*. New York: Anchor Books, 1961.

――. *The Firmament of Time*. New York: Atheneum, 1962.

――. *The Immense Journey*. London: Gollancz, 1958.

Farrington, Benjamin. *Greek Science*. Baltimore: Penguin, 1944.

Glutz, Melvin A. *The Manner of Demonstrating in Natural Philosophy*. River Forest, Ill.: Aquinas Institute, 1956.

Goudge, T. A. *The Ascent of Life*. Toronto: University of Toronto Press, 1961.

Grene, Marjorie. *A Portrait of Aristotle*. Chicago: University of Chicago Press, 1963.

Hanson, N. R. *Observation and Explanation*. New York: Harper and Row, 1971.

――. *Patterns of Discovery*. Cambridge: Oxford, 1958.

――. *Perception and Discovery*, ed. by W. C. Humphreys. San Francisco: Freeman, Cooper, and Co., 1969.

Heidegger, Martin. *An Introduction to Metaphysics*, trans. by Ralph Manheim. New Haven: Yale, 1959.

Hempel, C. *Aspects of Scientific Explanation*. New York: Free Press, 1965.

Henderson, L. *The Fitness of the Environment*. Boston: Beacon Press, 1958.

Husserl, Edmund. *Phenomenology and the Crisis of Philosophy*, trans. by Quentin Lauer: New York: Harper, 1965.

Huxley, Sir Julian. *Evolution in Action*. New York: Mentor, 1953.

Huxley, Julian S., ed. *The New Systematics*. Oxford: Oxford, 1940.

Jepsen, Glenn L., Simpson, George Gaylord, and Mayr, Ernst, eds. *Genetics, Paleontology, and Evolution*. New York: Atheneum, 1963.

Jonas, Hans. *The Phenomenon of Life*. New York: Harper and Row, 1966.

John of St. Thomas. *The Material Logic of John of St. Thomas,* trans. by Yves R. Simon, John J. Glanville, and G. Donald Hollenhorst. Chicago: University of Chicago Press, 1955.

Jones, W. T. *A History of Western Philosophy*. New York: Harcourt, Brace and World, 1952.

Kordig, Carl R. *The Justification of Scientific Change*. Dordrecht-Holland: D. Reidel, 1971.

Kuhn, Thomas. *The Structure of Scientific Revolutions*. 2nd ed., enlarged; Chicago: University of Chicago Press, 1970.

Lewin, Kurt A. *A Dynamic Theory of Personality: Selected Papers of Kurt Lewin*. New York: McGraw-Hill, 1935.

Linnaeus, C. *A Selection of the Correspondence of Linnaeus and Other Naturalists from the Original Manuscripts*, ed. by Sir James Edward Smith. London: Longman, Hurst, Rees, Orme, and Brown, 1821, Vol. II.

Litt, Thomas. *Les corps célestes dans l'univers de saint Thomas d'Aquin*. Paris: Nauwelaerts, 1963.

Lonergan, Bernard, J. F. *Insight*. New York: The Philosophical Library, 1965.

Lovejoy, Arthur O. *The Great Chain of Being*. New York: Harper, 1936.

MacIver, R. M. *Social Causation*. New York: Harper, 1964.

Maritain, Jacques. *Creative Intuition in Art and Poetry*. New York: Pantheon, 1953.

———. *The Degrees of Knowledge*, trans. from the 4th French ed. under the general supervision of Gerald B. Phelan. New York: Scribner's, 1959.

———. *The Peasant of the Garonne*, trans. by Michael Cuddihy and Elizabeth Hughes. New York: Holt, Rinehart and Winston, 1968.

———. *A Preface to Metaphysics*. New York: Mentor Books, 1962.

———. *The Range of Reasons*. New York: Scribner's, 1952.

———. *Science and Wisdom*. New York: Scribner's, 1940.

Mayr, Ernst. *Animal Species and Evolution*. Cambridge, Mass.: Harvard, 1963.

———, Linsley, E. G., and Usinger, R. L. *Methods and Principles of Systematic Zoology*. New York: McGraw-Hill, 1953.

———. *Systematics and the Origin of Species*. New York: Columbia, 1942.

Merleau-Ponty, Maurice. *The Primacy of Perception*. Evanston, Ill.: Northwestern University Press, 1964.

———. *The Structure of Behavior*. Boston: Beacon Press, 1963.

Newton, Sir Isaac. *Isaac Newton's Papers and Letters on Natural Philosophy*, ed., by I. Bernard Cohen. Cambridge, Mass.: Harvard, 1958.

Nogar, Raymond J. *The Wisdom of Evolution*. New York: Doubleday, 1963.

Nordenskiold, E. *The History of Biology.* New York: Tudor, 1935.

Plato. *Timaeus,* in *The Dialogues of Plato,* trans. by B. Jowett. New York: Random House, 1965, Vol. II, pp. 3–68.

Polanyi, Michael. *Science, Faith and Society.* Chicago: Phoenix, 1964.

Popper, Karl. *The Logic of Scientific Discovery.* New York: Basic Books, 1959.

Randall, John Herman. *Aristotle.* New York: Columbia, 1960.

Roe, Anne, and Simpson, George Gaylord. *Behavior and Evolution.* New Haven: Yale, 1958.

Sertillanges, A. D. *L'Idée de création et ses retentissements en philosophie.* Paris: Aubier, 1945.

Sikora, Joseph J. *The Scientific Knowledge of Sensible Nature.* Paris: Desclée, 1966.

Simon, Yves. *The Great Dialogue of Nature and Space.* New York: Magi, 1970.

———. *The Tradition of Natural Law.* New York: Fordham, 1965.

Simpson, G. G., Pittendrigh, C. S., and Tiffany, L. H., *Life,* New York: Harcourt, 1957.

———. *The Meaning of Evolution.* New Haven: Yale, 1949.

———. *Principles of Animal Taxonomy.* New York: Columbia, 1961.

———. *This View of Life.* New York: Harcourt, Brace and World, 1963.

Singer, C. *The Story of Living Things.* New York: Harper, 1931.

Tax, Sol, and Callender, Charles, eds., *Issues in Evolution,* Vol. III of *Evolution after Darwin.* Chicago: University of Chicago Press, 1960.

Teilhard de Chardin, Pierre. *The Future of Man,* trans. by Norman Denny. New York: Harper, 1964.

———. *How I Believe,* Teilhard's own redaction and expansion of *Comment je crois,* as delivered personally to George B. Barbour, from whom in turn I received a copy. On the authenticity of this document, see Barbour, George B., *In the Field with Teilhard de Chardin.* New York: Herder and Herder, 1965, p. 151.

———. *The Phenomenon of Man,* trans. by Bernard Wall. New York: Harper, 1959.

Toulmin, Stephen. *Foresight and Understanding.* New York: Harper and Row, 1961.

Waddington, C. H. *The Ethical Animal.* New York: Atheneum, 1961.

B.   Articles

Adler, Mortimer J. "The Next Twenty-five Years in Philosophy," *The New Scholasticism,* XXV (January, 1951), pp. 81–110.

Ashley, Benedict M. "Aristotle's Sluggish Earth, Part I: The Problematics of the *De caelo,*" *The New Scholasticism,* XXXII (January, 1958), pp. 1–31.

———. "Aristotle's Sluggish Earth, Part II: Media of Demonstration," *The New Scholasticism,* XXXII (April, 1958), pp. 202–234.

———. "Does Natural Science Attain Nature or Only the Phenomena?" in *The Philosophy of Physics,* ed. by V. E. Smith. Jamaica, New York: St. John's University Press, 1961, pp. 63–82.

Brodbeck, M. "Explanation, Prediction, and 'Imperfect' Knowledge," in *Scientific*

Explanation, Space, and Time, ed. by H. Feigl and G. Maxwell, Vol. III of *Minnesota Studies in the Philosophy of Science,* Minneapolis: University of Minnesota Press, 1962, pp. 231–272.

Case, E. C. "The Dilemma of the Paleontologist," in *Contributions from the Museum of Paleontology,* IX, No. 5. Ann Arbor: University of Michigan Press, 1951.

Crespy, Georges. "Evolution and Its Problems." Mimeographed text of 1965 lecture at the University of Chicago, sponsored by the Chicago Theological Seminary, January 16, 1965.

Davis, Philip J. "Number," *Scientific American,* 211 (September, 1964), pp. 50–59.

Deely, John N. "The Immateriality of the Intentional as such," *The New Scholasticism,* XLII (Spring, 1968), pp. 293–306.

——. "The Philosophical Dimensions of the Origin of Species," Parts I and II, *The Thomist,* XXXI (January and April, 1969), pp. 75–149 and 251–335.

De Koninck, Charles, "Abstraction from Matter," *Laval théologique et philosophique,* XIII (1957), pp. 133–197.

——. "Réflexious sur le problème de l'indéterminisme," *Revue Thomiste,* XLIII, Part I, (juillet–septembre, 1937), pp. 227–252; Part II (novembre–decembre, 1937), pp. 393–409.

——. "The Unity and Diversity of Natural Science," in *The Philosophy of Physics,* ed. by V. E. Smith, Jamaica, New York: St. John's University Press, 1961, pp. 5–24.

Donceel, Joseph. "Teilhard de Chardin: Scientist or Philosopher?" *International Philosophical Quarterly,* V (May, 1965), pp. 248–266.

Emmet, Dorothy M. "The Choice of a World-Outlook," *Philosophy,* 23 (1948), pp. 208–226.

Ewing, Franklin J. "Précis on Evolution," *Thought,* XXV (March, 1950), pp. 53–78.

Feyerabend, P. K. "Consolations for the Specialist," in *Criticism and the Growth of Knowledge,* ed. by I. Lakatos and A. Musgrave. Cambridge: Oxford, 1970, pp. 197–230.

——. "How to Be a Good Empiricist: A Plea for Tolerance in Matters Epistemological," in *The Deleware Seminar in Philosophy of Science,* ed. by B. Baumrin. New York: Interscience, 1963, Vol. II, pp. 3–39.

——. "Reply to Criticism," in *Boston Studies in the Philosophy of Science,* ed. by R. S. Cohen and M. W. Wartofsky. New York: Humanities, 1965, Vol. II, pp. 223–261.

Francoeur, Robert, "The God of Disorder II: A Response," *Continunm,* 4 (Summer, 1966), pp. 264–271.

Hayek, F. A. "The Theory of Complex Phenomena," in *The Critical Approach to Science and Philosophy,* ed. by Mario Bunge. Glencoe, Ill.: The Free Press, 1964.

Husserl, Edmund. "The *Lebenswelt* as forgotten foundation of meaning for natural science," in *Husserliana,* Band VI, p. 48. The Hague: Martinus Nijhoff, 1963.

Kordig, Carl R. "The Theory-Ladenness of Observation." *The Review of Metaphysics,* XXIV (March, 1971), pp. 448–484.

Kuhn, Thomas. "Logic of Discovery or Psychology of Research?" in *Criticism and the Growth of Knowledge,* ed. by I. Lakatos and A. Musgrave. Cambridge: Oxford, 1970, pp. 1–23.

——. "Reflections on my Critics," in *Criticism and the Growth of Knowledge,* ed. by I. Lakatos and A. Musgrave. Cambridge: Oxford, 1970, pp. 231–278.

Landesman, Charles. "Introduction" to *The Problem of Universals*. New York: Basic Books, 1971, pp. 3–17.

Langer, Susanne K. "Preface" to Ernst Cassirer, *Language and Myth*, trans. by S. K. Langer. New York: Dover, 1946.

Le Gros Clark, W. "The Crucial Evidence for Human Evolution," *American Scientist*, XLVII (1959).

Maritain, Jacques. "The Conflict of Methods at the End of the Middle Ages," *The Thomist*, III (October, 1941), pp. 527–538.

———. "Sign and Symbol," in *Redeeming the Time*. London: Geoffrey Bles, 1943, pp. 191–224.

———. "Teilhard de Chardin and Teilhardism," in *The Peasant of the Garonne*. trans. by Michael Cuddihy and Elizabeth Hughes. New York: Holt, Rinehart and Winston, 1968, pp. 116–126.

Mayr, Ernst, "Cause and Effect in Biology," *Science*, 134 (10 November 1961), pp. 1501–1506.

Nogar, Raymond J. "Evolution: Scientific and Philosophical Dimensions," in *Philosophy of Biology*, ed. by V. E. Smith. Jamaica, New York: St. John's University Press, 1962, pp. 23–66.

———. "From the Fact of Evolution to the Philosophy of Evolutionism," *The Thomist*, XXIV (April, July, and October, 1961), pp. 463–501.

———. "The God of Disorder," *Continuum*, 4 (Spring, 1966), pp. 102–113.

———. "The God of Disorder III: A Postscript," *Continuum*, 4 (Summer, 1966, pp. 264–275.

———. "The Mystery of Cosmic Epigenesis," in the *ACPA Proceedings*, XXIX (1965), pp. 112–124.

Ruse, Michael. "Narrative Explanation and the Theory of Evolution," *Canadian Journal of Philosophy*, I (September, 1971), pp. 59–74.

———. "The Revolution in Biology," *Theoria*, XXXVI (1971), pp. 1–22.

Scriven, Michael. "Explanation and Prediction," *Science*, 130 (26 August 1959), pp. 477–482.

Simon, Yves. "Maritain's Philosophy of the Sciences," *The Thomist*, V (January, 1943), pp. 85–102.

———. "The Nature and Process of Mathematical Abstraction," *The Thomist*. XXXIX (April, 1965), pp. 117–139.

Teilhard de Chardin, Pierre. "Contingence de l'univers et gout humain de survivre" (1953).

———. "Comment je vois" (1948).

———. "La lutte contre la multitide" (1917).

———. "Les noms de la matière" (1919).

———. "Note sur les modes de l'action divine dans l'univers" (1920).

———. "L'Union creatrice" (1917).

Toulmin, Stephen. "Does the Distinction Between Normal and Revolutionary Science Hold Water?" in *Criticism and the Growth of Knowledge*. ed. by I. Lakatos and A. Musgrave. Cambridge: Oxford, 1970, pp. 39–48.

———. "On Teilhard de Chardin," *Commentary*, 39 (March, 1965).

———. "Reply," in *Synthese*, 18 (1968), pp. 462–463.

Tresmontant, Claude. "Le Père Teilhard de Chardin et la Théologie," *Lettre*, 49–50 (septembre–octobre, 1962).

Weisheipl, James A. "The Celestial Movers in Medieval Physics," in *The Dignity of Science*, ed. by James A. Weisheipl. Washington, 1961, pp. 150–190.

Part II:

"Contemporary discussions":
in six sections

SECTION I—THE UNIQUENESS OF MAN

1. Adler, Mortimer J. *The Difference of Man and the Difference It Makes.* New York: Holt, Rinehart & Winston, 1967.
2. Aquinas, Thomas. *Aristotle's De Anima with the Commentary of Saint Thomas Aquinas.* New Haven, Connecticut: Yale University Press, 1951.
3. ————. *Treatise on Man.* Trans. James F. Anderson. Englewood Cliffs, New Jersey: Prentice-Hall, 1962.
4. Barnett, Anthony. *The Human Species.* Baltimore: Penguin, 1961.
5. Bates, Marston. *Man in Nature.* 2nd ed. Englewood Cliffs, New Jersey: Prentice-Hall, 1961.
6. Bergson, Henri. *Time and Free Will.* Trans. F. L. Pogson. New York: Harper Torchbooks TB 1021, 1960.
7. Boelen, Bernard J. *Symposium on Evolution.* Pittsburgh: Duquesne, 1959.
8. Boring, E. *The Physical Dimensions of Consciousness.* New York: Dover, 1933.
9. Buytendijk, F. J. *L'homme et l'animal. Essai de psychologie comparée.* Paris: Gallimard, 1965.
10. Carrington, Richard. *A Million Years of Man.* New York: Mentor Books, 1963.
11. Cassirer, E. *An Essay on Man.* New York: Doubleday, 1953.
12. Coon, C. S. *The Story of Man.* New York: Alfred A. Knopf, 1954.
13. Critchley, Macdonald. "The Evolution of Man's Capacity for Language," *Evolution after Darwin.* Ed. Sol Tax. Chicago: The University of Chicago Press, 1960. Vol. II, 289–308.
14. Darlington, C. D. *Genetics and Man.* Baltimore: Penguin, 1966. (This volume is a revised edition of the author's earlier book called *The Facts of Life.*)

15. Darwin, Charles. *The Origin of Species and the Descent of Man.* New York: Modern Library Giant, 1949.

16. Deely, John N. "The Emergence of Man," *The New Scholasticism*, XL (April, 1966), 141–176.

17. ———. "Animal Intelligence and Concept-Formation," *The Thomist*, XXXV (January, 1971), 43–93.

18. DeVore, Paul L., ed., *The Origin of Man*, restricted transcript of a symposium sponsored by the Wenner-Gren Foundation for Anthropological Research, Inc., convened by Sol Tax, held at the University of Chicago, April 2–4, 1965. Transcript distributed through *Current Anthropology* for the Wenner-Gren Foundation.

19. Dobzhansky, Theodosius. *The Biological Basis of Human Freedom.* New York: Columbia University Press, 1956.

20. ———. *Evolution, Genetics, and Man.* New York: John Wiley & Sons, Science Editions, 1963.

21. ———. *Genetics and the Origin of Species.* 3rd ed., rev. New York: Columbia University Press, 1951.

22. ———. *Heredity and the Nature of Man.* New York: Harcourt, Brace & World, 1964.

23. ———. *Mankind Evolving.* New Haven, Connecticut: Yale University Press, 1962.

24. Geertz, Clifford. "The Transition to Humanity," in *Horizons of Anthropology.* Ed. Sol. Tax. Chicago: Aldine Publication Company, 1964. Pp. 37–48.

25. Hallowell, A. Irving. "Self, Society, and Culture in Phylogenetic Perspective," in *Evolution after Darwin.* Ed. S. Tax. Chicago: Univ. of Chicago Press, 1960. Vol. II, 309–372.

26. Harrison, R. J. *Man the Peculiar Animal.* Baltimore: Penguin, 1957.

27. Herrick, J. *The Evolution of Human Nature.* New York: Harper Torchbooks, 1956.

28. Highet, G. *Man's Unconquerable Mind.* New York: Columbia University Press, 1954.

29. Hilgard. "Psychology after Darwin," in *Evolution after Darwin.* Ed. Sol Tax. Chicago: Univ. of Chicago Press, 1960. Vol. II, 269–288.

30. Hockett, C. F. and Ascher, R. "The Human Revolution," *Current Anthropology*, V (June, 1964), 135–168.

31. Howells, W. *Ideas on Human Evolution.* Cambridge, Massachusetts: Harvard University Press, 1962.

32. ———. *Back of History.* New York: Anchor, 1954.

33. Huxley, Sir Julian S. *The Uniqueness of Man.* London: Chatto and Windus, 1941.

34. ———. *Evolution in Action.* New York: Mentor Books MP 491, 1953.

35. ———. *Evolution: The Modern Synthesis.* New York: John Wiley & Sons, Science Editions, 1964.

36. Huxley, Thomas H. *Man's Place in Nature.* Ann Arbor: University of Michigan Press, 1959.

37. Le Gros Clark, W. E. *The Antecedents of Man.* New York: Harper Torchbooks, 1963.

38. ———. *History of the Primates.* Chicago: Phoenix Books, 1961.

39. Mayr, E. *Animal Species and Evolution*. Cambridge, Massachusetts: The Belknap Press of Harvard University Press, 1963.
40. Montagu, Ashley. *The Human Revolution*. New York: The World Publication Company, 1965.
41. Nogar, R. J. *The Wisdom of Evolution*. New York: Mentor-Omega, 1963. Esp. chapters 6 & 7.
42. North, Robert. *Teilhard and the Creation of the Soul*. Milwaukee: Bruce, 1967.
43. Oakley, Kenneth P. *Man, the Tool-Maker*. Chicago: Phoenix Books, 1959.
44. Rensch, B. *Evolution Above the Species Level*. New York: Science Editions, 1966.
45. Romer, A. S. *Man and the Vertebrates*. Baltimore, Maryland: Pelican Books, 1954. 2 Volumes.
46. Schenk, Gustav. *The History of Man*. New York: Chilton, 1961.
47. Sherrington, C. *Man on His Nature*. Cambridge: University Press, 1951.
48. Simpson and Roe, editors. *Behavior and Evolution*. New Haven, Connecticut: Yale University Press, 1958.
49. Sinnot, E. *Matter, Mind and Man*. New York: Atheneum, 1962.
50. Spuhler, James N., editor. *The Evolution of Man's Capacity for Culture*. Detroit: Wayne State University Press, 1959.
51. Tax, Sol. ed. *Anthropology Today*. Chicago: University of Chicago Press, 1962.
52. ———, editor. *The Evolution of Man*. Vol. II of *Evolution After Darwin*, Chicago: University of Chicago Press, 1960.
53. ———. *Horizons of Anthropology*. Chicago: Aldine, 1964.
54. Teilhard de Chardin, P. *The Appearance of Man*. New York: Harper, 1965.
55. ———. "The Idea of Fossil Man," *Anthropology Today*. Sol Tax, ed. Chicago: University of Chicago Press, 1962. Pp. 31–38.
56. ———. *Man's Place in Nature*. New York: Harper, 1966.
57. ———. *The Vision of the Past*. New York: Harper, 1966.
58. Thorpe, W. H. *Learning and Instinct in Animals*. 2nd ed. London: Methuen, 1963.
59. ———. *Biology and the Nature of Man*. London: Oxford University Press, 1962.
60. Von Bonin, Gerhard. *The Evolution of the Human Brain*. Chicago: University of Chicago Press, 1963.
61. Wendt, H. *In Search of Adam*. Boston: Houghton-Mifflin Company, 1956.
62. White, Leslie. *The Science of Culture*. New York: Grove Press, 1959. Chs. I and II, 22–48.

SECTION II—THE HUMANNESS OF MAN

1. Adler, Alfred. *Understanding Human Nature*. New York: Premier, 1954.
2. Aries, Philippe. *Centuries of Childhood, a Social History of Family Life*. New York: Vintage, 1965.
3. Barnouw, Victor. *Culture and Personality*. Homewood, Ill.: Dorsey Press, 1963.
4. Benedict, R. *Patterns of Culture*. New York: Mentor, 1946.

5. Bidney, David. *Theoretical Anthropology.* New York: Columbia University Press, 1953.
6. Childe, V. Gordon. *Man Makes Himself.* New York: Mentor, 1951.
7. ————. *What Happened In History.* Baltimore: Penguin, 1946.
8. Coon, Carleton S. and Hunt, Edward E. *Anthropology A to Z.* New York: Grosset & Dunlap, 1963.
9. Cornford, F. M. *From Religion to Philosophy.* New York: Harper Torchbooks, 1957.
10. Cumming, J. and Cumming, E. *Ego and Milieu.* New York: Atherton Press, 1966.
11. Dollard, J., and Miller, N. E. *Personality and Psychotherapy.* New York: McGraw-Hill, 1950.
12. Durkheim, Emile. *The Elementary Forms of Religious Life.* New York: Free Press, 1965.
13. Erikson, Erik H. *Childhood and Society.* 2nd ed., rev. New York: Horton, 1963.
14. Ferm, V., ed. *Ancient Religions.* New York: Citadel, 1965.
15. Geertz, Clifford. "The Growth of Culture and the Evolution of Mind." In *Theories of the Mind.* Ed. Jordan Scher. New York: Free Press, 1962. Pp. 713–740.
16. Henry, Jules. *Culture Against Man.* New York: Vintage, 1963.
17. *History of Mankind.*
    Volume I: Jacquetta Hawkes and Sir Leonard Woolley, *Prehistory and the Beginnings of Civilization;*
    Volume II: Luigi Pareti, *The Ancient World, 1200 BC to 500 AD.* New York: Harper and Row, 1966.
18. Holton, Gerald, ed. *Science and Culture.* Boston: Beacon Press, 1967.
19. Huntington, Elsworth. *Mainsprings of Civilization.* New York: Mentor, 1959.
20. Hutchins, Robert M., and Adler, Mortimer J., Editors. *The Great Ideas Today 1966. The Difference of Woman and the Difference it Makes. A Symposium.* Chicago: Britannica, 1966. Pp. 1–99.
21. Huxley, Julian S. "Evolution, Cultural and Biological." *Yearbook of Anthropology.* Ed. W. Thomas. New York: Wenner-Gren, 1955.
22. Kluckhohn, Clyde. *Culture and Behavior.* New York: Free Press, 1962.
23. ————. *Mirror for Man.* New York: Premier, 1965.
24. Kroeber, A. L. *Anthropology.* New York: Harcourt, 1948.
25. ————, and Kluckhohn, Clyde. *Culture, a Critical Review of Concepts and Definitions.* New York: Vintage, 1963.
26. Kroeber, A. L. "Evolution, History, and Culture," in *Evolution after Darwin.* Ed. Sol Tax. Chicago: Univ. of Chicago Press, 1960. Vol. II, 1–16.
27. Levi-Strauss, Claude. *Structural Anthropology.* Trans. Claire Jacobson and Brooke Grundfest Schoepf. New York: Basic Books, 1963.
28. Levy, G. Rachel. *Religious Conceptions of the Stone Age.* New York: Harper, 1963.
29. Lewin, Kurt. *Field Theory in Social Science.* New York: Harper Torchbooks, 1964.
30. ————. *A Dynamic Theory of Personality.* New York: McGraw-Hill, 1935.
31. Lowie, Robert H. *Primitive Religion.* New York: Grossett & Dunlap, 1952.
32. ————. *Primitive Society.* New York: Harper, 1961.
33. Malinowski, Bronislaw. *Magic, Science and Religion.* New York: Anchor, 1954.
34. ————. *A Scientific Theory of Culture.* New York: Galaxy, 1960.

35. Maritain, Jacques. *Integral Humanism*, trans. by Joseph Evans. New York: Scribner's, 1968.
36. McCord, W. *The Springtime of Freedom*. New York: Oxford University Press, 1965.
37. Mead, Margaret. *Male and Female*. New York: Mentor, 1955.
38. ———. *Cultural Patterns and Technical Change*. New York: Mentor, 1955.
39. ———. *Anthropology*. New York: Van Nostrand, 1964.
40. ———. *Continuities in Cultural Evolution*. New Haven, Connecticut: Yale University Press, 1964.
41. Menaker, E., and Menaker, W. *Ego in Evolution*. New York: Grove Press, 1965.
42. Montagu, Ashley. *Anthropology and Human Nature*. New York: McGraw-Hill, 1954.
43. ———. *Man's Most Dangerous Myth: The Fallacy of Race*. New York: Meridian, 1964.
44. Parsons, T., and Shils, E. *Toward a General Theory of Action*. New York: Harper, 1962.
45. Platt, John R., ed. *New Views on the Nature of Man*. Chicago: University of Chicago Press, 1965.
46. Radcliffe-Brown, A. R. "White's View of a Science of Culture," *American Anthropologist*, LI (1949), 503–512.
47. Radin, Paul. *Primitive Man as Philosopher*. New York: Dover, 1927.
48. ———. *The World of Primitive Man*. New York: Grove Press, 1960.
49. Redfield, Robert. *The Primitive World and its Transformations*. New York: Cornell University Press, 1953.
50. Sahlins, M. D., and Service, E. R. *Evolution and Culture*. Ann Arbor: University of Michigan Press, 1960.
51. Shapiro, Harry L., ed. *Man, Culture, and Society*. New York: Galaxy, 1960.
52. Sorokin, P. A. *Basic Trends of our Times*. New Haven: College & University Press, 1964.
53. ———. *Crisis of our Age: The Social & Cultural Outlook*. New York: Dutton, 1941.
54. ———. *Fads and Foibles in Modern Sociology*. Chicago: Gateway, 1956.
55. ———. *Modern Historical & Social Philosophies*. New York: Dover, 1964.
56. ———. *Social and Cultural Dynamics*. New York: American Book Co., 1937. 4 Vols. (One vol. edition available from Boston: Sargent Porter, Inc.)
57. ———. *Sociocultural Causality, Space, Time*. New York: Russell and Russell, 1943.
58. ———. *Sociological Theories of Today*. New York: Harper, 1966.
59. Spindler, G. D. *Education and Culture*. New York: Holt, 1963.
60. Steward, Julian H. *Theory of Culture Change*. Urbana: University of Illinois Press, 1955.
61. Summer, W. G. *Folkways*. New York: Mentor, 1960.
62. Tax, Sol, ed. *Anthropology Today*. Chicago: University of Chicago Press, 1962.
63. ———. *Horizons of Anthropology*. Chicago: Aldine, 1964.
64. Teilhard de Chardin, P. "The Antiquity and World Expansion of Human Culture," in *Man's Role in Changing the Face of the Earth*. International Symposium edited by William L. Thomas. Chicago: University of Chicago Press, 1956. Pp. 103–112.

65. Van den Berg, J. H. *The Changing Nature of Man*. New York: Delta, 1964.
66. Washburn, S. L., and Howell, F. Clark. "Human Evolution and Culture," in *Evolution after Darwin*. Ed. Sol Tax. Chicago: Univ. of Chicago Press, 1960. Vol. II, 33–56.
67. White, Leslie. *The Evolution of Culture*. New York: McGraw-Hill, 1959.
68. ———. "The Individual and the Culture Process," *Centennial Proceedings*. Washington: American Association of Advanced Science, 1950.
69. ———. *The Science of Culture*. New York: Grove Press, 1949.

SECTION III—THE MORAL ISSUES

1. Adler, Mortimer J. *A Dialectic of Morals*. New York: Ungar, 1941.
2. ———. *The Idea of Freedom*. 2 Volumes. New York: Doubleday, Volume I, 1958; Volume II, 1961.
3. ———. *The Time of Our Lives*. New York: Holt, Rinehart & Winston, 1970.
4. ———. *What Man Has Made of Man*. New York: Ungar, 1937.
5. Anscombe, G. E. M. "Modern Moral Philosophy." *Philosophy* XXXIII (1958), 1–19.
6. Anshen, Ruth Nanda, ed. *Moral Principles of Action : Man's Ethical Imperative*. New York: Harper, 1952.
7. Baker, Wesley C. *The Open End of Christian Morals*. Philadelphia: The Westminster Press, 1967.
8. Beis, Richard H. "Some Contributions of Anthropology to Ethics." *The Thomist*, XXVIII (April, 1964), 174–224.
9. Bennett, John C., *et al. Storm Over Ethics*. United Church Press, 1967.
10. Bird, Otto. *The Idea of Justice*. New York: Praeger, 1967.
11. Bodenheimer, Edgar. *Jurisprudence*. Cambridge, Mass.: Harvard, 1962.
12. Bourke, Vernon J. *History of Ethics*. New York: Doubleday, 1968.
13. Broad, C. D. "Symposium on the Relation Between Science and Ethics." *Proceedings of the Aristotelian Society*. XLII (1941–42), 65–100H.
14. Bronowski, J. *Science and Human Values*. New York: Harper, 1965.
15. Cogley, John, ed. *Natural Law and Modern Society*. New York: Meridian, 1966.
16. Cox, Harvey, ed. *The Situation Ethics Debate*. Philadelphia: Westminster, 1968.
17. Daly, Cahal B. *Natural Law Morality Today*. London: Burns and Oates, 1965.
18. D'Arcy, Eric. *Conscience and Its Right to Freedom*. New York: Sheed and Ward, 1961.
19. de Beauvoir, Simone. *The Ethics of Ambiguity*. New York: Citadel, 1967.
20. De Koninck, Charles. "General Standards and Particular Situations in Relation to the Natural Law." *ACPA Proceedings*, XXIV (1950), 28–32.
21. Dentrèves, A. P. *Natural Law. An Historical Survey*. New York: Harper, 1965.
22. Dewey, John. *Human Nature and Conduct*. New York: Holt, 1922.
23. Dubarle, Dominique. "Does Man's Manner of Determining his Own Destiny Constitute a Threat to his Humanity?" *Concilium*, VI (June, 1966), 41–46.
24. Emmet, Dorothy. *Rules, Roles and Relations*. New York: St. Martin's Press, 1966.

25. Evans, Illtud, ed. *Light on the Natural Law*. Baltimore: Helicon, 1965.
26. Ewing, A. C. "Symposium on the Relations Between Science and Ethics." *Proceedings of the Aristotelian Society*, XLII (1941–42), 65–100H.
27. Fletcher, Joseph. *Moral Responsibility. Situation Ethics at Work*. Philadelphia: Westminster, 1967.
28. ———. *Situation Ethics*. Philadelphia: Westminster, 1966.
29. Gerharz, George E. "Natural Law and Self-Realization." *Listening*, II (Autumn, 1967), 184–193.
30. Gilkey, Langdon. "Evolutionary Science and the Dilemma of Freedom and Determinism." *The Christian Century*, LXXXIV (March 15, 1967), 339–343.
31. Gilson, Etienne. *Moral Values and the Moral Life*. Hamden, Conn.: Shoe String Press, 1961.
32. Hook, Sidney, ed. *Law and Philosophy*. New York: New York University, 1964.
33. Huxley, Julian S. *Evolutionary Ethics*. London: Oxford, 1943.
34. ———. *The Human Crisis*. Seattle: University of Washington Press, 1963.
35. ———. *Knowledge, Morality, and Destiny*. New York: Mentor, 1957.
36. ———. *Man Stands Alone*. New York: Harper, 1941.
37. Huxley, J. S., and Huxley, Thomas H. *Touchstone For Ethics*. New York: Harper, 1947.
38. Jenkens, Iredell. "The Present Status of the Value Problem," *Review of Metaphysics*, IV (September 1950), 85–110.
39. Kluckhohn, Clyde. "Ethical Relativism, Sic et Non," *Journal of Philosophy*, LII (November, 1955), 666–677.
40. ———. "The Scientific Study of Values." *Three Lectures*. Toronto: University of Toronto Press, 1958.
41. Köhler, Wolfgang. *The Place of Values in a World of Facts*. New York: Mentor, 1966.
42. Lachance, Louis. *Le Concept de droit selon Aristote et S. Thomas*. 2nd ed., rev. Montreal: Editions du Levrier, 1948.
43. Leake, C. D. "Ethicogenesis." *Scientific Monthly*, LX (1945), 245–253.
44. ———, and Romanell, P. *Can We Agree? A Scientist and a Philosopher Argue About Ethics*. Austin: University of Texas Press, 1950.
45. Leclerque, Jacques. *La philosophie morale de saint Thomas devant la pensée contemporaine*. Paris: Vrin, 1955.
46. Lepp, Ignace. *The Authentic Morality*. New York: Macmillan, 1965.
47. Lottin, Odon. "Natural Law, Natural Right and Natural Reason." *Philosophy Today*, III (1959), 10–18.
48. Luijpen, W. A. *Phenomenology of Natural Law*. Pittsburgh: Duquesne, 1967.
49. Margeneau, Henry. *Ethics and Science*. New York: Van Nostrand, 1964.
50. Maritain, Jacques. *Moral Philosophy*. New York: Scribner's, 1964.
51. ———. *Neuf leçons sur les notions premières de la philosophie morale*. Paris: Tequi, 1951.
52. McAllister, Joseph B. "Psychoanalysis and Morality." *The New Scholasticism*, XXX (1956), 310–329.
53. McGill, V. J. *The Idea of Happiness*. New York: Praeger, 1967.
54. Messner, J. *Ethics and Facts*. St. Louis: Herder, 1952.
55. ———. *Social Ethics*. Trans. J. J. Doherty. Rev. ed. St. Louis: Herder, 1965.
56. Mouroux, Jean. *The Meaning of Man*. New York: Image, 1961.

57. Muller, H. J. "Human Values in Relation to Evolution." *Science*, CCXXVII (1958), 625–629.

58. Myers, W. H. *Human Personality and Its Survival of Bodily Death.* New York: Longmans, 1954. 2 Vols.

59. Northrop, F. S. C. *The Complexity of Legal and Ethical Experience.* Boston: Little, Brown, and Co., 1959.

60. Pinchaers, S. *Le renouveau de la morale.* Paris: Casterman, 1964.

61. Platt, John. "The New Biology and the Shaping of the Future." in *The Great Ideas Today 1968.* Ed. Robert M. Hutchins and Mortimer J. Adler. Chicago: Britannica, 1968, pp. 120–169.

62. Ramsey, Paul. *Nine Modern Moralists.* Englewood-Cliffs, N.J.: Prentice-Hall, 1962.

63. Rommen, Heinrich A. *The Natural Law.* New York: Vail-Ballou Press, 1947.

64. ———. "Natural Law and War Crimes Guilt." *ACPA Proceedings*, XXIV (1950), 40–57.

65. Schleichert, Herbert. "The Communicable Content of the Conventional Bases for the Natural Law." *Philosophy Today*, VII (1963), 33–38.

66. Sertillanges, A.-D. *La philosophie des Lois.* Paris: Alsatia, 1946.

67. Simon, Yves R. *Freedom and Community.* New York: Fordham, 1968.

68. ———. *Freedom of Choice.* New York: Fordham, 1969.

69. Sonneborn, T. M. "Implications of the New Genetics for Biology and Man." *Bulletin of the American Institute of Biological Sciences* (April, 1963), 26 ff.

70. Stevenson, C. L. *Facts and Values.* New Haven, Conn.: Yale, 1963.

71. Stone, Julius. *Human Law and Human Justice.* Stanford: Stanford University Press, 1966.

72. Strauss, L. *Natural Right and History.* Chicago: Univ. of Chicago Press, 1953.

73. Teilhard de Chardin, P. "Some Reflections on the Rights of Man." In UNESCO Symposium on *Human Rights.* New York: Columbia, 1949.

74. Tillich, Paul. *Morality and Beyond.* New York: Harper, 1963.

75. Toulmin, Stephen. *Reason in Ethics.* Cambridge: The University Press, 1964.

76. Waddington, C. H. *The Ethical Animal.* New York: Atheneum, 1961.

77. ———. "The Relations Between Science and Ethics." *Nature*, CXLVIII (1941), 270–274; Comments, 274–280.

78. ———. *Science and Ethics.* London: Allen and Unwin, 1942.

79. ———. "Science-Ethics-Religion." *World Review* (July, August, September, 1946).

80. ———, Ewing, A. C., and Broad, C. D. "Symposium on the Relation Between Science and Ethics." *Proceedings of the Aristotelian Society*, XLII (1941–42), 65–100H.

SECTION IV—THE METAPHYSICAL ISSUES

As was mentioned in the Table of Contents and in the "Rationale" for the readings in this Section above, by "metaphysical issues" here is understood whatever pertains to things insofar as they are considered but not produced by human reason. In the Thematic Remarks at the beginning of this book, we observed that the task peculiar to philosophical reflection is the achievement of an understanding of the past in its interarticulation with and relevance to the present. In that essay, however, limitations of space restricted us to illumining the idea of evolution chiefly within noetic or epistemological perspectives; while in our readings we have been limited to issues in areas proximately bearing on humanistic thought.

Now, it is a fact that in the whole matter of evolution, the mass of data which scientific research has uncovered and attested has had a cumulative effect in making an ancient and traditional problem about the nature and origin of species a currently insistent one, and one which, in the contemporary mind, serves to exemplify in a clear and striking manner the incompatibility of modern science and traditional wisdom. Accordingly, if this book were to adequately discharge philosophy's peda-gogical obligation in treating of the idea of evolution, it would be necessary, in addition to our survey of the noetic landscape of Darwin's world, to essay a similar survey of its ontological landscape. Such an essay, to achieve its end, would have to show first of all that evolutionary science has not altered the metaphysical structure of the species problematic, but has on the contrary specified that problematic so as to make its options clearer and their alternatives more definite. In showing that, the survey in question would on the one side clear away the morass of philosophical perplexities in post-Darwinian thought due not to the accumulation of evolu-tionary data (as is commonly supposed) but primarily and directly to those ambigu-ities and uncertainties latent in Classical Antiquity's notion itself of species, which features the labor of evolutionary research has merely forced to the fore; while on the other side, the forthright acknowledgment and philosophical resolution of these no longer latent ambiguities and uncertainties would render the evolutionary data themselves more intelligible in their own line of explanation which is not mathe-matical (species are not numbers), but that of natural philosophy, wherein are assigned reasons for the changes that never cease around us.

Though this problem of natural species is not *at first glance* as proximate to humanistic thought as are other of the evolutionary issues, its historical importance and fundamental character will force anyone who seeks to grasp the final thrust of the idea of evolution to come to terms with the issue of the number and interrelation of essentially distinct kinds of being in the world of natures and their transforming interactions.

In line with this central exigency of the evolutionary problematic, and in keep-ing with the nature of this book as a guide for personal research, we will divide our selected bibliography on the "metaphysical" issues into two parts, one part listing key philosophical works influenced or dominated by the realization that the whole of nature is not just in motion but in process, and that the time of the world is the measure not only of its changes but of its genesis; another part listing the repre-sentative materials, both philosophical and scientific, which hold the key to the solution of the problem of specific natures. For extensive assistance in the compila-tion of sources on the question of species, I am indebted to an old friend of student days, Richard J. Woods of the Aquinas Institute.

THE METAPHYSICAL ISSUES, PART A: BIBLIOGRAPHY ON THE PROBLEM OF NATURAL SPECIES

1. Abercrombie, M., Hickman, C., and Johnson, N. *A Dictionary of Biology.* Baltimore: Penguin Books, 1951.

2. Adler, Mortimer. *The Problem of Species*. New York: Sheed and Ward, 1940.

3. ———. "Solution of the Problem of Species." *The Thomist*, III (April, 1941), 279–379.

4. ———. "The Hierarchy of Essences." *The Review of Metaphysics*, VI (September, 1952), 3–30.

5. ———. "The Philosophers Give All the Answers and Establish None," in *The Difference of Man and the Difference It Makes*. New York: Holt, 1967. Ch. 4, 51–65.

6. Arendt, Hannah. *The Human Condition*. Garden City, New York: Doubleday Anchor Books, 1959.

7. Aristotle. "On the Generation of Animals," in *The Basic Works of Aristotle*. Ed. R. McKeon. New York: Random House, 1941.

8. ———. *History of Animals*.

9. ———. *Metaphysics*.

10. ———. *On the Parts of Animals*.

11. ———. *Physics*.

12. Ashley, Benedict M. "Does Natural Science Attain Nature or Only the Phenomena?" in *The Philosophy of Physics*. Ed. Vincent E. Smith. Jamaica, New York: St. John's University Press, 1961. Pp. 63–82.

13. Beaudry, Jean. "The Species Concept: Its Evolution and Present Status." *Review of Canadian Biology*, XIX (1960).

14. Bergson, Henri. *Creative Evolution*. Trans. Arthur Mitchell. New York: Modern Library ed., 1944.

15. Burma, Benjamin H. "The Species Concept: A Semantic Review," *Evolution*, III (1949), 369–370.

16. Burtt, E. A. *The Metaphysical Foundations of Modern Science*. New York: Doubleday Anchor Books, 1954.

17. Cailleux, André. "How Many Species?" *Evolution*, VIII (1960).

18. Cain, A. J. *Animal Species and Their Evolution*. New York: Harper, 1960.

19. Clark, W. E. Le Gros. *The Antecedents of Man*, New York: Harper Torchbooks, 1963.

20. Coleman, William. *George Cuvier : Zoologist. A Study in the History of Evolution Theory*. Cambridge, Massachusetts: Harvard University Press, 1964.

21. Collingwood, R. G. *The Idea of Nature*. New York: Oxford University Press Galaxy ed., 1960.

22. Commoner, Barry. "Is DNA the 'Secret of Life'?" *Clinical Pharmacology and Therapeutics*, VI (1965), 273–278.

23. Coon, Carleton. *The Origin of Races*. New York: Alfred A. Knopf, 1962.

24. Coonen, L. P. "Evolution of Method in Biology." *Philosophy of Biology*. Ed. Vincent E. Smith. New York: St. John's Univ. Press, 1962.

25. Darwin, Charles. *The Origin of Species*. New York: Modern Library ed. (n.d.).

26. Deely, John N. *Evolution : Concept and Content*. Reprint booklet of "Evolution: Concept and Content," Part I, *Listening*, Volume 0, No. 0 (Autumn, 1965), 27–50, and Part II, *Listening*, Volume I, No. 1 (Winter, 1966), 38–66 (2nd ed.; Dubuque, 1967).

27. ———. "The Philosophical Dimensions of the Origin of Species." *The Thomist*, XXXIII (January and April, 1969), 75–149 (Part I) and 251–342 (Part II).

28. Dobzhansky, Theodosius. *Genetics and the Origin of Species*. 3rd rev. ed. New York: Columbia University Press, 1951.
29. ————. *Mankind Evolving. The Evolution of the Human Species*. New Haven: Yale University Press, 1962.
30. Dodson, E. O. *Evolution: Process and Product*. New York: Reinhold Publishing Co., 1960.
31. Fernandez, Aniceto. "Science and Philosophy According to St. Albert The Great." *Angelicum*, XIII (1936), 24–59.
32. Fothergill, Philip G. *Historical Aspects of Organic Evolution*. New York: Philosophical Library, 1953.
33. Gilson, Etienne. *The Philosopher and Theology*. Trans. Cecile Gilson. New York: Random House, 1962.
34. Grant, Verne. "The Plant Species in Theory and Practice," in Mayr, *The Species Problem*. Washington, D.C.: The American Association for the Advancement of Science, 1957.
35. Grant, W. F. "The Categories of Classical and Experimental Taxonomy and the Species Concept." *Review of Canadian Biology*, XIX (1960).
36. Grene, Marjorie. *A Portrait of Aristotle*. Chicago: University of Chicago Press, 1963.
37. ————. "Biology and the Problem of Levels of Reality." *The New Scholasticism*, XLI (Autumn, 1967), 427–449.
38. Guyenot, Emile. *The Origin of Species*. New York: Walker, 1964.
39. Heidegger, Martin. *An Introduction to Metaphysics*. New Haven, Yale: 1959.
40. Huxley, Julian. *Evolution: the Modern Synthesis*. New York: Harper and Bros., 1942.
41. ————. *Evolution in Action*. New York: Mentor Books, 1953.
42. ————, ed. *The New Systematics*. Oxford: The University Press, 1940.
43. Jolivet, Régis. *La notion de substance.* Paris: Beauchesne, 1929.
44. Leakey, L.S.B. *Adam's Ancestors*. London: Methuen and Co., 1934.
45. Lovejoy, Arthur O. *Essays in the History of Ideas*. New York: George Braziller, Inc., 1955.
46. ————. *The Great Chain of Being*. New York: Harper, 1960.
47. Maritain, J. *Bergsonian Philosophy and Thomism*. New York: Philosophical Library, 1951.
48. ————. *Philosophy of Nature*. New York: Philosophical Library, 1951.
49. ————. *The Range of Reason*. New York: Charles Scribner's Sons, 1942. Ch. IV, 33–38.
50. Mayr, Ernst. *Animal Species and Evolution*. Cambridge, Massachusetts: Harvard University Press, 1963.
51. ————. *The Species Problem*. Washington, D.C.: The American Association for the Advancement of Science, 1957.
52. ————. "The Species Concept: Semantics versus Semantics." *Evolution*, III (December, 1949).
53. ————. *Systematics and the Origin of Species*. New York: Dover Publications, Inc., 1964.
54. Mayr, E., Linsley, E. G., and Usinger, R. L. *Methods and Principles of Systematic Zoology*. New York: McGraw-Hill, 1953.
55. Murray, Desmond. *Species Revalued*. London: Blackfriars' Publications, 1955.

56. Nogar, R. J. "Evolution: Scientific and Philosophical Dimensions." *Philosophy of Biology*. Ed. Vincent E. Smith. Jamaica, New York: St. John's University Press, 1962. Pp. 23–66.

57. ———. *The Wisdom of Evolution*. Garden City, New York: Doubleday and Co., Inc., 1963.

58. Nordenskiold, E. *The History of Biology*. New York: Tudor, 1963.

59. Oparin, A. J. *The Origin of Life*. 2nd ed. New York: Dover, 1953.

60. Perelman, Chaim. "Some Reflections on Classification." *Philosophy Today*, IX (1965), 268–272.

61. Pieper, Josef. *Guide to Thomas Aquinas*. New York: Mentor Books, 1964.

62. Reisser, Oliver L. "The Concept of Evolution in Philosophy." *A Book that Shook the World*. Pittsburgh, Pa.: Univ. of Pittsburgh Press, 1961.

63. Renoirte, F. *Cosmology : Elements of a Critique of the Sciences and of Cosmology*. New York: Joseph Wagner, 1950.

64. Rensch, B. *Evolution Above the Species Level*. New York: Science Editions, 1961.

65. Roe, Anne, and Simpson, G. G., eds. *Behavior and Evolution*. New Haven: Yale University Press, 1958.

66. Ross, Herbert A. *A Synthesis of Evolutionary Theory*. Englewood Cliffs, New Jersey: Prentice-Hall, Inc., 1962.

67. Russell, John L. "The Concept of Natural Law." *Heythrop Journal*, VI (1965), 434–446.

68. Schrödinger, E. *What Is Life?* New York: Anchor, 1956.

69. Scientific American Reader. *The Physics and Chemistry of Life*. 1955.

70. Shapley, Harlow, "On the Evidence of Inorganic Evolution," in *Evolution after Darwin*. Ed. Sol Tax. Chicago: University of Chicago Press, 1960. Vol. I, 23–38.

71. Simpson, G. G. "The Species Concept." *Evolution*, V (Dec. 1951). 285–298.

72. ———. "The History of Life," in *Evolution after Darwin*. Ed. Sol Tax. Chicago: University of Chicago Press, 1960. Vol. I, 117–180.

73. ———. *Life of the Past : An Introduction to Paleontology*. New Haven: Yale University Press, 1953.

74. ———. *The Major Features of Evolution*. New York: Columbia, 1953.

75. ———. *Principles of Animal Taxonomy*. New York: Columbia, 1961.

76. Simpson, G. G., Jepsen, G. L., and Mayr, E., eds. *Genetics, Paleontology, and Evolution*. New York: Atheneum, 1963.

77. Solmsen, F. *Aristotle's System of the Physical World*. Ithaca, N.Y.: Cornell Univ. Press, 1960.

78. Sonneborn, T. M. "Breeding Systems, Reproductive Methods, and Species Problems in Protozoa," in *The Species Problem*. Washington, D.C.: The American Association for the Advancement of Science, 1957.

79. Thomas Aquinas, St. *Commentary on the Metaphysics of Aristotle*. Trans. John P. Rowan. Chicago: Henry Regnery Co., 1961.

80. Verdenius, W. J., and Waszink, J. H. *Aristotle : On Coming to Be and Passing Away*. Leiden, Holland: E. J. Brill, 1946.

81. Weisheipl, J. A. "The Concept of Nature." *The New Scholasticism*, XXVIII (1954), 377–408.

82. ———. *Nature and Gravitation*. River Forest, Ill.: Aquinas Institute, 1955.

83. Weizäcker, C. F. *The History of Nature*. Chicago: The University of Chicago Press, 1959.

84. Whitehead, Alfred North. *The Concept of Nature*. London: Cambridge University, 1955.
85. ———. *Process and Reality*. New York: Harper, 1941.
86. Williams, George C. *Adaptation and Natural Selection*. Princeton: The University Press, 1966.

THE METAPHYSICAL ISSUES, PART B:
GENERAL BIBLIOGRAPHY

1. Alexander, Samuel. *Space, Time and Deity*. New York: Dover, 1916–18. 2 Vols.
2. Barthélemy-Madaule, Madeleine. *Bergson et Teilhard de Chardin*. Paris: Editions du Seuil, 1963.
3. Bergson, Henri. *Creative Evolution*. Authorized translation by Arthur Mitchell. New York: the Modern Library, 1944.
4. ———. *The Creative Mind*. Trans. Mabelle L. Andison. New York: The Philosophical Library, 1946.
5. ———. *The Two Sources of Morality and Religion*. New York: Anchor, 1955.
6. Bunge, Mario, ed. *The Critical Approach to Science and Philosophy*. New York: Free Press, 1964.
7. Cassirer, E. *An Essay on Man*. New York: Doubleday, 1953.
8. Collingwood, R. G. *The Idea of Nature*. New York: Galaxy, 1960.
9. ———. *The Idea of History*. New York: Galaxy, 1956.
10. Deely, John N. "Finitude, Negativity, and Transcendence: The Problematic of Metaphysical Knowledge." *Philosophy Today*, XI (Fall, 1967), 184–206.
11. Dewey, John. *Intelligence in the Modern World*. Ed. Joseph Ratner. New York: Modern Library, 1939.
12. Eiseley, Loren. *Darwin's Century*. New York: Anchor, 1958.
13. Eliade, Mircea. *Cosmos and History*. New York: Harper, 1959.
14. Elsasser, Walter M. *Atom and Organism, A New Approach to Theoretical Biology*. Princeton: The University Press, 1966.
15. Fackenheim, Emil L. *Metaphysics and Historicity*. Milwaukee: Marquette University Press, 1961.
16. Fraser, J. T. *The Voices of Time*. New York: George Braziller, 1966.
17. Geiger, L. B. "Métaphysique et relativité historique," in *Philosophie et Spiritualité*. Vol. II, 97–144.
18. Goudge, T. A. *The Ascent of Life*. Toronto: University of Toronto Press, 1961.
19. Grene, Marjorie. *A Portrait of Aristotle*. Chicago: University of Chicago Press, 1963.
20. Hayek, F. A. *The Counter-Revolution of Science*. New York: Free Press, 1952.
21. Henderson, Lawrence J. *The Fitness of the Environment. An inquiry into the biological significance of the properties of matter*. Boston: Beacon Press, 1913.
22. Jonas, Hans. *The Phenomenon of Life*. New York: Harper, 1966.
23. Koyré, Alexander. *From the Closed World to the Infinite Universe*. New York: Harper, 1958.

24. ———. *On the Philosophy of History*. New York: Scribner's, 1957.

25. ———. *Science and Wisdom*. New York: Scribner's, 1940.

26. Maritain, Jacques. "Vers une idée thomiste de l'evolution," *Nova et Vetera*, XLII (avril-juin, 1967), pp. 87–136.

27. Morgan, C. Lloyd. *Emergent Evolution*. London: Williams and Norgate, 1923.

28. Randall, John H. *Aristotle*. New York: Columbia University Press, 1960.

29. Ross, Sir David. *Aristotle*. New York: Barnes and Noble, 1964.

30. Scheler, M. *Man's Place in Nature*. Trans. Hans Meyerhoff. Boston: Beacon Press, 1961.

31. Simpson, George Gaylord. *The Meaning of Evolution*. New Haven, Connecticut: Yale University Press, 1949.

32. Sinnot, E. *Biology of the Spirit*. Viking: Compass, 1955.

33. Smuts, J. C. *Holism and Evolution*. New York: The Macmillan Co., 1926.

34. Spencer, Herbert. *First Principles*. 6th ed. New York: Appleton, 1903.

35. Teilhard de Chardin, Pierre. *The Phenomenon of Man*. New York: Harper & Row, 1959.

36. Van Laer, Henry. *Philosophico-Scientific Problems*. Pittsburgh: Duquesne University Press, 1953.

37. Voeglin, Eric. *Israel and Revelation. Order and History*. Baton Rouge: Louisiana State University Press, 1956. Vol. I.

38. ———. *The World of the Polis. Order and History*. Baton Rouge: Louisiana State University Press, 1961. Vol. II.

39. ———. *Plato and Aristotle. Order and History*. Baton Rouge: Louisiana State University Press, 1966. Vol. III.

40. Ward, Leo. *God and World Order. A Study of Ends in Nature*. St. Louis: Herder, 1961.

41. Whitehead, Alfred North. *Adventures of Ideas*. New York: Free Press, 1961.

42. ———. *The Concept of Nature*. London: Cambridge University Press, 1955.

43. ———. *The Function of Reason*. New York: Beacon, 1958.

44. ———. *Process and Reality*. New York: Harper Tb, 1929.

45. ———. *Science and the Modern World*. New York: Free Press, 1953.

46. Whitrow, G. J. *The Structure and Evolution of the Universe*. New York: Harper Tb, 1959.

47. Whyte, Lancelot Law. *Essay on Atomism : From Democritus to 1960*. New York: Harper, 1963.

48. ———. *The Unconscious Before Freud*. New York: Basic Books, 1960.

SECTION V—THE IMPACT OF EVOLUTION ON CHRISTIAN THOUGHT

1. Albright, W. F. *New Horizons in Biblical Research*. London: Oxford University Press, 1966.

2. Alszeghi, Zoltan. "Development in the Doctrinal Formulations of the Church Concerning the Theory of Evolution." *Concilium*, VI (June, 1967), 14–17.

3. Alszeghy, L., and Flick, M. "Il peccato originale in prospettiva evoluzionistica." *Gregorianum*, XLVII (1966), 201–205; and XLVI (1965), 705.

4. Barbour, Ian G. *Issues in Science and Religion*. Englewood Cliffs, New Jersey: Prentice-Hall, 1966.

5. Baum, Gregory. *The Future of Belief Debate*. New York: Herder and Herder, 1967.

6. Beauchamp, E. *La Bible et le sens religieux de l'univers*. Paris: Cerf, 1959.

7. Birch, C. *Nature and God*. London: SCM Press, 1965.

8. Bone, P. E. "Un siècle d'anthropologie préhistorique: Compatibilité ou incompatibilité scientifique du monogénisme." *Nouvelle Revue Théologique*, LXXXIV (1962), 709–734.

9. ———. "'Homo habilis', nouveau venu de la paléoanthropologie." *Nouvelle Revue Théologique*, LXXXVI (1964), 619–632.

10. Bröker, W. "Aspects of Evolution." *Concilium*, VI (June, 1967), 5–13.

11. Brunner, E. *The Christian Doctrine of Creation and Redemption*. New York: 1952.

12. Bultmann, R. *History and Eschatology*. Edinburgh: The University Press, 1957.

13. Chifflot, T. G. *Approaches to a Theology of History*. New York: Desclée, 1965.

14. Cobb, John B. *A Christian Natural Theology: Based on the Thought of Alfred North Whitehead*. Philadelphia: Westminster Press, 1965.

15. Collin, R. *Evolution*. New York: Hawthorne, 1959.

16. Comstock, W. Richard. "Theology After the 'Death of God'." *Cross Currents*, XVI (Summer, 1966), 265–305.

17. Connor, James L. "Original Sin: Contemporary Approaches," *Theological Studies*, XXIX (June, 1968), 215–240.

18. Corte, N. *The Origins of Man*. New York: Hawthorne, 1958.

19. Corvez, M. "Création et évolution du monde." *Revue Thomiste*, LXIV (1964), 549–568.

20. Curran, Charles. *Moral Absolutes in Theology*. Washington: Corpus, 1968.

21. Deely, John N. "The Vision of Man in Teilhard de Chardin." *Listening* (Autumn, 1966), 201–209.

22. de Fraine, Jean. *Adam and the Family of Man*. New York: Alba House, 1965.

23. ———. *The Bible and the Origin of Man*. New York: Desclée, 1962.

24. de Lavalette, H. "Le péché originel." *Recherches des sciences religieuses*, LV (avril, 1967), 227–242.

25. de Solages, Bruno. "Christianity and Evolution." *Cross Currents*, IV (Summer, 1951), 26–37.

26. Dewart, Leslie. *The Future of Belief*. New York: Herder & Herder, 1966.

27. Dobzhansky, T. *The Biology of Ultimate Concern*. New York: The New American Library, 1966.

28. ———. *Mankind Evolving*. New Haven: Yale, 1962.

29. Dolch, Heimo. "Sin in an Evolutive World." *Concilium*, VI (June, 1967), 36–40.

30. Dondeyne, A. *Contemporary European Thought and Christian Faith*. Pittsburgh: Duquesne University Press, 1963.

31. Dubarle, A. M. *The Biblical Doctrine of Original Sin*. New York: Herder, 1964.

32. ———. "Le gémissement des créatures dans l'ordre divin du cosmos." *Revue des sciences philosophiques et théologiques*, XXXVIII (1954), 445–465.

33. Dumont, C. "La prédication du péché originel." *Nouvelle Revue Théologique*, LXXXIII (1961), 113–134.
34. Eliade, Mircea. *Patterns in Comparative Religion*. New York: Meridian, 1966.
35. Flannagan, D. "Some Theological Implications of Evolution." *Irish Theological Quarterly*, XXXIII (1966), 265–270.
36. Fontinell, Eugene. "Reflections on Faith and Metaphysics." *Cross Currents*, XVI (Winter, 1966), 15–44.
37. Fothergill, P. G. *Evolution and Christians*. London: Longmans, Green & Company, 1961.
38. Francoeur, Robert T. *Perspectives in Evolution*. Baltimore: Helicon, 1965.
39. Gilkey, Langdon. "The Concept of Providence in Contemporary Theology." *Journal of Religion*, XLIII (1963), 171–192.
40. ———. "Cosmology, Ontology, and the Travail of Biblical Language." *Journal of Religion*, XLI (1961), 194–205.
41. ———. "Evolutionary Science and the Dilemma of Freedom and Determinism." *The Christian Century*, LXXXIV (March 15, 1967), 339–343.
42. Gleason, R. W. "Theology and Evolution." *Thought*, XXXIV (1959), 249–257.
43. Greene, J. C. *Darwin and the Modern World-View*. New York: Mentor, 1961.
44. Grelot, P. *Réflexions sur le problème du péché originel*. Paris: Casterman, 1968.
45. ———. "Faut-il croire au péché originel?" *Etudes*, CCCXXVII (1967), 231–251.
46. ———. "Péché originel et rédemption dans l'Epître aux Romains." *Nouvelle Revue Théologique*, XC (1968), 449–478.
47. Gunther, Ludwig. *Le chrétien et la vision scientifique du monde*. Paris: Cerf, 1965.
48. Haag, H. *Bible et évolution*. Barcelona: Herder, 1965.
49. Hamman, A. "La foi chrétienne au Dieu de la création." *Nouvelle Revue Théologique*, LXXXVI (1964), 1049–1057.
50. Hauret, Charles. *Beginnings*, 2nd revised ed. Chicago: The Priory Press, 1964.
51. Heim, Karl. *Christian Faith and Natural Science*. New York: Harper, 1953.
52. Hibbert, Giles. *Man, Culture, and Christianity*. London: Sheed and Ward, 1967.
53. Hutchins, Robert M. and Adler, Mortimer J. *The Great Ideas Today 1967: Should Christianity Be Secularised? A Symposium*. Chicago: Britannica, 1967, 1–81.
54. Jacques de Bivort de la Sandée. *God, Man and the Universe*. New York: P. J. Kennedy, 1953.
55. Kogan, B. R., ed. *Darwin and his Critics*. Belmont, California: Wadsworth Publishing Company, 1960.
56. Labourdette, M. M. *Le péché originel et les origines de l'homme*. Paris: Alsatia, 1953.
57. Lack, D. *Evolutionary Theory and Christian Belief*. London: Methuen, 1957.
58. Lang, G. O. and Vollert, Cyril. *Symposium on Evolution*. Pittsburgh: Duquesne University Press, 1959.
59. Lavocat, R. "Réflexions d'un paléontologiste sur l'état originel de l'humanité et le péché originel." *Nouvelle Revue Théologique*, LXXXIX (1967), 582–600.
60. Leclerc, J. *Le chrétien durant la planétorisation du monde*. Paris: Fayard, 1958.
61. Levie, Jean. *The Bible: Word of God in Words of Men*. New York: Kennedy, 1961.
62. Ligier, L. *Péché d'Adam et péché du monde*. Paris: Aubier, 1960–1961. 2 Vols.
63. Lyonnet, Stanislas, S. J. *De Peccato et Redemptione*, Vol. I, "E Notione Peccati,"

Vol. II, "De Vocabularia Redemptiones." Rome: Pontifical Biblical Institute, 1957 & 1960, respec.

64. ———. *Les épîtres de saint Paul aux Galates, aux Romains.* (Bible de Jérusalem), 1ère éd., 1953; 2e éd., 1959. Notes sur Rom. 5, 12.

65. ———. *Quaestiones in epistulam ad Romanos.* Prima series, Romae 1955, 182–234. Ces pages seront reprises, entièrement remaniées, dans une 2e édition à paraître (—Quaestiones).

66. ———. "Le sens de EPHO en Rom 5, 12 et l'exégèse des Pères grecs, *Biblica,* 36 (1955), 436–456.

67. ———. "Le péché originel et l'exégèse de Rom. 5, 12, *Recherches des sciences religieuse,* XLIV (1956), 63–84. (L'article a été reproduit, légèrement remanié et allégé, dans J. Huby, *Saint Paul, Epître aux Romains* [Verbum salutis], 2e éd. 1957. Pp. 521–557).

68. ———. *Le Péché Originel en Rom 5, 12 et le Concile de Trente.* Rome: Institute Biblique Pontifical, 1961.

69. ———. "Le Sens de Perazein en Sap 2, 24 et la doctrine du péché originel," *Biblica,* XXXIX (1958), 27–36, notamment pp. 32–36.

70. Macquarrie, John. *An Existentialist Theology.* New York: Harper, 1965.

71. ———. *The Scope of Demythologizing.* New York: Harper, 1966.

72. Mascall, E. "The Scientific Outlook and the Christian Message." *Concilium,* VI (June, 1967), 60–63.

73. Messenger, E. C., ed. *Theology and Evolution.* London: Sands & Company, 1949.

74. Mixter, R. L. *Evolution and Christian Thought Today.* Grand Rapids, Michigan: William B. Eerdmans Publishing Company, 1959.

75. Moretti, J. *Biologie et réflexion chrétienne.* Paris: Fayard, 1967.

76. Motherway, T. J. "Adam and the Theologians." *Chicago Studies* (Fall, 1962), 115–132.

77. Murray, John Courtney. *The Problem of Religious Freedom.* Westminster, Md: Newman, 1965.

78. Nogar, R. J. *The Lord of the Absurd.* St. Louis: B. Herder, 1966.

79. ———. *The Wisdom of Evolution.* New York: Doubleday, 1963. Esp. Ch. 14.

80. O'Brien, John A. *Evolution and Religion.* New York: Century, 1932.

81. Ong, Walter, ed. *Darwin's Vision and Christian Perspectives.* New York: Macmillan, 1960.

82. Paul VI. *Address to the Symposium on Original Sin.* Convoked at Rome, July 19, 1966. National Catholic Welfare Conference Documentary Service.

83. Piault, B. *La création et le péché originel.* Paris: Spes, 1960.

84. Pius XII. *Humani Generis.* New York: Paulist Press, 1950.

85. Rabut, O. *God in an Evolving Universe.* St. Louis: Herder and Herder, 1966.

86. Rahner, Karl. "Thoughts on the Possibility of Belief Today," in *Theological Investigations.* Trans. Karl H. Kruger. Baltimore: Helicon, 1966. Vol. V, 3–22.

87. ———, et al. *The Bible in a New Age.* New York: Sheed and Ward, 1965.

88. ———. "Christology Within an Evolutionary View of the World," in *Theological Investigations.* Vol. V, 157–172.

89. ———. *The Church After the Council.* New York: Herder, 1966.

90. ———. "Evolution and Original Sin." *Concilium,* VI (June, 1967), 30–35.

91. ———. "History of the World and Salvation History," in *Theological Investigations,* Vol. V, 97–114.

92. ———. *Hominisation. The Evolutionary Origin of Man as a Theological Problem.* St. Louis: Herder & Herder, 1965, 229–296.

93. ———. "Science as a Confession," in *Theological Investigations.* Vol. III, 385–400.

94. ———. *Science, évolution et pensée chrétienne.* Paris: Desclée, 1967.

95. ———. "Theological Reflexions on Monogenism," in *Theological Investigations,* Vol. I, 229–296.

96. Renkins, H. *La Bible et les origines du monde.* Paris: Desclée, 1964.

97. Ricoeur, P. *Finitude et Culpabilité.* Paris: Aubier, 1960.

98. Rideau, E. "Humanisme chrétien et humanisme païen." *Nouvelle Revue Théologique,* LXXV (1953), 141–160.

99. Schoonenberg, Piet. *God's World in the Making.* Pittsburgh: Duquesne University Press, 1964.

100. ———. *Man and Sin.* Notre Dame, Indiana: Notre Dame University Press, 1965.

101. Sertillanges, A.-D. *L'univers et l'âme.* Paris: Ouvrières, 1965.

102. Smulders, Piet. *The Design of Teilhard de Chardin.* Westminster, Md.: Newman, 1967.

103. Smyth, Kevin, translator. *A New Catechism.* New York: Herder and Herder, 1967.

104. Stebbins, G. Ledyard. "Why I Gave Up Traditional Christianity," with a "Response" by Matthew Fox, *Listening,* I (Spring, 1966), 150–161.

105. Teilhard de Chardin, Pierre. *The Divine Milieu.* New York: Harper & Row, 1960.

106. ———. *Oeuvres.* 10 Vols. Paris: Seuil (English translations, New York: Harper).

107. *Theology Digest,* XV (Autumn, 1966): Cross-section of current research on the *quid sit* of Original Sin.

108. Tillich, Paul. *Theology of Culture.* New York: Galaxy, 1964.

109. Vanhengel, M. C., and Peters, J. "Death and Afterlife." *Concilium,* VI (June, 1967), 75–83.

110. Van Onna, Ben. "The State of Paradise and Evolution." *Concilium,* VI (June, 1967), 64–68.

111. Verrielle, A. *Le Surnaturel en nous et le péché originel.* Paris: Bloud et Gay, 1934.

112. von Balthasar, Hans Urs. *Science, Religion and Christianity.* London: Burns and Oates, 1958.

113. ———. *A Theological Anthropology.* New York: Sheed and Ward, 1967.

114. ———. *A Theology of History.* New York: Sheed and Ward, 1963.

115. Ward, L. R. *God and the World Order.* St. Louis: B. Herder, 1961.

116. Zimmerman, P. A. *Darwin, Evolution, and Creation.* St. Louis: Concordia Publishing House, 1961.

SECTION VI—TOWARD AN EVOLUTIONARY WORLD-VIEW

1. Arendt, H. *The Human Condition.* Chicago: University Press, 1959.

2. Bell, Daniel, and the American Academy Commission on the Year 2000. "Toward the Year 2000." *Daedalus,* XCVI (Summer, 1967), 639–679.

3. Bloom, Benjamin S. *Stability and Change in Human Characteristics*. New York: John Wiley & Sons, 1964.

4. Brown, Harrison. *The Challenge of Man's Future*. New York: Compass, 1954.

5. Campbell, Joseph. *The Masks of God*, New York: Viking Press. Four Volumes:
   I. *Primitive Mythology* (1959);
   II. *Oriental Mythology* (1962);
   III. *Occidental Mythology* (1964);
   IV. *Creative Mythology* (1968).

6. Clarke, Arthur C. *Profiles of the Future*. New York: Bantam, 1962.

7. Cox, Harvey. "Evolutionary Progress and Christian Promise." *Concilium*, VI (June, 1967), 18–23.

8. ———. *The Secular City*. New York: Macmillan, 1965.

9. Dilthey, Wilhelm. *The Essence of Philosophy*. Chapel Hill: University of North Carolina Press, 1961.

10. ———. *Pattern and Meaning in History*. Ed. H. P. Rickman. New York: Harper Torchbooks, 1961.

11. Dubarle, Dominique. *Scientific Humanism and Christian Thought*. New York: Philosophical Library, 1956.

12. ———. "Does Man's Manner of Determining His Own Destiny Constitute a Threat to His Humanity?" *Concilium*, VI (June, 1967), 41–46.

13. Eiseley, L. *The Firmament of Time*. New York: Atheneum, 1962.

14. ———. *The Immense Journey*. New York: Random House, 1957.

15. Ellul, Jacques. "The Technological Revolution: Its Moral and Political Consequences." *Concilium*, VI (June, 1967), 47–51.

16. Elsasser, Walter M. *Atom and Organism, A New Approach to Theoretical Biology*. Princeton: The University Press, 1966.

17. Emmet, D. M. "The Choice of a World Outlook," *Philosophy*, XXIII (1948), 208–227.

18. Endres, Joseph. *Man as the Ontological Mean*. New York: Desclée, 1965.

19. Erikson, E. *Insight and Responsibility*. New York: Norton, 1964.

20. Ewald, William R. *Environment for Man, The Next 50 Years*. Bloomington: Indiana University Press, 1967.

21. Fessard, Gaston. "The Theological Structure of Marxist Atheism." *Concilium*, VI (June, 1966), 5–13.

22. Fetscher, Irving. "Developments in the Marxist Critique of Religion." *Concilium*, VI (June, 1966), 56–67.

23. Greene, John C. *Darwin and the Modern World-View*. New York: Mentor, 1963.

24. ———. *The Death of Adam*. New York: Mentor, 1961.

25. Hoffer, Eric. *The True Believer: Thoughts on the Nature of Mass Movements*. New York: Perennial, 1966.

26. Hoyle, Fred. *Of Men and Galaxies*. Seattle: University of Washington Press, 1966.

27. Hutchins, Robert M. and Adler, Mortimer J. "An Essay on Time," in *The Great Ideas Today*. 1963. Pp. 83–134.

28. ———. *The Great Ideas Today 1964. New Europe and the U.S.A.* Chicago: Britannica, 1964. Pp. ix–179.

29. ———. *The Great Ideas Today 1963. A Symposium on Space*. Chicago: Britannica, 1963. Pp. 1–82.

30. ———. *The Great Ideas Today 1965. Work, Wealth, and Leisure.* Chicago: Britannica, 1965. Pp. 1–105.
31. Huxley, Julian. *Essays of a Humanist.* New York: Harper, 1964.
32. ———, ed. *The Humanist Frame.* London: Allen & Unwin, 1961.
33. ———. *Man in the Modern World.* New York: Mentor, 1955.
34. ———. *Religion Without Revelation.* New York: Mentor, 1955.
35. Jackson, F. L., and Moore, P. A. *Life in the Universe.* New York: Norton, 1962.
36. Jaspers, Karl. *Man in the Modern Age.* New York: Anchor, 1957.
37. ———. *The Future of Mankind.* Chicago: University of Chicago Press, 1959.
38. Kahn, Herman, and Wiener, Anthony J. *The Year 2000.* New York: Macmillan, 1967.
39. Kaplan, A. "Freud and Modern Philosophy," in *Freud and the Twentieth Century.* Ed. B. Nelson. Gloucester, Massachusetts: Peter Smith Company, 1958.
40. Kardiner, A., and Preble, E. *They Studied Man.* New York: Mentor, 1961
41. Maritain, Jacques. *The Degrees of Knowledge.* Trans. from the 4th French edition under the direction of Gerald Phelan. New York: Scribner's, 1958.
42. ———. *Integral Humanism.* Trans. by J. Evans. New York: Scribner's, 1968.
43. Mayer, Milton. "The New Man," in *The Great Ideas Today 1966.* Eds. Robert M. Hutchins and Mortimer J. Adler. Chicago: Britannica, 1966. Pp. 100–144.
44. Medawar, P. B. *The Future of Man.* New York: Mentor, 1959.
45. Meier, Richard L. *Developmental Planning.* New York: McGraw-Hill, 1965.
46. Mesthene, E. "Religious Values in the Age of Technology," *Concilium,* VI (June, 1967), 52–59.
47. Meyerhoff, Hans, ed. *The Philosophy of History in our Time.* New York: Anchor, 1959.
48. Mills, C. Wright, ed. *Images of Man.* New York: George Braziller, 1960.
49. Moltmann, Jürgen. "Hope Without Faith: An Eschatological Humanism Without God." *Concilium,* VI (June, 1966), 14–21.
50. Nogar, Raymond J. *Evolutionism: Its Power and Limits.* Washington: Compact Studies, 1964.
51. ———. "The God of Disorder." *Continuum,* IV (Spring, 1966), 102–113.
52. Nogar, R., and Francoeur, R. "The God of Disorder II: A Response," (Francoeur), and "The God of Disorder III: A Postscript," (Nogar). *Continuum,* IV (Summer, 1966), 264–275.
53. Northrop, F. S. C., ed. *Ideological Differences and World Order.* New Haven: Yale, 1963.
54. Parsons, Talcott, and Shils, Edward, eds. *Toward a General Theory of Action.* New York: Harper Tb, 1951.
55. Pepper, Stephen. *World Hypotheses, A Study in Evidence.* Los Angeles: University of California Press, 1966.
56. Pieper, Joseph. "Death and Immortality." *Philosophy Today,* VI (1962), 34–44.
57. Platt, John R. "The New Biology and the Shaping of the Future," in *The Great Ideas Today 1968.* Eds. Robert M. Hutchins and Mortimer J. Adler. Chicago: Britannica, 1968, pp. 120–169.
58. ———. *The Step to Man.* New York: John Wiley and Sons, 1966.
59. Popper, K. R. "On the Sources of Knowledge and of Ignorance," *Proceedings of the British Academy* (1960), 39–71.

60. ———. *The Poverty of Historicism.* Boston: Beacon Press, 1957.
61. Rabut, Olivier. *God in an Evolving World.* St. Louis: Herder, 1966.
62. Rahner, Karl. *The Christian of the Future.* New York: Herder and Herder, 1967.
63. ———. "Christianity and Ideology." *Concilium,* VI (June, 1965), 23–31.
64. Roberts, M. *The Behavior of Nations.* London: Dent, 1941.
65. Roheim, G. *Psycho-analysis and Anthropology.* New York: International Universities Press, 1950.
66. Rokeach, Milton. *The Open and Closed Mind.* New York: Basic Books, 1960.
67. Schiffers, N. "Physics Questions Theology." *Concilium,* VI (June, 1967), 69–74.
68. Schlette, H. R. "The Problem of Ideology and Christian Belief." *Concilium,* VI (June, 1965), 56–67.
69. Schroedinger, E. *Science and Humanism.* New York: Cambridge University Press, 1951.
70. Sertillanges, A.-D. *Dieu ou Rien.* Foreword by Teilhard de Chardin. Paris: Flammarion, 1965.
71. ———. *Le Problème du Mal.* Paris: Aubier, Vol. I: "L'Histoire" (1948); Vol. II: "La Solution" (1951).
72. Shapley, H. *Of Stars and Men.* London: Elek, 1958.
73. ———. *The View From a Distant Star.* New York: Delta, 1964.
74. Shklovski, I. S., and Sagan, Carl. *Intelligent Life in the Universe.* New York: Delta, 1966.
75. Siegmund, Georg. *God on Trial. A Brief History of Atheism.* New York: Desclée, 1967.
76. Simpson, George Gaylord. "Biological Sciences," in *The Great Ideas Today 1965.* Eds. Robert M. Hutchins and Mortimer J. Adler. Chicago: Britannica, 1965. Pp. 286–319.
77. Singh, Jagjit. *Great Ideas and Theories of Modern Cosmology.* New York: Dover, 1961.
78. Teilhard de Chardin, P. "The Psychological Conditions of Human Unification." *Cross Currents,* III (1952), 1–5.
79. ———. "Building the Earth." *Cross Currents,* IX (1959), 315–330.
80. ———. *The Future of Man.* New York: Harper, 1964.
81. Thomas, William L., ed. *Man's Role in Changing the Face of the Earth.* International Symposium. Chicago: University of Chicago Press, 1956.
82. Toulmin, Stephen. "Contemporary Scientific Mythology," in *Metaphysical Beliefs.* London: SCM Press, 1957. Pp. 13–81.
83. Vahanian, G. *The Death of God.* New York: George Braziller, 1961.
84. Van Doren, Charles. *The Idea of Progress.* New York: Praeger, 1967.
85. von Balthasar, Hans Urs. *A Theology of History.* New York: Sheed and Ward, 1963.
86. Watson, James D. *The Double Helix.* New York: Atheneum, 1968.
87. White, L. A. "Ethnological Theory," in *Philosophy for the Future.* New York: Macmillan Company, 1943.
88. Whitehead, A. N. *An Enquiry Concerning the Principles of Natural Knowledge.* New York: Macmillan Company, 1925.
89. Wiener, Norbert. *God & Golem, Inc.* Cambridge, Mass.: MIT Press, 1966.
90. Wolstenholme, G., ed. *Man and His Future.* Boston: Little, Brown and Company, 1963.

91. Zaehner, R. C. *Matter and Spirit*. New York: Harper, 1963.
92. Zirkle, Conway. *Evolution, Marxian Biology, and the Social Scene*. Philadelphia: University of Philadelphia, 1959.

Retrospect

This book was as long in the printing as it was in the writing: a somewhat curious fate, and a tale that bears telling. It will give me a chance, at the same time, to add some critical qualifications to themes the work develops.

Raymond Nogar was a teacher of mine in college and graduate school, a most generous and inspired teacher. I had the good fortune to be his student, and eventually to work with him closely, in what was the most fruitful period of his work, from about 1962 until his death in 1967. His first book, *The Wisdom of Evolution*, was published in 1963. The preparation of proofs, the choice of cover design, the advertising plans—these were excitements in the life of a writer and teacher that he would share with his class in ways that made us understand and feel the weight of the subject, the depth of the questions, the commitments and intensity of a mind that questions, studies, and contemplates the mysteries of the past and future.

Between 1963 and the publication in 1966 of *The Lord of the Absurd*, Nogar plunged into his work as fully as any man could. He gave more and more of his time, besides his classes, articles, and book manuscript, to lecture tours, tours in which his audiences were carried entranced into the far reaches of space and time that were more and more the haunt of Raymond Nogar's anxious and sometimes fearful reflections on "the meaning of it all."

At the Aquinas Institute in River Forest an annual celebration was held in honor of St. Thomas Aquinas, the high point of which was the reading of a formal academic paper to an assemblage of faculty and

437

students from several Chicago area colleges, followed by an open discussion and finally a social evening. In 1965 I was chosen by the President of the Institute to write and present the formal paper, and—as a result of two years' work constantly inspired by classes and conversations with Nogar—I selected as my topic a discussion of human origins, titling my paper "The Emergence of Man." Naturally, Fr. Nogar was assigned to direct my writing, an office he filled with a delicacy and concern for the independent growth of another's thought that is as impressive in memory as it was in life.

The evening was a great success, but the astonishing upshot came the next morning, when Fr. Nogar asked me to his room to discuss several points I had made in the paper that seemed to him novel and promising—but which he had no recollection of having made himself, in passing, in certain of his class lectures! That was one of the forms the spontaneity of the spoken word took in Fr. Nogar's lectures: he would often seize on an insight and develop it *ex tempore*, so carried away by the discourse that he frequently had no recollection of particular points, no matter how striking, that he had mentioned along the way. He was, above all, a spirit of the spoken word.

At Fr. Nogar's urging, I revised "The Emergence of Man" for publication. It appeared in *The New Scholasticism* for 1966, and is reprinted—again with Nogar's concurrence—as a reading in this volume.

It was early in 1967, while I was in the midst of work on a book on the philosophy of Martin Heidegger—since published by Martinus Nijhoff under the title, *The Tradition Via Heidegger*—that Fr. Nogar offered me the chance to collaborate with him in preparing a volume on evolution for Appleton-Century-Crofts. Naturally, I was at once flattered and honored. To work closely with Nogar was more than intellectually exciting: It was a spiritual adventure. I accepted his invitation without hesitation, despite the complications it would cause in completing my work on Heidegger.

My wife and I drove to Chicago for dinner with Nogar—he had not yet met Simone, and "wanted to see this team" he "would be working with." Simone was as charmed with him as he was with her; we launched the work with a clink of wine-filled glasses.

By the 4th of July we were well on our way, and Fr. Nogar wrote to our editor:

> This morning I posted our doubly-signed contract for *The Problem of Evolution* to Appleton-Century-Crofts, and I have long wanted to express my gratitude for your patience in working me into your important series, Current Problems in Philosophy. It often seemed almost absurd (someone ought to write a book about absurdity!) to me how difficult it was to get anything done. I had fifty public lectures between last October and May 1st.

And in early June, I landed in the hospital again for seven days.

But Jack has been a gem, and I think it was very good for him, in the creative sense. I made a two-hour tape for him of what I wanted in the book and, while I was in the hospital, he lived in my room in River Forest with all my library and notes at his disposal. When I returned we had a three day series of meetings, revampings, and mullings-over (is that a word?). I like the general pattern, and I hope that you do, too.

I will have Jack's final form of the readings in late August. Meantime, I will have begun to draft my sixty pages [of introductory essay]. When I do the final editing (from our side) and add my introductory part which will tie in closely with the readings, and touch it up as I see fit, we will send it off to you—as A.-C.-C. (as they affectionately refer to themselves) advised. That should be very early in September.

On my way back from New Mexico to New Brunswick in late August, I dropped off the selected readings as scheduled, but Fr. Nogar had not yet returned from California. I did not know it at the time, but the sickness that was to take his life on the 17th of November had already set upon him. My meeting with Fr. Nogar in June was to be the last time I ever saw him. His parting words had been: "This won't be the last time we'll be working together." So much for the plans of men.

Upon learning of Fr. Nogar's death, our publisher asked me to complete the book on my own. We agreed to proceed under the original contract. I flew to Chicago to pick up the manuscript, and I took as a reminder from my dead friend's room a black glass ashtray with the skeletons of man's evolutionary ancestors in white on the bottom.

Between December, 1967 and November, 1968 I worked on the manuscript. Of course Fr. Nogar's death had been a heavy and unexpected blow. No one so close to me had died before, and I brooded much over the classes, conversations, work, and laughter we had shared. I decided to make of the volume a fitting final memorial to this empathic friend and impassioned thinker.

As I look back now, I can better sympathize with what the editors at A.-C.-C. must have felt when they finally received the completed manuscript—300% over the size stipulated in the original contract, considerably more technical and far more comprehensive than they had planned for. With the intransigence and naiveté of youth, I insisted that the book be published as it was, allowing for some editorial changes, principally concerning the overly extensive footnotes of Part I.

These changes were drawn out over a number of months. Meanwhile, I was wrestling with yet another question I had first been awakened to by Fr. Nogar: the problem of species. The results of this work tied in closely with many sections of the A.-C.-C. manuscript, and the several changes and cross-references I sent in as a result of the publication by *The Thomist* of the species work in early 1969—as a long, two-part article called "The Philosophical Dimensions of the Origin of Species"—not only made for an untidiness in references all around, but yet further delayed the preparation of the book.

In the spring of 1969 I left the Faculty of Philosophy at the University of Ottawa to take a position offered me at the Institute for Philosophical Research in Chicago. By this time all the many revisions of the evolution manuscript had been finalized, and I turned to my new work—research for a book on the philosophy of language—with a sense of relief and delight in the prospect of the publication at last of *The Problem of Evolution*.

´Now at last the book is finally to appear. I hope it will help fill the conspicuous gap in evolutionary studies that Fr. Nogar and I intended it to fit into, by providing a work that is philosophically sophisticated as well as scientifically informed and historically accurate. The work has its faults, to be sure. There is a certain youthful turgidity and uncompromising bombast of style in parts, for which tone I must bear the blame—one finds none of that from Nogar's hand. In organizing Sections 1 and 3 of Part II, I was perhaps a bit less qualificatory than I should have been concerning the facile "dialectical scheme" of Adler's work (fn. 2, p. 86 above notwithstanding). Still, nothing there is indefensible, with the right qualifications in mind, except for one point discussion of which I reserve for the last paragraphs of these remarks in retrospect, because of its importance. There are, however, several more strictly technical points that, though perhaps not indefensible as they stand, do have qualifications that demand a more explicit remark, and these I may mention here.

A point of particular importance that needs a more careful statement is expressed in my remark in n. 53, p. 33 above that "if one prescinds from the role mathematics plays, the explanations of modern physics and biology alike approximate closely, as to an ideal, to the 'covering-law model' of explanation." In order to envision any such approximation, I now see, one must also prescind from the logical empiricists' notion (after Hume) of causality as constant conjunction (replacing the classical view of causality as a dependence in being), which makes it impossible for the covering-law model *à la* Hempel *et al.* to accommodate the difference between what the Latin Aristotelians termed demonstration *quia* ("proof of the fact") vs. demonstration *propter quid* ("proof of the reason for the fact"). Of these two types of "demonstration," only the latter is demonstration strictly speaking, i.e., a knowledge of the reasoned fact, an explanation in the philosophical sense. (Cf. Aristotle's *Posterior Analytics*, Book I, ch. 13, esp. 78a28–78b3, for the classical *locus* of this distinction.) The point is effectively shown, and without any recourse to Latin sources, by B. A. Brody in his remarkable essay, "Toward an Aristotelian Theory of Scientific Explanation," *Philosophy of Science,* 39 (March, 1972), pp. 20–31, for the knowledge of which article—knowledge which came too late, unfortunately, for correcting the galleys of this book—I am grateful to W. A. Wallace.

Another critical point that needs not so much added qualification as expansion into an article in its own right is the account of the ontological foundations for the prospective and retrospective characters, respectively, of the Platonic and Aristotelian Explanatory Modes, as sketched briefly in fn. 64, p. 36 above.

I could add to this list, but these express the reservations that seem most vital at the moment, and no doubt the addition of other points may be confidently left to reviewers. Suffice it to say here that the essential lines of the work as a whole still seem to me sound, and an answer to a genuine need in the evolutionary literature for a philosophically structured interdisciplinary overview of mankind evolving.

Over the past year or two, it has not been possible for me to follow the current books and articles on evolution as closely as I did before taking up my researches on language. Hence it seemed wise to make one last check of the evolutionary literature for quite recent work of singular significance that may have escaped my notice. As is always the case, the quantitative proliferation of books and articles seemed to have well exceeded the increase of qualitative insight. To doublecheck this impression, I decided to canvass as many as I could contact of the authors included in Part II of the book, to ask them whether in their judgment any work of singular significance had appeared during the last two to three years, one that had either changed their own thinking significantly or developed such a cogent presentation of alternative views that it should be drawn to the particular attention of all interested readers.

Among the authors included in this book, I was able to contact Leslie White (now moved from the University of Michigan to the University of California at Santa Barbara), David Bidney, Mortimer Adler, Francisco J. Ayala (now moved with Theodosius Dobzhansky from the Rockefeller University in New York to the University of California at Davis), Benedict Ashley, and Loren Eiseley. To a man, they maintained their views as represented in their essays in this volume, and noted that, while many good and readable studies fill the journals and bookstores, there seems for the moment to be nothing in the offing of revolutionary significance to bring about shifts in the basic outlines of the general problems. In connection with Part I of this book, however, I might call the reader's attention to a further and somewhat different line of development in my own views of myth, "The Myth as Integral Objectivity," *ACPA Proceedings,* XLV (1971), pp. 67–76; to William A. Wallace's two-volume historical-philosophical study of *Causality and Scientific Explanation,* published by the University of Michigan Press; and to

Henry Veatch's *Two Logics* (Evanston: Northwestern University Press, 1969). Veatch develops a distinction between what he calls "What-logic" and "Relating-logic" in a way that provides many analogues helpful in understanding the rather different Platonic/Aristotelian Explanatory Mode distinction as developed in this book.

"A philosophical work," it has been said, "must stand the changing scene that develops around it, and does not need to be recast every season." We may hope that the present volume is such a book; if so, it will stand also as a lasting memorial to the man whose spirit and generosity made it possible—Raymond J. Nogar. To this end, may I close with mention of one major point in this book that, when the needed qualifications are added, far from becoming convincing, proves itself untenable from the start. Here I follow the suggestion of Aristotle in the *Nicomachean Ethics*, 1096a14–16, to the effect that "it would perhaps be thought to be better, indeed to be our duty, for the sake of maintaining the truth even to destroy what touches us closely, especially as we are philosophers or lovers of wisdom; for, while both are dear, piety requires us to honor truth above our friends."

The point, then, that seems to me beyond the frontiers of the defensible, is the one that Mortimer Adler proposes on p. 233 above, in insisting that "we do not *know* whether man's difference in kind [from animals] is superficial or radical; we do not *know* whether the materialist or immaterialist hypothesis is nearer the truth." When I pressed him recently concerning his over-all argument on this point, he would say only that dissatisfactions with it, as far as he is concerned, show only that "a lot of people don't want to face up to the fact that their good impulses are based on theories that might be questioned by science." However, properly philosophical reflection need no more wait on science than science need wait on it. Hence, it does not require "Turing's game" (see p. 235, n. 2a above) to articulate and vindicate the natural sanity of the intelligence that serves notice that, however we may understand the nature of man, no such understanding can be sound or complete which leaves out of account or positively excludes the maturing individual's capacity for self-determination, i.e., his responsibility for what in his world he finally accepts and embraces as truly expressive of what he is and would be. Yet, as Adler strongly argues before floating his disclaimer of knowledge, the only alternative to the radical difference of the "immaterialist hypothesis" (not an especially felicitous phrase) is a view of man that precludes the freedom that begets moral responsibility. Such an alternative is an alternative in words only; for, given, with all its limitations, the common and primary experience of personal responsibility—a *datum* in the sense defined above (pp. 53 and 67) if ever there was one—it is certain in advance that any theoretical view of man that explains this experience away altogether proves in the very effort its bankruptcy. Surely it is a fallacy of misplaced emphasis to entertain as theoretically possible a view that renders the whole of man's moral experience one gigantic illusion.

It is characteristic of the time, however, not merely to entertain views of man irreconcilable with the deepest capacities of human nature, but to endorse and disseminate such views—which, to tell the truth, could be ascertained (if at all) only by the most subtle philosophy—as certified by experimental science. This great charade will one day pass. But in the meantime, it bears witness to the extreme state of philosophical dissociation which holds sway in contemporary culture and in which the modern mind exists, to the extent that it educates men in its institutions to adopt views logically antinomious, not only among themselves, but vis-à-vis the moral experience of mankind. Such a state cannot long endure; but in its passing, it may take with it the civilization in whose bosom the antinomies were nurtured. Thought, by mistaking itself for the instrument for overcoming everything, has at this point come close to engendering conditions under which its own existence is imperiled.

Chicago, Illinois
September 18, 1972

Index

This is an index of proper names, subjects, and terms combined. It covers the book only up to p. 404, that is to say, there are items — mostly names — that occur both in the bibliography and in the "Retrospect" that are not entered in the index.

All the main entries are ordered alphabetically. For the subentries, however, two different systems of ordering are used. Subentries under main entries that are subjects or terms are ordered alphabetically according to the first word of import. Subentries under proper name main entries, by contrast, are ordered "chronologically," i.e., in the sequence in which they occur in the book.

With regard to the proper name entries, prior to the subentries, if any, and provided more than a reduplication of page numbers for subentries is involved, the main entry is followed by a list of all the pages whereon the person in question plays a role. Three distinct typographical devices are employed to achieve this. Take, for example, the main entry, "Dobzhansky, Theodosius." Simple numerals occurring after the name — e.g., 12, 15, 20, etc. — indicate pages where the name is explicitly mentioned in the main text. Numerals followed by the letter n, or by nn — e.g., 60 n, 140–145 nn, 201 n — indicate that the the name is mentioned on the page indicated by the numeral, but in footnotes rather than in main texts. Numerals followed by a parenthesis enclosing an n or nn together with another numeral — e.g., 121 (n.5), 129 (nn.38–40, 42), 137 (nn.102, 104) — indicate that the person in question, though not mentioned explicitly in the main text on the page indicated, holds an opinion involved in the main text immediately preceding the footnote number or numbers given within the parenthesis.

The index is as complete and accurate, according to the above explained plan, as it was possible to make it within the very limited time available. My thanks go to Ms. Frances Plass for her assistance in preparing the index according to the twelfth edition of the University of Chicago's *Manual of Style*. It would not have been possible to do the work at all adequately without her enormous labor.

Non-physical factor in man. *See* Freedom; Human behavior; Humanization; Humanness, essence of; *Humanum, the;* Immaterial factor in man; Intellect

Noösphere, 346–47

Nordenskiold, E., 44 n

Norm of reaction. *See* Genotype, norm of reaction of

Northrop, F. S. C., 154

Not-being, 290–91

Obligation, nature of, 189. *See also* Ethics; Morality; Natural law

Observation, order of, 27, 67. *See also* Explanation, order of; Observation and explanation

Observation and explanation: anti-classical position on, 28 n.43

as applied to evolutionary notions, 27–29

as basis for understanding the transition from Classical Antiquity to Darwin's World, 29, 31, 52–53, 64

classical view of, 28 n.43

as formulated by Maritain, 35, 35 n.61

as involved in the distinction between science and philosophy, 52 n.114

as knowledge within the perspective of fact vs. the perspective of reasoned fact, 27–28

as mutually conditioning the growth of scientific understanding, 66

proof of their distinction, 52 n.114, 67 n.168. *See also* Contradiction, principle of; Dualism; Explanatory dualisms

Ogonik, M. J., 250 n

Onan, 177 n

Ontological explanation. *See* Aristotelian Explanatory Mode; Metaphysics

Ontological gap, 122. *See* Humanness, essence of; *Humanum, the;* Immaterial factor in man; Intellect

Openmindedness: 26

to divine communication, 335–37, 369–74

Opportunism in evolution, 25, 25 n.37, 98. *See also* Natural selection

Opposition between sciences and humanities, 280–81

Order, 269, 289, 294, 396 n.5. *See also* Chance; Natural selection

Organism: classification of by patterns of behavior, 109

dependence on environment, 112

viewed in evolutionary perspective, 137. *See also* Behavior; Natural unit; Substance

Organization: patterns of in evolution 342

Origin, human. *See* Genetics; Human behavior; Human nature; Humanization; Humanness

Origin of justice, 188, 191, 202 n.15. *See also* Justice

Origin of life. *See* Life

Origin of species: causality of, 132. *See also* Ecology; Genetics; Natural selection

Original sin: problem of, 309–10, 313

in what the doctrine of consists, 200. *See also* Fall of man

Ortega y Gasset, José, 189, 201 n

Orthogenesis, definition of, 25, 100

Osborn, Frederick, 239 n

Overpopulation, 239–40, 239 n.6, 240 n.8, 245, 350–51, 382

Paleolithic man, 361, table III (pp. 18–19)

Pantheism, 24, 25 n.38. *See also* Monism

Pardies, Fr., 53 n, 72 n

Parmenides, 269, 277, 292

Parsons, T., 135 (n.89)

Pascal, 299 n

Patterns of behavior, 105–109. *See also* Behavior; Minding

Patterns of culture, 154–55. *See also* Cultural evolution

Patterns in language, 96. *See also* Language

Paul, Saint: on original sin, 313–15, 318–19

Pavlov, experiment in behavior, 105, 112

Péguy, C., 321 n

Perceptual intelligence. *See* Intelligence

Perinoetic Intellection: as critiqued by Ashley, 33 n.53

as critiqued by Deely, 70–71

Period of man's rise to dominance, tables II and III (pp. 16–19)

Person. *See* Personality

Personality, concept of, 195–96, 224, 235 n.9, 387. *See also* Freedom; Human nature; Humanization; Humanness; *Humanum,* the

Phenomena, as basis of knowledge, 73. *See also* Fact; Knowledge; Observation and explanation

Phenomenology, 56 n

Phenomenology vs. ontology of evolution, 27

Phenotype, 129. *See also* Genetics; Genotype

Philosophical Explanatory Mode, 61. *See also* Aristotelian Explanatory Mode